Springer Polar Sciences

Springer Polar Sciences

Springer Polar Sciences is an interdisciplinary book series that is dedicated to research in the Arctic, sub-Arctic regions, and the Antarctic. In recent years, the polar regions have received increased scientific and public interest. Both the Arctic and Antarctic have been recognized as key regions in the regulation of the global climate, and polar ecosystems have been identified to be particularly susceptible to the ongoing environmental changes. Consequently, the international efforts in polar research have been enhanced considerably, and a wealth of new findings is being produced at a growing rate by the international community of polar researchers.

Springer Polar Sciences aims to present a broad platform that will include state-of-the-art research, bringing together both science and humanities to facilitate an exchange of knowledge between the various polar science communities. The Series offers an outlet to publish contributions, monographs, edited works, conference proceedings, etc. Topics and perspectives will be broad and will include, but not be limited to: climate change impacts, environmental change, polar ecology, governance, health, economics, indigenous populations, tourism and resource extraction activities. Books published in the series will appeal to scientists, students, polar researchers and policy makers.

More information about this series at http://www.springer.com/series/15180

Susana Castro-Sowinski
Editor

The Ecological Role of Micro-organisms in the Antarctic Environment

Springer

Editor
Susana Castro-Sowinski
Biochemistry and Molecular Biology, Faculty of Sciences
Universidad de la República
Montevideo, Uruguay

ISSN 2510-0475 ISSN 2510-0483 (electronic)
Springer Polar Sciences
ISBN 978-3-030-02785-8 ISBN 978-3-030-02786-5 (eBook)
https://doi.org/10.1007/978-3-030-02786-5

Library of Congress Control Number: 2018967764

This Springer imprint is published by the registered company Springer Nature Switzerland AG
The registered company address is: Gewerbestrasse 11, 6330 Cham, Switzerland

Preface

Antarctica, officially a desert and technically the largest reservoir of freshwater, is the world's most arid place, with very low humidity and temperature, strong winds, and UV irradiation, and an example of an oligotrophic environment. Among the seven continents, Antarctica is the least populous one (up to 4000 researchers and personnel during summer and 1000 in the winter), but it is in fifth place when ranked by size (ca. 5,500,000 mi^2, 14,245,000 km^2), followed by Europe (3,997,929 mi^2, 10,354,636 km^2) and Australia/Oceania (2,967,909 mi^2, 7,686,884 km^2). This continent is a great white silence place (*The Great White Silence*, a documentary from the filmmaker Herbert Ponting that recorded an expedition in Terra Nova, 1910–1913) protected by the Antarctic Treaty System (opened for signature in Washington, on December 1, 1959; https://www.ats.aq/e/ats.htm) that established that this is a place of international cooperation, peace, and scientific finding.

Despite the harsh environmental conditions found in this continent, molecular phylogenetic and physiological surveys have revealed a vast array of microbial groups within. Antarctica has been considered as a pristine environment, but it continuously receives microbial intruders, a fact mainly supported by the high frequency of apparently cosmopolitan species found in snow, ice, and geothermal sites, among others. These microorganisms (native and nonnative ones) have major roles in this oligotrophic environment. They contribute to nutrient cycling and manage food webs and water/soil/air quality, among others. As a whole, they have a role as regulators of ecosystem processes.

This book volume deals with the description of microbial groups and the environments that they inhabit, with emphasis in their physiological traits and the strategies they use to shape the Antarctic environment. Through the different book chapters, the readers will find the description of microbes in terms of their physiology and metabolism and their ecological roles in the Antarctic environment. The role of bacteria/cyanobacteria, fungi, yeasts, and viruses on different ecosystems, such as rocks, water, soil, plants, glacial environments, and animals, has been introduced in a didactic way. However, a few chapters describe the Antarctic microbes in terms of potential genetic source for the development of biotechnological products.

Contributions from researchers from Argentina, Brazil, Germany, Italy, Japan, New Zealand, Singapore, South Africa, the United States, and Uruguay have shaped this book. These countries are representing six continents. But, what do they have in common? They have permanent or summer working scientific Antarctic stations that contribute to the knowledge on the Antarctic environment.

Montevideo, Uruguay Susana Castro-Sowinski

About the Book

This book describes how microbes face the harsh conditions found in Antarctica and how they colonize all possible niches. Their presence, abundance, and physiology, with a focus on their roles in the Antarctic environment, is overviewed in this book. A multidisciplinary picture of the microbial diversity in a changing world with a focus on the environmental sustainability of Antarctica has been captured. Through the different book chapters, the readers will find the description of the role of the microbial communities found in Antarctic soil, water, air, plants, rocks and invertebrates, and how they shape the biogeochemical cycles of carbon and nitrogen.

Contents

Contributors

Diego Alem Biochemistry and Molecular Biology, Faculty of Sciences, Universidad de la República, Montevideo, Uruguay

Epigenetics and Genomics Instability Laboratory, Instituto de Investigaciones Biológicas Clemente Estable, Montevideo, Uruguay

Vanesa Amarelle Department of Microbial Biochemistry and Genomics, Biological Research Institute Clemente Estable, Montevideo, Uruguay

Stephen David James Archer Institute for Applied Ecology New Zealand, Auckland University of Technology, Auckland, New Zealand

Yale-NUS College, National University of Singapore, Singapore, Singapore

Maurizio Azzaro Institute for the Biological Resources and Marine Biotechnologies, National Research Council, U.O.S. Messina, Messina, Italy

Gastón Azziz Unidad Microbiología Molecular, Instituto de Investigaciones Biológicas Clemente Estable, Montevideo, Uruguay

Silvia Batista Unidad Microbiología Molecular, Instituto de Investigaciones Biológicas Clemente Estable, Montevideo, Uruguay

Valentina Carrasco Department of Microbial Biochemistry and Genomics, Biological Research Institute Clemente Estable, Montevideo, Uruguay

Susana Castro-Sowinski Biochemistry and Molecular Biology, Faculty of Sciences, Universidad de la República, Montevideo, Uruguay

Ivana Cavello Research and Development Center for Industrial Fermentations, CINDEFI (CONICET-La Plata, UNLP), La Plata, Argentina

María Fernanda Cerdá Laboratory of Biomaterials, Faculty of Sciences, Universidad de la República, Montevideo, Uruguay

Don A. Cowan Centre for Microbial Ecology and Genomics, Department of Biochemistry, Genetics and Microbiology, University of Pretoria, Pretoria, South Africa

Juan Cristina Laboratorio de Virología Molecular, CIN, Facultad de Ciencias, Universidad de la República, Montevideo, Uruguay

Maria T. de Paula Departamento de Microbiologia, Universidade Federal de Minas Gerais, Belo Horizonte, MG, Brazil

Elena Fabiano Department of Microbial Biochemistry and Genomics, Biological Research Institute Clemente Estable, Montevideo, Uruguay

Cecilia G. Flocco Julius Kühn-Institut (JKI), Federal Research Centre for Cultivated Plants, Braunschweig, Germany

Instituto Antártico Argentino (IAA), Buenos Aires, Argentina

Instituto de Nanobiotecnología (NANOBIOTEC), Universidad de Buenos Aires-Consejo Nacional de Investigaciones Científicas y Técnicas (CONICET), Buenos Aires, Argentina

César X. García-Laviña Biochemistry and Molecular Biology, Faculty of Sciences, Universidad de la República, Montevideo, Uruguay

Gabriela Garmendia Laboratory of Biotechnology, Microbiology Area, Department of Bioscience, Faculty of Chemistry, Universidad de la República, Montevideo, Uruguay

Paul R. Gill Nevada City Biolabs, Nevada City, CA, USA

Matías Giménez Unidad Microbiología Molecular, Instituto de Investigaciones Biológicas Clemente Estable, Montevideo, Uruguay

Satoshi Imura National Institute of Polar Research (NIPR), Tachikawa, Tokyo, Japan

Department of Polar Science, SOKENDAI (The Graduate University for Advanced Studies), Tachikawa, Tokyo, Japan

Katie King-Miaow Institute for Applied Ecology New Zealand, Auckland University of Technology, Auckland, New Zealand

Donnabella LaCap-Bugler Institute for Applied Ecology New Zealand, Auckland University of Technology, Auckland, New Zealand

Kevin Lee Institute for Applied Ecology New Zealand, Auckland University of Technology, Auckland, New Zealand

Angelina Lo Giudice Institute for the Biological Resources and Marine Biotechnologies, National Research Council, U.O.S. Messina, Messina, Italy

Walter P. Mac Cormack Instituto Antártico Argentino (IAA), Buenos Aires, Argentina

Instituto de Nanobiotecnología (NANOBIOTEC), Universidad de Buenos Aires-Consejo Nacional de Investigaciones Científicas y Técnicas (CONICET), Buenos Aires, Argentina

Teruya Maki Department of Chemical Engineering, Kanazawa University, Kanazawa, Japan

Juan José Marizcurrena Biochemistry and Molecular Biology, Faculty of Sciences, Universidad de la República, Montevideo, Uruguay

Adalgisa Martínez-Silveira Laboratory of Biotechnology, Microbiology Area, Department of Bioscience, Faculty of Chemistry, Universidad de la República, Montevideo, Uruguay

Takeshi Naganuma Graduate School of Biosphere Science, Hiroshima University, Higashihiroshima, Japan

Ryosuke Nakai Applied Molecular Microbiology Research Group, Bioproduction Research Institute, National Institute of Advanced Industrial Science and Technology (AIST), Sapporo, Japan

Mayara B. Ogaki Departamento de Microbiologia, Universidade Federal de Minas Gerais, Belo Horizonte, MG, Brazil

Franciane M. Pellizzari Department of Biological Sciences, Paraná State University, Paranaguá, Brazil

Luiz H. Rosa Departamento de Microbiologia, Universidade Federal de Minas Gerais, Belo Horizonte, MG, Brazil

Alexandre Soares Rosado Laboratory of Molecular Microbial Ecology, Institute of Microbiology Paulo de Góes, Federal University of Rio de Janeiro, Rio de Janeiro, Brazil

Daniele Ruas Departamento de Microbiologia, Universidade Federal de Minas Gerais, Belo Horizonte, MG, Brazil

Stefano Schiaparelli Department of Earth, Environment and Life Sciences, University of Genoa, Genoa, Italy

Italian National Antarctic Museum (MNA, Section of Genoa), University of Genoa, Genoa, Italy

Júnia Schultz Laboratory of Molecular Microbial Ecology, Institute of Microbiology Paulo de Góes, Federal University of Rio de Janeiro, Rio de Janeiro, Brazil

Kornelia Smalla Julius Kühn-Institut (JKI), Federal Research Centre for Cultivated Plants, Braunschweig, Germany

Marc W. Van Goethem Centre for Microbial Ecology and Genomics, Department of Biochemistry, Genetics and Microbiology, University of Pretoria, Pretoria, South Africa

Silvana Vero Laboratory of Biotechnology, Microbiology Area, Department of Bioscience, Faculty of Chemistry, Universidad de la República, Montevideo, Uruguay

Michael Wisniewski USDA-ARS Appalachian Fruit Research Station, Kearneysville, WV, USA

About the Editor

Dr. Susana Castro-Sowinski is Ph.D. in Chemistry and has completed her academic training at the Department of Plant Pathology and Microbiology, Robert Smith Faculty of Agriculture, Food and Environment in Rehovot, Israel. Currently, she is Associate Professor at the Biochemistry and Molecular Biology Section, Faculty of Sciences, Universidad de la República, in Montevideo, Uruguay, and member of the Researcher National System (SNI), the Program for Developing Basic Sciences (PEDECIBA) and the Editorial Board of Symbiosis (Springer) and Environmental Sustainability (Springer). Her specific area of research has been the molecular-plant microbe interaction, but currently, she is exploring a new area of research, focusing her work on the microbial ecology of Antarctic systems and the development of biotechnological products using the genetic material of Antarctic microorganisms.

Part I
The Role of Microorganisms in the Biogeochemical Cycles in Antarctica

Chapter 1
Role of *Cyanobacteria* in the Ecology of Polar Environments

Marc W. Van Goethem and Don A. Cowan

Abstract *Cyanobacteria* are the dominant living features of Antarctic terrestrial environments. They have the capacity to directly influence components of the cryosphere including nutrient acquisition, soil stabilisation and driving soil community structure. This book chapter incorporates recent literature to discuss how gradients of cyanobacterial abundance and diversity across Antarctic soil and lithic biotopes influence local biogeochemical cycling regimes, drive community structure and enhance primary productivity. Most recent studies have gleaned novel insights into the ecological importance of Antarctic cyanobacteria by applying so-called multi-'omics' technologies. While these breakthroughs have undoubtedly improved our understanding of metabolic potential in polar niches; cultivation-based analyses of cyanobacteria should be leveraged to gain perspectives into actual physiological attributes and morphological variation within Antarctica. Combined, these studies show that members of the cyanobacteria are critical carbon and nitrogen regulators and are essential for making nutrients available to associated community members.

Keywords Cyanobacteria · Carbon biogeochemical cycling · Nitrogen biogeochemical cycling · Primary production · Interactions in cryptic niches

1.1 Introduction

Some of the most dramatic increases in surface air temperature in the Southern Hemisphere have been recorded on the Antarctic continent (Turner et al. 2005). The effects of warming across the continent are thought to be exacerbated by changes to numerous connected processes including ozone depletion and the loss of local sea ice (Turner et al. 2016). Combined, these effects have imparted substantial changes

M. W. Van Goethem · D. A. Cowan (✉)
Centre for Microbial Ecology and Genomics, Department of Biochemistry, Genetics and Microbiology, University of Pretoria, Pretoria, South Africa
e-mail: Don.Cowan@up.ac.za

S. Castro-Sowinski (ed.), *The Ecological Role of Micro-organisms in the Antarctic Environment*, Springer Polar Sciences,
https://doi.org/10.1007/978-3-030-02786-5_1

to regional geomorphology and biodiversity; leading to both the 'greening' of previously arid soils and glacial front retreat in the Antarctic Peninsula (Cook et al. 2005). Moreover, anticipated changes in water distribution across the continent will impact both soil carbon and nitrogen cycling through associated perturbations in soil properties such as freeze-thaw cycling and soil moisture content (Cowan and Ah Tow 2004). Climate models have predicted increases in the average surface temperature by as much as 4.8 °C by the year 2100 (Kirtman et al. 2013), and these rates of soil warming will cause dramatic changes to the availability of water across continental and maritime Antarctic soil ecosystems (Bromwich et al. 2013).

Climate warming will lead to changes in sea-ice cover through increases in ocean temperatures (Bintanja et al. 2013) with associated variations in large-scale weather circulation patterns in the Southern Hemisphere (Fyfe and Saenko 2006). A warmer atmosphere associated with increased near-surface air temperatures will undoubtedly lead to higher rates of precipitation, which has the potential to add 30–70 mm sea level equivalent by 2100 (Ligtenberg et al. 2013). Due to the enhanced poleward transport of moisture, there is great potential for change to regional climate in continental Antarctica. The effects of such climatic changes on Antarctic soil microbial biodiversity remain unclear even though recent studies have highlighted the rapid responses of soil communities to local wetting events (Niederberger et al. 2012, 2015a, b) and temperature variations (de Scally et al. 2016; Benhua et al. 2014). Importantly, any observed alterations to species composition or membership may not reflect variations in community activity (Benhua et al. 2014), which is often dependent on the degree to which conditions are altered (Hopkins et al. 2006).

Even though the combined effects of anticipated temperature changes remain unclear, substantial changes to microbial diversity and functionality are predicted across many continental Antarctic cryptic and refuge niches, including hypoliths (Cowan et al. 2010; Makhalanyane et al. 2013a; De los Ríos et al. 2014a), endoliths (De la Torre et al. 2003; Yung et al. 2014; Wei et al. 2015a), surface soils (Niederberger et al. 2015a; van Horn et al. 2013; Makhalanyane et al. 2016) and microbial mat communities (Stal 1995; Varin et al. 2012; Zhang et al. 2015; Van Goethem et al. 2016). *Cyanobacteria* serve as keystone taxa (also referred to as 'cornerstone' taxa (Mills et al. 1993)) for most of these cryptic Antarctic niches as their contributions to local nutrient cycling typically exceeds their anticipated importance inferred from relative abundance data (Makhalanyane et al. 2015). The unique responses of cyanobacterial lineages to altered temperature and moisture regimes are unknown despite recent advances in our understanding of the ecology and physiology of the cyanobacteria across the Antarctic continent (Makhalanyane et al. 2015). These advances are driven by the application of shotgun metagenomic sequencing to resolve questions relating to cyanobacterial physiology and metabolism (Cowan et al. 2015). Gauging how cyanobacteria alter their functional traits in response to increasing temperatures, and concomitant upshifts in moisture availability and nutrient mobilisation from deep soil, will be fundamental to improving our understanding of Antarctic ecology within the context of future variations in climate. In addition, further research is required to investigate how altered environmental conditions will result in changes to biogeochemical cycling strategies and the mechanisms through which cyanobacteria share resources with associated microorganisms.

The phylum *Cyanobacteria* comprises a wide diversity of functionally and morphologically distinct Gram-negative bacteria (Paerl et al. 2000). *Cyanobacteria* are central regulators of both carbon and nitrogen cycling in depauperate environments and can be viewed as 'ecosystem engineers' (Chrismas et al. 2018). The contributions of this key bacterial phylum to biogeochemical cycling, and particularly Carbon (C) and nitrogen (N), are exemplified in diverse cold desert ecosystems, including the Tibetan Plateau soils (Wong et al. 2010) and Arctic permafrost (Hultman et al. 2015), where soil microbial communities mediate core functional processes relating to soil nutrient turnover (Barrow 1992; Delgado-Baquerizo et al. 2018). In Antarctica, the absence of vascular plants places more importance on the cyanobacteria in shaping the ecology of the continent at local scales (Cary et al. 2010), and on the key functional roles mediated by soil-based photoautotrophs.

The unique physiologies used by the cyanobacteria to drive nutrient cycling are beginning to be understood through improved sequencing technologies and direct observations, and those novel insights are collated and reviewed in this chapter. We also explore how recent applications of modern technologies have bridged gaps in our understanding of this critical functional guild and discuss the potential for novel methodologies to contribute to our understanding of *Cyanobacteria* in cryptic Antarctic niches.

1.2 Physiological Adaptations for Nutrient Cycling in Polar Environments

Cyanobacteria are thought to have influenced the Earth's biosphere for at least the last 2.5 billion years (Gya) (Schirrmeister et al. 2016) and may have existed for more than 2.9 Gya (Blank and Sanchez-Baracaldo 2010). Contemporary and historical surveys of continental Antarctica supply strong evidence that *Cyanobacteria* are central to nutrient cycling processes (Friedmann and Ocampo 1976; Friedmann et al. 1988; Cowan et al. 2011a) and are essential for numerous ecosystem processes (Wynn-Williams 1996). Members of the cyanobacteria are thought to be critical providers of ecosystem services in Antarctica (Wynn-Williams 1996) and are key determinants in guiding microbial community structure in cold, nutrient-starved soils (Cary et al. 2010). Recent research has expanded our understanding of the contributions of cyanobacteria to nutrient cycling in Antarctica by coupling measurements and observations of nutrient and energy fluxes, to assessments of the physiological underpinnings of these processes by applying metagenomic and genomic sequencing techniques.

Regional warming in Antarctica has the potential to increase water availability through ice melt which will in turn mobilise soil nutrients from deep soil layers

(Cowan and Ah Tow 2004). These and other perturbations, such as prolonged freeze-thaw events, may impart substantial changes on Antarctic microbial communities. The extent to which these anticipated changes will alter cyanobacterial functionality is currently unknown, owing, in part, to contradictory results that have suggested either an increase in microbial activity (Benhua et al. 2014) or a reduction in functionality through keystone species loss resulting from soil warming (Philippot et al. 2013). Regardless of the direct implications of warming on Antarctic soil community composition, consolidating direct observations with metagenomic (and genomic) evidence of cyanobacterial functionality and physiology in continental Antarctica will be essential to predict how climatic variations will impact regional biogeochemical cycling rates.

Incremental rises in temperature and water availability are thought to increase soil microbial activity (Benhua et al. 2014), while the release of stored soil carbon may drive increases in microbial abundance (Chen et al. 2015). *Cyanobacteria* are known to use hydrotaxis in response to moisture events in deserts (Pringault and Garcia-Pichel 2004) and seemingly contribute to soil moisture retention in soil (Belnap and Gardner 1993). Their mobility and increased activity resulting from exposure to moisture will likely have profound impacts on the rates of nitrogen fixation and carbon cycling at local scales.

Investigations into microbial physiological traits is tractable in Antarctica due to the limited biodiversity of microorganisms and microinvertebrates which is the result of extreme environmental stressors, including sub-zero temperatures, prolonged periods of darkness and hyperoligotropy (Pointing et al. 2009). The microbiology of Antarctica seemingly departs from latitudinal gradients of species richness, with biodiversity instead exhibiting very strong endemic signals (Bahl et al. 2011) linked to cryptic niche availability (Makhalanyane et al. 2015; Pointing et al. 2014) and the availability of photosynthetically active radiation (Zhang et al. 2015). Cyanobacteria harbour numerous physiological adaptations to contend with climatic stress and are capable of colonising, adapting and growing in broad environmental ranges across geographical space and time. The Antarctic cryosphere is thus a model system for studies aimed at exploring the diverse and multifaceted interactions between cyanobacteria and associated microorganisms (heterotrophic bacteria, archaea, fungi, microinvertebrates) that shapes their ecology at the cold limits of life.

1.2.1 Nitrogen Cycling

The nitrogen content of Antarctic soils is typically very low and is often at the accurate detection limits (Cowan and Ah Tow 2004). The nitrate concentrations (NO_3^--N) of Antarctic mineral soils are driven by atmospheric precipitation (Wada et al. 1981) and are the lowest of any terrestrial soil system – ranging from 0 to 1250 μg g^{-1} (Vishniac 1993; Hopkins et al. 2009). Consequently, nitrogen scavenging mechanisms and microbe-driven nitrogen input into Antarctic soils are essential strategies for contending with hyperoligotropy. N_2 fixation offers a viable mechanism for directly supplying the nitrogen-based molecules required for life.

Nitrogen fixation, also referred to as diazotrophy, is the key elemental conversion step of the nitrogen cycle. Essentially inert atmospheric N_2 gas is reduced into ammonia (NH_3) or more commonly into ammonium in neutral pH soils (NH_4^+). This irreversible reaction is mediated by free-living, aerobic *Cyanobacteria* that produce the nitrogenase enzyme complex (Burgess and Lowe 1996). In the presence of oxygen, nitrogenase activity is inhibited (Howard and Rees 1996). Heterocystous (nitrogen-fixing) cyanobacteria overcome this limitation by physically separating nitrogen fixation and oxygen production through the production of heterocysts; specialised cells formed during nitrogen starvation (Bergman et al. 1997). By contrast, non-heterocystous *Cyanobacteria* decouple oxygen production and nitrogen fixation temporally, through manipulations of their circadian rhythm (Golden et al. 1998). Not all cyanobacteria are N_2 fixers as the high energetic cost of the process (16 ATPs per molecule of fixed N_2 (Scherer et al. 1988)) and evidence of repeated independent nitrogenase gene (*nifH*) loss from cyanobacterial genomes (Latysheva et al. 2012) suggests that the nitrogen fixation is not always the most viable mechanism of nitrogen acquisition from the environment.

Heterocystous cyanobacteria, predominantly *Nostoc commune*, have the capacity to drive nitrogen fixation in Antarctic soils and lithic niches (Makhalanyane et al. 2014). Recently, maritime-specific N_2-fixing cyanobacteria on Ulu Peninsula, James Ross Island, have been identified using traditional culturing approaches (Komárek et al. 2015). Using a polyphasic approach, 11 species of cyanobacteria were identified, of which 4 were novel. These appear to be dominant and endemic components of the microflora in the Peninsula as most of the species occur only on Antarctic islands and not on the continent. Although *Nostoc commune* isolates were the dominant soil colonists, members of the *Calothrix*, *Dichothrix*, *Nodularia* and *Hydrocoryne* were shown to be important and potentially novel heterocystous cyanobacteria capable of fixing elemental nitrogen (Komárek et al. 2015). In addition to physical observations of nitrogen fixation, acetylene (C_2H_2) reduction assays are a useful proxy for measuring in situ nitrogenase activity (Cowan et al. 2011a), while metagenomic surveys targeting the key nitrogenase gene *nifH* also provide insights into the potential for communities to transform atmospheric N_2 into organic substrates (Hutchins and Miller 2017).

Diazotrophy is also an important biogeochemical process in the McMurdo Dry Valleys of Eastern Antarctica. Hypolithic communities, cryptic assemblages below translucent rocks, have been shown to fix atmospheric nitrogen at rates between 0.02 and 0.174 nmol N g^{-1} h^{-1} in Miers Valley (Cowan et al. 2011a), while ephemerally wetted mineral soils from the same region show very high in situ N_2 fixation rates ranging from 0.04 to 5.8 nmol N cm^{-3} h^{-1} compared to arid soils that did not receive an exogenous moisture input (0–0.5 nmol N cm^{-3} h^{-1}) (Niederberger et al. 2012). Here, the availability of moisture has a profound impact on both microbial community composition and functionality (Niederberger et al. 2015a). The low-lying Miers Valley is hyperoligotrophic (very low N and C concentration), and discrete patches of microbial activity, such as hypolithic communities, may offer a viable strategy for maintaining regional nutrient budgets when moisture availability is low.

Functional gene surveys using the universal *nifH* gene primers, PolF and PolR (Poly et al. 2001), and shotgun metagenomic sequencing have highlighted the prevalence of nitrogenase genes across distinct Antarctic niches. Dry Valley endoliths, microbial communities within rocks, and hypoliths have been shown to be richer in *nifH* gene signatures than exposed desert soils (Chan et al. 2013; Wei et al. 2015b), which correlates strongly with the higher abundance of heterocystous cyanobacteria in cryptic niche communities compared to surface soils (Van Goethem et al. 2016; Cowan et al. 2015; Chan et al. 2012). This is broadly consistent with the producer to consumer ratio of hypoliths (P:C = 3.28) compared to soil (P:C = 1.04) (Wei et al. 2015b) in which primary producers (phototrophs) greatly outnumber heterotrophs. Hypoliths are typically dominated by *Cyanobacteria* (Cockell and Stokes 2004) with *Synechococcus*-like taxa dominant across the Dry Valleys and *Microcoleus*, *Phormidium* and *Oscillatoria* species less common (Wei et al. 2015b).

Cyanobacterial key genes involved in downstream components of the nitrogen cycle have also been found. The potential for cyanobacteria to perform nitrification (the conversion of ammonium (NH_4^+) to nitrate (NO_2^-) through oxidation) has been shown in multiple metagenomic studies (Wei et al. 2015a; Chan et al. 2013). Ammonia-oxidising taxa such as *Nitrosospira*- and *Nitrosomonas*-like phylotypes have also been implicated in driving nitrification in continental Antarctic soils from Upper Wright Valley and Battleship Promontory (Magalhães et al. 2014). A recent microarray-based study has indicated that microbial communities in hypoliths and endoliths from McKelvey Valley are important sources of nitrogen input in hyperoligotrophic niches, with nitrification primarily driven by *Cyanobacteria* and denitrification (the reduction of nitrate to N_2 gas) mediated by *Deltaproteobacteria* and *Bacteroidetes* (Chan et al. 2013). Combined, these results support the concept that cooperativeness between cyanobacteria and associated functional microorganisms is essential to complete the nitrogen cycle in nutrient-poor Antarctic habitats.

Complete cyanobacterial genomes of cultured *Chroococcidiopsis* spp. from Victoria Valley hypoliths have revealed the capacity for Antarctic cyanobacteria to transport nitrate, reduce nitrite (*nit*) and assimilate ammonia, possibly by coupling glutamate synthetase with NADPH oxidation (MWVG, unpublished genomic data). However, both *Chroococcidiopsis* genomes lacked the key *nifH* gene for nitrogen fixation, despite reported nitrogenase activity for members of the genus found in endolithic communities (Boison et al. 2004; Banerjee and Verma 2009). *Cyanobacteria* from the Antarctic Peninsula showed a strong correlation with *norB* genes (Yergeau et al. 2009) which encode the enzyme nitric oxide reductase that catalyses the reduction of nitric oxide (NO) to nitrous oxide (N_2O). Non-denitrifying cyanobacteria such as *Synechocystis* have been shown to harbour the *norB* gene within their genomic architecture (Büsch et al. 2002). Moreover, microarray data show that *Proteobacteria* and *Chloroflexi* have relationships in principal coordinate analyses with *nifH*, *amoA* (ammonia monooxygenase) and *narGH* (nitrate reductase) genes, indicating that distinct microbial guilds are required to complete different components of the nitrogen cycle (Yergeau et al. 2009).

1.2.2 Carbon Cycling

Antarctic permafrost, surface and subsurface soils are characterised by extremely low nutrient availability compared to other desert soils and temperate soil ecosystems (Fierer et al. 2012). Organic carbon content ranges from between 0.02% and 0.16% across the Dry Valleys, which is orders of magnitude lower than rainforests and arctic tundra soils (Fierer et al. 2012). The paucity of organic carbon in Antarctic soils, coupled with extremely low temperatures, places constraints on microbial metabolism, growth rates and biogeochemical cycling regimes. Cyanobacteria are important sources of carbon input to soils and lithic niche ecosystems as higher photoautotrophs are absent owing to the extreme environmental conditions and the absence of water (Cameron et al. 1970).

Carbon fixation is a critical process in hyperoligotrophic soils as carbon scarcity challenges microbial metabolism. The direct acquisition of inorganic carbon from the environment is central to providing the molecular building blocks for cellular growth and metabolism (i.e. the organic biosphere). Members of the cyanobacteria are often the sole primary producers in Antarctic soils, and their contribution to carbon fixation and the production of complex sugars cannot be understated. *Cyanobacteria* prefer CO_2 as their primary carbon source, and exclusively utilise the photosynthetic carbon reduction cycle (also known as the Calvin-Benson-Bassham cycle, or simply the Calvin cycle) to metabolise and reduce atmospheric carbon dioxide through either photo- or chemoheterotrophic metabolism. That is, phototrophic metabolism is initiated when cyanobacteria are irradiated either by white light or ultraviolet B (UV-B) radiation (Huang et al. 2002), while chemoheterotrophic metabolism proceeds in low-light conditions.

Antarctic niches are essentially exposed to 24-h of sunlight during the austral summer, and 24 h of darkness throughout the winter months (Cowan et al. 2014). Although unilluminated niches such as permafrost and subsurface soils populate the Antarctic continent (see Romanovsky, et al. 2010 for a recent review on Northern Hemisphere permafrost), they have not been extensively studied within the context of cyanobacterial physiology. Consequently, we have focussed this chapter on the illuminated soils, lithic niches and cryptogamic mat consortia that occur across Antarctica and have been the subject of extensive polyphasic characterisation studies.

Metagenomic surveys have revealed the distribution of genes encoding phosphoribulokinase (*prk*) in Antarctic soils and lithic niches (Fierer et al. 2012) as well as in marine Antarctic ecosystems (Williams et al. 2013). The phosphoribulokinase enzyme catalyses the final reaction in the Calvin cycle and converts D-ribulose 5-phosphate to D-ribulose 1,5-bisphosphate (RuBP) using ATP as phosphoryl group donor. The regenerated RuBP can then be used as an electron acceptor that undergoes carboxylation by ribulose bisphosphate carboxylase/oxygenase (RuBisCO) which leads to the fixation of atmospheric CO_2. *Prk* genes appear to be common features of members of the order *Nostocales*, with both *Chlorogloeopsis fritschii* and *Anabaena cylindrica* encoding the complete phosphoribulokinase gene in their

Fig. 1.1 Cryoconite holes in Antarctic glacial ice. (Photo, Jarishma Gokul, with permission)

genomes (Marsden and Codd 1984; Serra et al. 1989), while *Synechococcus* sp. PCC 7942 (Synechoccales) also harbours the *prk* gene (Kobayashi et al. 2003). Similarly, *Nostocales* lineages from the Sør Rondane Mountains, Dronning Maud Land, were identified as primary producers based on clone libraries of RuBisCO genes (*cbbLM*) (Tahon et al. 2016).

The CO_2-fixing enzyme RuBisCO is spatially segregated from the phosphoribulokinase enzyme which occurs in the cytoplasm (Marsden et al. 1984). Within subcellular carboxysomes, RuBisCO facilitates the incorporation of CO_2 into organic molecules via the Calvin-Benson-Bassham (CBB) cycle. The RuBisCO enzyme comprises two protein subunits, namely, the large chain (*rbcL*) and the small chain (*rbcS*). The active substrate binding sites for the substrate RuBP are located on the large chain of the enzyme. Chemosynthetic carbon fixation is supported in Robison Ridge (Wilkes Land) and Adams Flat soils where *Actinobacteria* dominate and possess type IE RuBisCO genes (*rbcL1E*) (Ji et al. 2017). Here soils characterised as high *Actinobacteria*-low *Cyanobacteria* communities use nutrient scavenging strategies to contend with low nutrient status in the absence of cyanobacterial photoautotrophs. Mechanisms of utilising dependable energy sources from the atmosphere, such as H_2 gas and carbon monoxide, may be more widespread than previously appreciated (Cowan and Makhalanyane 2017) and have implications for microbial survival and energy acquisition in Antarctic soils from which cyanobacteria are absent or inactive.

Cyanobacteria found in cryoconite holes (Fig. 1.1) in glaciers and ice sheets have been shown to sequester inorganic carbon dioxide and are essential components of the microbial food chain by driving in situ primary production and respiration (Anesio et al. 2009; Stibal et al. 2012). Microorganisms aggregate on Antarctic glaciers and ice sheet surfaces, which enhance local supraglacial ice melt and the production of glacial streams (Fortner et al. 2005). The enhanced water bioavailability, coupled with organic carbon break down and CO_2 fixation, has implications for biogeochemical cycling in downstream ecosystems such as hyperarid surface soils (Niederberger et al. 2012, 2015a). The processes involved in primary productivity may be susceptible to changes in available moisture content, as relative cyanobacterial abundance (GpI and GpIV) has been shown to increase with wetting in Miers Valley (Niederberger et al. 2015b), while carbon fixation also increased from ~12 nmol C/cc/h to up 750 nmol C/cc/h in wetter soils (Niederberger et al. 2015a).

1.2.3 Primary Production

While metagenomic studies have elucidated the prevalence of *rbcLS* genes across Antarctic soils, and more predominantly in lithic communities, direct measurements of primary productivity have shown the importance of cyanobacteria in driving carbon accumulation (Novis et al. 2007). Direct measurements of CO_2 fixation have shown that cryptoendolithic cyanobacteria photosynthesise only at high water potentials (>−6.9 MPa; 90% relative humidity) (Palmer and Friedmann 1990). Within endoliths, snowfall permeates porous sandstone rocks with water which is retained in the rock during dry seasons, thus supporting microbial metabolism and biomass production (Palmer and Friedmann 1990).

Hypoliths from the Dry Valleys can be as productive as nearby lichens, bryophytes and plants (Thomas and Nielsen 2005). Here, a sufficient quantity of light penetrates translucent quartz rocks to support photosynthetic carbon assimilation (Cockell and Stokes 2004). Net primary productivity (NPP) yields, defined as the difference between photosynthesis and respiration, of 1–20 g C m^{-2} $year^{-1}$ can occur at 0.1% irradiance of incident light, temperatures less than −30 °C and with minimal water availability (Novis et al. 2007; Thomas and Nielsen 2005; Schlesinger et al. 2003). Consistent with these very low estimates of cyanobacteria-driven productivity, carbon assimilation in the Dry Valleys is thought to occur over centuries or millennia (Matsumoto et al. 1990). For example, radiocarbon dating estimates of organic matter in endolithic horizons from the coastal Larsemann Hills (Prydz Bay) can reach 480 ± 25 years (Mergelov et al. 2012).

Despite extremely low rates of C-fixation by photoautotrophic community members, the contributions by cyanobacteria to primary productivity remain important sources of organic matter for associated heterotrophic bacteria (Pointing et al. 2009), and are central food sources for grazing nematodes and protozoans, and thus are the basis for survival of whole ecosystems in these extreme polar environments (Freckman and Virginia 1997). Interestingly, Victoria Land microinvertebrate communities are thought to be the least diverse on Earth, with fewer than five species identified (Adams et al. 2006). Nematodes are thought to graze on heterotrophic bacteria and protozoan consumers that had acquired organic matter directly from cyanobacteria (endogenous inputs from photoautotrophs) or from exogenous inputs (i.e. lacustrine sediment deposition) (Cary et al. 2010). The aeolian redistribution of cyanobacterial mats can serve as contemporary sources of carbon input in the Dry Valleys (Hopkins et al. 2009). This process also has the potential to inoculate surface soils with cyanobacterial cells (Wood et al. 2008) that may ultimately stimulate local primary productivity (Geyer et al. 2017).

In cyanobacteria-poor soils (average relative abundance of 0.28%), oxygenic photosynthesis capacity attributed to cyanobacterial lineages has still been documented (Ji et al. 2017), even though the majority of primary productivity could be attributed to atmospheric H_2, CO_2 and CO consumption by the phyla *Actinobacteria*, AD3 and WPS-2. Here, microbial biomass production can be envisaged by H_2-driven CO_2 fixation in microorganisms containing group 1 h [NiFe]-hydrogenase genes as well as type IE RuBisCO genes (Ji et al. 2017).

1.3 Cyanobacterial Interactions in Cryptic Niches

1.3.1 Hypoliths

Hypolithic communities colonise the undersides of quartz stones and other transparent rocks (Fig. 1.2) where they experience dramatically improved climatic conditions compared to exposed soils (Pointing 2016). Notably, the hypolithic community, also referred to as a hypolithon, is protected from physical wind abrasion, extreme temperature fluctuations and intense ultraviolet irradiation receipt (Cowan et al. 2011b), while moisture is trapped under the rock and provides bioavailable moisture

Fig. 1.2 Antarctic quartz hypolithic community. (Photo, Leo Sancho, with permission)

to the hypolithon (Cowan and Ah Tow 2004). These refuge niches are common features of the Dry Valley landscape in Eastern Antarctica (Cowan et al. 2010; De Los Ríos et al. 2014b) and have also been reported in maritime locations of the continent, including Mars Oasis on Alexander Island (Cockell and Stokes 2004).

Cyanobacteria-dominated hypoliths are important reservoirs of microorganisms in Antarctica (Cowan et al. 2011b) and serve as 'hotspots' of microbial activity and nutrient acquisition in drylands across the globe (Chan et al. 2012). In the Antarctic Dry Valleys, three distinct hypolithic morphotypes have been identified (Cowan et al. 2010) and characterised as successional stages of microbial community development (Makhalanyane et al. 2013a). Type I hypoliths are dominated by cyanobacterial lineages and serve as the basal community stage that recruits bacterial members from surrounding soils (Makhalanyane et al. 2013b). Type I hypoliths typically comprise cyanobacteria that co-occur with heterotrophic bacteria and archaea and develop into Type II hypolithic communities after fungal species are recruited. Fungal species dominate Type II hypoliths and subsequently develop into Type III hypoliths when moss-phylotypes become the dominant community members (Makhalanyane et al. 2013a).

The development of hypoliths hinges upon the salinity and porosity of soil beneath the quartz rock (Pointing et al. 2009) and also on rock transmittance (Cowan et al. 2011b). For example, the rock transmittance of Type I hypoliths is 50% lower than that of fungal- and moss-dominated hypoliths (Cowan et al. 2011b), suggesting that polar *Cyanobacteria* have mechanisms of tolerating low-light conditions (Laybourn-Parry et al. 2012). A 16S rRNA gene-based analysis of lithobiontic niches from Victoria Valley showed that *Leptolyngbya* and *Phormidium* (both *Oscillatoriales*) were the dominant cyanobacterial members of Type I hypoliths, although, unusually, *Actinobacteria* were the most abundant bacterial members in the hypolithon (Van Goethem et al. 2016). Clone libraries have shown a dominance of *Oscillatoriales* and *Nostocales* in Miers Valley hypoliths (Khan et al. 2011), while microscopic analyses of quartz rocks from McKelvey Valley indicated a predominance of *Leptolyngbya* morphotypes (Pointing et al. 2009).

1.3.2 Endoliths

Cryptoendoliths are microbial communities that inhabit the pores and interstices between mineral grains of sandstone and granite rocks (Fig. 1.3). By contrast chasmoendoliths colonise cracks and fissures within rocks (Wierzchos et al. 2013); thus, the mode of endolithic colonisation is delineated by the structural properties of the rock (Makhalanyane et al. 2014). Moreover, the translucence of the mineral substrate determines the depth of cryptoendolithic colonisation (Golubic et al. 1981; Büdel et al. 2009). Typically dominated by cyanobacteria, but occasionally found in association with lichens and algae, cryptoendolithic communities inhabit the subsurface zone of lithic substrates and are exposed to a favourable microclimate (Friedmann 1982). The beneficial microclimatic conditions include reduced

Fig. 1.3 Cryptoendolithic microbial community in Antarctic sandstone. (Photo, Don Cowan)

incident UV-radiation receipt (De Los Ríos et al. 2007), improved moisture retention compared to exposed surface soils (Palmer and Friedmann 1990; Büdel et al. 2008) and protection from desiccation, physical wind scouring and exposure to extreme temperature fluctuations (Wierzchos et al. 2013; Hughes and Lawley 2003).

In addition to altering their mode of growth, endolithic cyanobacteria also harbour multiple physiological adaptations to contend with extreme abiotic stressors, such as DNA repair mechanisms (Baqué et al. 2013). In Antarctica, endolithic communities are generally dominated by members of the cyanobacteria (Yung et al. 2014; Van Goethem et al. 2016), although communities dominated by eukaryotic fungal and algal phylotypes have been observed in East Beacon using FT-Raman spectroscopic analysis (Russell et al. 1998) and in upper Miers Valley using LTSEM (low-temperature SEM) (De Los Ríos et al. 2014b). Endoliths from Larsemann Hills are composed of cyanobacteria, green algae and lichens which together serve as primary producers in the rock subsurface and aid in rock destruction (Mergelov et al. 2012). *Cyanobacteria* in these crystalline granite rocks are essential biogenic components of the rock subsurface layer as the production of organic matter supports endolithic heterotrophs such as bacteria and micromycetes (Mergelov et al. 2012).

Microscopic analyses have been useful in visualising how *Cyanobacteria* support diverse populations of fungi, heterotrophic bacteria (Siebert et al. 1996) and algal taxa (Büdel et al. 2008) which utilise metabolites from cyanobacteria as carbon sources (Hughes and Lawley 2003). Endolithic cyanobacteria are also very important in the ecology of Antarctic soil communities by serving as sources of inocula that can be dispersed over long ranges. These propagules are important for colonising local soils during wetting periods (Wynn-Williams 1991). Endoliths occur much more extensively in the McMurdo Dry Valleys than in the Antarctic Islands, despite reports of unique translucent gypsum crust endoliths inhabiting the surface of sandstone boulders on Alexander Island, West of the Antarctic Peninsula (Hughes and Lawley 2003). However these *Chlorogloea*-dominated endoliths appear to support very low levels of primary productivity (Hughes and Lawley 2003).

Dry Valley endoliths are colonised by a diverse group of cyanobacterial taxa, while community membership and species richness can vary dramatically between distinct valleys, rock types and climatic conditions (Yung et al. 2014; Wierzchos et al. 2013; Friedmann 1982). The most ubiquitous endolithic colonists include *Chroococcidiopsis* (*Chroococcidiopsis*), *Gloeocapsa* and *Hormathonema* (both *Chroococcales*) species (Friedmann et al. 1988). Microscopic techniques coupled with live/dead microbial fluorescence staining have shown that granitic endoliths from Mont Falconer and carbonate rocks from Goldman Glacier comprise live photosynthesising cyanobacterial lineages, predominantly of the genus *Synechocystis* (*Chroococcales*) (de los Ríos et al. 2004). Phylogenetic surveys of endoliths from the Dry Valleys have shown *Chroococcidiopsis* to dominate in coarse-grained marbles from Taylor Valley (Büdel et al. 2008, 2009), sandstone substrates in McKelvey Valley (Pointing et al. 2009) and granite rocks in Miers Valley (Yung et al. 2014). *Synechococcus* phylotypes appear to favour warmer and wetter slopes compared to *Chroococcidiopsis* species which dominate on colder, drier slopes (Yung et al. 2014). Interestingly, sandstone endoliths from Victoria Valley mainly comprised *Acaryochloris*, *Leptolyngbya*, *Phormidium* and *Synechococcus* lineages (Van Goethem et al. 2016), which may have been due to their proximity to meltwater streams. Meltwater streams appear to be hotspots of microbial activity and diversity in the austral summer, and the cyanobacteria-dominated communities within these habitats are thought play significant roles in structuring surrounding soil communities (Wood et al. 2008).

1.3.3 Microbial Mat Communities

Lacustrine mat communities proliferate during the austral summer when microclimatic conditions are favourable for the development of photosynthetic biofilms. Subsequently, microbial mat communities are found along melt water streams and ponds in the Dry Valleys where organic matter averages ~257 g C/m^2 (Moorhead et al. 2003). Molecular evidence suggests that cyanobacteria are redistributed from lakes to exposed edaphic soil or lithic niches (Pointing et al. 2009; Wood et al. 2008). This is consistent with a survey of aerosols in the Dry Valleys indicating widespread dissemination of bacteria cells via aeolian transport (Bottos et al. 2014). *Synechococcus* are dominant in cryptogamic mats as *R*-selected species grow quickly in the summer at the expense of slow-growing *Chroococcidiopsis* (*K*-strategists) (Pointing 2016). Benthic microbial mats dominated by *Phormidium* and *Nostoc* spp. dominate carbon cycling and ecosystem processes with NPP between 10.1 and 24.6 g C m^{-2} year^{-1} (Moorhead et al. 2003). *Nostoc commune* drive most functional processes related to C and N cycles in Antarctica (Makhalanyane et al. 2016) and in many Arctic habitats (Rhodes et al. 2013).

Benthic mats in meromictic, ice-covered lakes and shallow seasonally frozen ponds (Fig. 1.4) have been shown to harbour high proportions of *Oscillatoriales* and *Chroococcales* (Zhang et al. 2015). The absence of layer mixing within a lake can

Fig. 1.4 Shallow seasonally-frozen pond in the McMurdo Dry Valleys (Antarctica) showing cyanobacterial mat structure. (Photo, Don Cowan)

lead to the formation of benthic mats with stable stratification. Laminated mats are also stratified and have been shown to possess layers of sulphur-reducing and methanogenic bacteria between cyanobacterial guilds (Stal 1995). Despite physical separation of unique microbial guilds, the metabolic interdependencies of different taxa in cyanobacteria-dominated communities have been highlighted recently. *Microcystis* spp. share synergistic relationships with heterotrophic bacteria in cyanobacterial blooms, and the reliance on metabolite sharing between photoautotrophs and heterotrophs certainly occurs in Antarctic mat communities as well. In freshwater blooms *Microcystis* supply heterotrophic bacteria with carbon and energy, while associated bacteria synthesise Vitamin B12 that is essential for the growth of *Microcystis* (Xie et al. 2016).

1.4 Conclusions and Perspectives

The Antarctic continent is characterised by extreme chemical and climatic features, which has led to the formation of soil biotopes of extremely low microbial biomass and productivity (van Horn et al. 2013). Despite severe constraints on cellular metabolism and phylogenetic diversity, Antarctic soil communities maintain important functional processes including carbon and nitrogen turnover. Central to nutrient cycling in Antarctic soils and lithic niches are members of the *Cyanobacteria*. Within cryptic niche communities of the Antarctic (i.e. hypoliths, crypto- and chasmoendoliths and microbial mats) phototrophic guilds drive most of the primary production, as well as carbon and nitrogen fixation. The distribution of cyanobacterial populations across these niches is critical to provide ecosystem services in cold, polar deserts.

There has been considerable recent research aimed at exploring the physiological attributes of cold-adapted *Cyanobacteria* from Antarctica. Combining metagenomic and/or genomic sequence data with direct measurements of functional processes and high-resolution microscopy has paved the way toward novel insights into the functional importance of the cyanobacteria in Antarctic desert soils. Even though we still lack information about the spectrum of interactions between cyanobacteria and associated microbial community members across all niches, we now know how cyanobacteria shape Antarctic microbial ecology through their unique physiological adaptations.

Cyanobacteria are critical role players in Antarctic soil biotopes due to the absence of higher eukaryotic photoautotrophs (Cary et al. 2010). Metagenomic data has revealed the prevalence of cyanobacterial lineages across a range of distinct niche types. Some of the most common cyanobacteria include members of the orders *Oscillatoriales*, *Nostocales* and *Chroococcales*. These cold-adapted cyanobacteria are capable of colonising, and even thriving, in freezing habitats that are made more challenging by high incident UV irradiation, low-light penetration and frequent freeze-thaw events (Makhalanyane et al. 2015). Moreover, their symbiotic competence with fungi has been reported (Meeks et al. 2001), while network structure analyses have shown the importance of cyanobacteria in maintaining hypolithic community in hot deserts (Valverde et al. 2015; Van Goethem et al. 2017). Recent research has shown the genomic adaptations harboured by these groups through the direct genomic sequencing of pure isolates and through the reconstruction of metagenome-assembled genomes (MAGs) (Chrismas et al. 2018; Parks et al. 2017).

One of the curiosities of recent sequenced-based studies has been the absence of genes for nitrogen fixation (*nifH*, *nifD* and *nifK*) from many Antarctic soil metagenomes and cyanobacterial genomes. Fixed nitrogen from the atmosphere is crucial to support the cellular biosynthesis of nitrogenous compounds. While acetylene reduction assays are indicative of actual nitrogen fixation, nitrogen isotopic data ($\delta^{15}N$ depletion) and cyanobacterial MAGs often do not corroborate these findings. Antarctic soils are very nitrate depleted, which is driven be the absence of water as nitrate accumulates in the soil (Walvoord et al. 2003). This suggests that alternative nitrogen sources, such as trace gases, may be more important for supplying nitrogen to the community than previously anticipated (Ji et al. 2017). While many cyanobacteria harbour the capacity to fix nitrogen in many cryptic niches, the low relative abundance of cyanobacteria in exposed polar soil surfaces poses questions as to potential alternative sources of nitrogen in these desert soils.

Both nitrogen and carbon cycling appear to be susceptible to variations in moisture content in Antarctic soils (Niederberger et al. 2012, 2015a). Wetter soils have also been shown to harbour a higher relative abundance of cyanobacteria-assigned sequences compared to drier soils (Niederberger et al. 2015b) which has the potential to drive changes in the functional attributes of the soil community. The application of metatranscriptomic (analysis of actual gene transcripts) and metaproteomic (translated products of gene transcripts) techniques will answer broad questions about actual microbial functionality in polar soils, and how shifts in climatic conditions drive changes in the allocation of functional traits.

Acknowledgements The authors thank the University of Pretoria and the South African National Research Foundation (NRF) for continued financial support of our research programmes.

References

Adams, B. J., et al. (2006). Diversity and distribution of Victoria Land biota. *Soil Biology and Biochemistry, 38*, 3003–3018. https://doi.org/10.1016/j.soilbio.2006.04.030.

Anesio, A. M., Hodson, A. J., Fritz, A., Psenner, R., & Sattler, B. (2009). High microbial activity on glaciers: Importance to the global carbon cycle. *Global Change Biology, 15*, 955–960.

Bahl, J., et al. (2011). Ancient origins determine global biogeography of hot and cold desert cyanobacteria. *Nature Communications, 2*, 163. https://doi.org/10.1038/ncomms1167.

Banerjee, M., & Verma, V. (2009). Nitrogen fixation in endolithic cyanobacterial communities of the McMurdo Dry Valley, Antarctica. *Science Asia, 35*, 215–219.

Baqué, M., Viaggiu, E., Scalzi, G., & Billi, D. (2013). Endurance of the endolithic desert cyanobacterium Chroococcidiopsis under UVC radiation. *Extremophiles: Life Under Extreme Conditions, 17*, 161–169.

Barrow, C. (1992). *World atlas of desertification (United Nations Environment Program)*. London: Edward Arnold.

Belnap, J., & Gardner, J. S. (1993). Soil microstructure in soils of the Colorado plateau: The role of the cyanobacterium *Microcoleus vaginatus*. *The Great Basin Naturalist, 53*, 40–47.

Benhua, S., et al. (2014). Biogeochemical responses to nutrient, moisture and temperature manipulations of soil from Signy Island, South Orkney Islands in the Maritime Antarctic. *Antarctic Science, 26*, 513–520.

Bergman, B., Gallon, J., Rai, A., & Stal, L. (1997). N2 fixation by non-heterocystous cyanobacteria1. *FEMS Microbiology Reviews, 19*, 139–185.

Bintanja, R., van Oldenborgh, G., Drijfhout, S., Wouters, B., & Katsman, C. (2013). Important role for ocean warming and increased ice-shelf melt in Antarctic sea-ice expansion. *Nature Geoscience, 6*, 376.

Blank, C., & Sanchez-Baracaldo, P. (2010). Timing of morphological and ecological innovations in the cyanobacteria–a key to understanding the rise in atmospheric oxygen. *Geobiology, 8*, 1–23.

Boison, G., Mergel, A., Jolkver, H., & Bothe, H. (2004). Bacterial life and dinitrogen fixation at a gypsum rock. *Applied and Environmental Microbiology, 70*, 7070–7077.

Bottos, E. M., Woo, A. C., Zawar-Reza, P., Pointing, S. B., & Cary, S. C. (2014). Airborne bacterial populations above desert soils of the McMurdo Dry Valleys, Antarctica. *Microbial Ecology, 67*, 120–128. https://doi.org/10.1007/s00248-013-0296-y.

Bromwich, D. H., et al. (2013). Central West Antarctica among the most rapidly warming regions on earth. *Nature Geoscience, 6*, 139.

Büdel, B., Bendix, J., Bicker, F. R., & Allan Green, T. (2008). Dewfall as a water source frequently activates the endolithic cyanobacterial communities in the granites of Taylor Valley, Antarctica. *Journal of Phycology, 44*, 1415–1424.

Büdel, B., Schulz, B., Reichenberger, H., Bicker, F., & Green, T. (2009). Cryptoendolithic cyanobacteria from calcite marble rock ridges, Taylor Valley, Antarctica. *Algological Studies, 129*, 61–69.

Burgess, B. K., & Lowe, D. J. (1996). Mechanism of molybdenum nitrogenase. *Chemical Reviews, 96*, 2983–3012.

Büsch, A., Friedrich, B., & Cramm, R. (2002). Characterization of the norB gene, encoding nitric oxide reductase, in the nondenitrifying cyanobacterium Synechocystis sp. strain PCC6803. *Applied and Environmental Microbiology, 68*, 668–672.

Cameron, R. E., King, J., & David, C. N. (1970). *Microbiology, ecology and microclimatology of soil sites in Dry Valleys of Southern Victoria Land, Antarctica*.

Cary, S. C., McDonald, I. R., Barrett, J. E., & Cowan, D. A. (2010). On the rocks: The microbiology of Antarctic Dry Valley soils. *Nature Reviews Microbiology, 8*, 129–138. https://doi.org/10.1038/nrmicro2281.
Chan, Y., et al. (2012). Hypolithic microbial communities: Between a rock and a hard place. *Environmental Microbiology, 14*, 2272–2282. https://doi.org/10.1111/j.1462-2920.2012.02821.x.
Chan, Y., van Nostrand, J. D., Zhou, J., Pointing, S. B., & Farrell, R. L. (2013). Functional ecology of an Antarctic dry valley. *Proceedings of the National Academy of Sciences, 110*, 8990–8995.
Chen, J., et al. (2015). Stronger warming effects on microbial abundances in colder regions. *Scientific Reports, 5*, 18032.
Chrismas, N. A., Anesio, A. M., & Sánchez-Baracaldo, P. (2018). The future of genomics in polar and alpine cyanobacteria. *FEMS Microbiology Ecology, 94*, fiy032.
Cockell, C. S., & Stokes, M. D. (2004). Ecology: Widespread colonization by polar hypoliths. *Nature, 431*, 414–414.
Cook, A., Fox, A., Vaughan, D., & Ferrigno, J. (2005). Retreating glacier fronts on the Antarctic peninsula over the past half-century. *Science, 308*, 541–544.
Cowan, D. A., & Ah Tow, L. (2004). Endangered antarctic environments. *Annual Review of Microbiology, 58*, 649–690. https://doi.org/10.1146/annurev.micro.57.030502.090811.
Cowan, D. A., & Makhalanyane, T. P. (2017). Energy from thin air. *Nature, 552*, 336.
Cowan, D. A., Khan, N., Pointing, S. B., & Cary, S. C. (2010). Diverse hypolithic refuge communities in the McMurdo Dry Valleys. *Antarctic Science, 22*, 714–720.
Cowan, D. A., et al. (2011a). Hypolithic communities: Important nitrogen sources in Antarctic desert soils. *Environmental Microbiology Reports, 3*, 581–586.
Cowan, D. A., et al. (2011b). Distribution and abiotic influences on hypolithic microbial communities in an Antarctic Dry Valley. *Polar Biology, 34*, 307–311.
Cowan, D. A., Makhalanyane, T. P., Dennis, P. G., & Hopkins, D. W. (2014). Microbial ecology and biogeochemistry of continental Antarctic soils. *Frontiers in Microbiology, 5*, 154. https://doi.org/10.3389/fmicb.2014.00154.
Cowan, D., Ramond, J., Makhalanyane, T., & De Maayer, P. (2015). Metagenomics of extreme environments. *Current Opinion in Microbiology, 25*, 97–102.
De la Torre, J. R., Goebel, B. M., Friedmann, E. I., & Pace, N. R. (2003). Microbial diversity of Cryptoendolithic communities from the McMurdo Dry Valleys, Antarctica. *Applied and Environmental Microbiology, 69*, 3858–3867. https://doi.org/10.1128/aem.69.7.3858-3867.2003.
de los Ríos, A., Wierzchos, J., Sancho, L. G., & Ascaso, C. (2004). Exploring the physiological state of continental Antarctic endolithic microorganisms by microscopy. *FEMS Microbiology Ecology, 50*, 143–152.
De Los Ríos, A., Grube, M., Sancho, L. G., & Ascaso, C. (2007). Ultrastructural and genetic characteristics of endolithic cyanobacterial biofilms colonizing Antarctic granite rocks. *FEMS Microbiology Ecology, 59*, 386–395.
De los Ríos, A., Cary, C., & Cowan, D. (2014a). The spatial structures of hypolithic communities in the Dry Valleys of East Antarctica. *Polar Biology*. https://doi.org/10.1007/s00300-014-1564-0.
De Los Ríos, A., Wierzchos, J., & Ascaso, C. (2014b). The lithic microbial ecosystems of Antarctica's McMurdo Dry Valleys. *Antarctic Science, 26*, 459–477.
de Scally, S., Makhalanyane, T., Frossard, A., Hogg, I., & Cowan, D. (2016). Antarctic microbial communities are functionally redundant, adapted and resistant to short term temperature perturbations. *Soil Biology and Biochemistry, 103*, 160–170.
Delgado-Baquerizo, M., et al. (2018). A global atlas of the dominant bacteria found in soil. *Science, 359*, 320–325.
Fierer, N., et al. (2012). Cross-biome metagenomic analyses of soil microbial communities and their functional attributes. *Proceedings of the National Academy of Sciences, 109*, 21390–21395.
Fortner, S. K., Tranter, M., Fountain, A., Lyons, W. B., & Welch, K. A. (2005). The geochemistry of supraglacial streams of Canada Glacier, Taylor Valley (Antarctica), and their evolution into proglacial waters. *Aquatic Geochemistry, 11*, 391–412.

Freckman, D. W., & Virginia, R. A. (1997). Low-diversity antarctic soil nematode communities: Distribution and responce to disturbance. *Ecology, 78*, 363–369.
Friedmann, E. I. (1982). Endolithic microorganisms in the Antarctic cold desert. *Science, 215*, 1045–1053.
Friedmann, E. I., & Ocampo, R. (1976). Endolithic blue-green algae in the dry valleys: Primary producers in the Antarctic desert ecosystem. *Science, 193*, 1247–1249.
Friedmann, E. I., Hua, M., & Ocampo-Friedmann, R. (1988). 3.6 Cryptoendolithic lichen and cyanobacterial communities of the Ross Desert, Antarctica. *Polarforschung, 58*, 251–259.
Fyfe, J. C., & Saenko, O. A. (2006). Simulated changes in the extratropical Southern Hemisphere winds and currents. *Geophysical Research Letters, 33*, L06701.
Geyer, K. M., Takacs-Vesbach, C. D., Gooseff, M. N., & Barrett, J. E. (2017). Primary productivity as a control over soil microbial diversity along environmental gradients in a polar desert ecosystem. *PeerJ, 5*, e3377.
Golden, S. S., Johnson, C. H., & Kondo, T. (1998). The cyanobacterial circadian system: A clock apart. *Current Opinion in Microbiology, 1*, 669–673.
Golubic, S., Friedmann, I., & Schneider, J. (1981). The lithobiontic ecological niche, with special reference to microorganisms. *Journal of Sedimentary Research, 51*.
Hopkins, D., et al. (2006). Carbon, nitrogen and temperature controls on microbial activity in soils from an Antarctic dry valley. *Soil Biology and Biochemistry, 38*, 3130–3140.
Hopkins, D. W., et al. (2009). Isotopic evidence for the provenance and turnover of organic carbon by soil microorganisms in the Antarctic dry valleys. *Environmental Microbiology, 11*, 597–608. https://doi.org/10.1111/j.1462-2920.2008.01830.x.
Howard, J. B., & Rees, D. C. (1996). Structural basis of biological nitrogen fixation. *Chemical Reviews, 96*, 2965–2982.
Huang, L., McCluskey, M. P., Ni, H., & Larossa, R. A. (2002). Global gene expression profiles of the cyanobacterium Synechocystis sp. strain PCC 6803 in response to irradiation with UV-B and white light. *Journal of Bacteriology, 184*, 6845–6858.
Hughes, K. A., & Lawley, B. (2003). A novel Antarctic microbial endolithic community within gypsum crusts. *Environmental Microbiology, 5*, 555–565.
Hultman, J., et al. (2015). Multi-omics of permafrost, active layer and thermokarst bog soil microbiomes. *Nature, 521*, 208–212.
Hutchins, P. R., & Miller, S. R. (2017). Genomics of variation in nitrogen fixation activity in a population of the thermophilic cyanobacterium Mastigocladus laminosus. *The ISME Journal, 11*, 78–86.
Ji, M., et al. (2017). Atmospheric trace gases support primary production in Antarctic desert surface soil. *Nature, 552*, 400.
Khan, N., et al. (2011). Hypolithic microbial communities of quartz rocks from Miers Valley, McMurdo Dry Valleys, Antarctica. *Polar Biology, 34*, 1657–1668. https://doi.org/10.1007/s00300-011-1061-7.
Kirtman, B., et al. (2013). *Near-term climate change: Projections and predictability*.
Kobayashi, D., Tamoi, M., Iwaki, T., Shigeoka, S., & Wadano, A. (2003). Molecular characterization and redox regulation of phosphoribulokinase from the cyanobacterium Synechococcus sp. PCC 7942. *Plant and Cell Physiology, 44*, 269–276.
Komárek, J., Genuário, D. B., Fiore, M. F., & Elster, J. (2015). Heterocytous cyanobacteria of the Ulu Peninsula, James Ross Island, Antarctica. *Polar Biology, 38*, 475–492.
Latysheva, N., Junker, V. L., Palmer, W. J., Codd, G. A., & Barker, D. (2012). The evolution of nitrogen fixation in cyanobacteria. *Bioinformatics, 28*, 603–606.
Laybourn-Parry, J., Tranter, M., & Hodson, A. J. (2012). *The ecology of snow and ice environments*. Oxford: Oxford University Press.
Ligtenberg, S., van de Berg, W., van den Broeke, M., Rae, J., & van Meijgaard, E. (2013). Future surface mass balance of the Antarctic ice sheet and its influence on sea level change, simulated by a regional atmospheric climate model. *Climate Dynamics, 41*, 867–884.

Magalhães, C. M., Machado, A., Frank-Fahle, B., Lee, C. K., & Cary, S. C. (2014). The ecological dichotomy of ammonia-oxidizing archaea and bacteria in the hyper-arid soils of the Antarctic Dry Valleys. *Frontiers in Microbiology, 5*, 515.

Makhalanyane, T. P., et al. (2013a). Evidence for successional development in Antarctic hypolithic bacterial communities. *The ISME Journal, 7*, 2080–2090.

Makhalanyane, T. P., et al. (2013b). Evidence of species recruitment and development of hot desert hypolithic communities. *Environmental Microbiology Reports, 5*, 219–224. https://doi.org/10.1111/1758-2229.12003.

Makhalanyane, T. P., Pointing, S. B., & Cowan, D. A. (2014). *Antarctic terrestrial microbiology* (pp. 163–179). Berlin: Springer.

Makhalanyane, T. P., et al. (2015). Ecology and biogeochemistry of cyanobacteria in soils, permafrost, aquatic and cryptic polar habitats. *Biodiversity and Conservation, 24*, 819–840. https://doi.org/10.1007/s10531-015-0902-z.

Makhalanyane, T. P., van Goethem, M. W., & Cowan, D. A. (2016). Microbial diversity and functional capacity in polar soils. *Current Opinion in Biotechnology, 38*, 159–166. https://doi.org/10.1016/j.copbio.2016.01.011.

Marsden, W., & Codd, G. (1984). Purification and molecular and catalytic properties of phosphoribulokinase from the cyanobacterium Chlorogloeopsis fritschii. *Microbiology, 130*, 999–1006.

Marsden, W., Lanaras, T., & Codd, G. (1984). Subcellular segregation of phosphoribulokinase and ribulose-1, 5-bisphosphate carboxylase/oxygenase in the cyanobacterium Chlorogloeopsis fritschii. *Microbiology, 130*, 2089–2093.

Matsumoto, G. I., Hirai, A., Hirota, K., & Watanuki, K. (1990). Organic geochemistry of the McMurdo dry valleys soil, Antarctica. *Organic Geochemistry, 16*, 781–791.

Meeks, J. C., et al. (2001). An overview of the genome of Nostoc punctiforme, a multicellular, symbiotic cyanobacterium. *Photosynthesis Research, 70*, 85–106.

Mergelov, N., Goryachkin, S., Shorkunov, I., Zazovskaya, E., & Cherkinsky, A. (2012). Endolithic pedogenesis and rock varnish on massive crystalline rocks in East Antarctica. *Eurasian Soil Science, 45*, 901–917.

Mills, L. S., Soulé, M. E., & Doak, D. F. (1993). The keystone-species concept in ecology and conservation. *Bioscience, 43*, 219–224.

Moorhead, D. L., Barrett, J. E., Virginia, R. A., Wall, D. H., & Porazinska, D. (2003). Organic matter and soil biota of upland wetlands in Taylor Valley, Antarctica. *Polar Biology, 26*, 567–576.

Niederberger, T. D., et al. (2012). Diverse and highly active diazotrophic assemblages inhabit ephemerally wetted soils of the Antarctic Dry Valleys. *FEMS Microbiology Ecology, 82*, 376–390.

Niederberger, T. D., et al. (2015a). Carbon-fixation rates and associated microbial communities residing in arid and ephemerally wet Antarctic Dry Valley soils. *Frontiers in Microbiology, 6*, 9.

Niederberger, T. D., et al. (2015b). Microbial community composition of transiently wetted Antarctic Dry Valley soils. *Frontiers in Microbiology, 6*, 9.

Novis, P. M., et al. (2007). Annual carbon fixation in terrestrial populations of Nostoc commune (Cyanobacteria) from an Antarctic dry valley is driven by temperature regime. *Global Change Biology, 13*, 1224–1237.

Paerl, H. W., Pinckney, J. L., & Steppe, T. F. (2000). Cyanobacterial–bacterial mat consortia: Examining the functional unit of microbial survival and growth in extreme environments. *Environmental Microbiology, 2*, 11–26.

Palmer, R. J., & Friedmann, E. I. (1990). Water relations and photosynthesis in the cryptoendolithic microbial habitat of hot and cold deserts. *Microbial Ecology, 19*, 111–118.

Parks, D. H., et al. (2017). Recovery of nearly 8,000 metagenome-assembled genomes substantially expands the tree of life. *Nature Microbiology, 2*, 1533.

Philippot, L., et al. (2013). Loss in microbial diversity affects nitrogen cycling in soil. *The ISME Journal, 7*, 1609.

Pointing, S. B. (2016). *Biological soil crusts: An organizing principle in drylands* (pp. 199–213). Cham: Springer.

Pointing, S. B., et al. (2009). Highly specialized microbial diversity in hyper-arid polar desert. *Proceedings of the National Academy of Sciences of the United States of America, 106*, 19964–19969. https://doi.org/10.1073/pnas.0908274106.

Pointing, S. B., Bollard-Breen, B., & Gillman, L. N. (2014). Diverse cryptic refuges for life during glaciation. *Proceedings of the National Academy of Sciences, 111*, 5452–5453.

Poly, F., Monrozier, L. J., & Bally, R. (2001). Improvement in the RFLP procedure for studying the diversity of nifH genes in communities of nitrogen fixers in soil. *Research in Microbiology, 152*, 95–103.

Pringault, O., & Garcia-Pichel, F. (2004). Hydrotaxis of cyanobacteria in desert crusts. *Microbial Ecology, 47*, 366–373.

Rhodes, M., et al. (2013). *The prokaryotes* (pp. 43–55). Berlin: Springer.

Romanovsky, V. E., Smith, S. L., & Christiansen, H. H. (2010). Permafrost thermal state in the polar Northern Hemisphere during the international polar year 2007–2009: A synthesis. *Permafrost and Periglacial Processes, 21*, 106–116.

Russell, N., Edwards, H., & Wynn-Williams, D. (1998). FT-Raman spectroscopic analysis of endolithic microbial communities from Beacon sandstone in Victoria Land, Antarctica. *Antarctic Science, 10*, 63–74.

Scherer, S., Almon, H., & Böger, P. (1988). Interaction of photosynthesis, respiration and nitrogen fixation in cyanobacteria. *Photosynthesis Research, 15*, 95–114.

Schirrmeister, B. E., Sanchez-Baracaldo, P., & Wacey, D. (2016). Cyanobacterial evolution during the Precambrian. *International Journal of Astrobiology, 15*, 187–204.

Schlesinger, W. H., et al. (2003). Community composition and photosynthesis by photoautotrophs under quartz pebbles, southern Mojave Desert. *Ecology, 84*, 3222–3231.

Serra, J. L., Llama, M. J., Rowell, P., & Stewart, W. D. (1989). Purification and characterization of phosphoribulokinase from the N2-fixing cyanobacterium Anabaena cylindrica. *Plant Science, 59*, 1–9.

Siebert, J., et al. (1996). Cryptoendolithic microorganisms from Antarctic sandstone of Linnaeus Terrace (Asgard Range): Diversity, properties and interactions. *Biodiversity and Conservation, 5*, 1337–1363.

Stal, L. J. (1995). Physiological ecology of cyanobacteria in microbial mats and other communities. *New Phytologist, 131*, 1–32.

Stibal, M., Šabacká, M., & Žárský, J. (2012). Biological processes on glacier and ice sheet surfaces. *Nature Geoscience, 5*, 771.

Tahon, G., Tytgat, B., Stragier, P., & Willems, A. (2016). Analysis of cbbL, nifH, and pufLM in soils from the Sør Rondane Mountains, Antarctica, reveals a large diversity of autotrophic and phototrophic bacteria. *Microbial Ecology, 71*, 131–149.

Thomas, C. M., & Nielsen, K. M. (2005). Mechanisms of, and barriers to, horizontal gene transfer between bacteria. *Nature Reviews Microbiology, 3*, 711–721.

Turner, J., et al. (2005). Antarctic climate change during the last 50 years. *International Journal of Climatology, 25*, 279–294.

Turner, J., et al. (2016). Absence of 21st century warming on Antarctic Peninsula consistent with natural variability. *Nature, 535*, 411.

Valverde, A., Makhalanyane, T. P., Seely, M., & Cowan, D. A. (2015). Cyanobacteria drive community composition and functionality in rock–soil interface communities. *Molecular Ecology, 24*, 812–821.

Van Goethem, M. W., Makhalanyane, T. P., Valverde, A., Cary, S. C., & Cowan, D. A. (2016). Characterization of bacterial communities in lithobionts and soil niches from Victoria Valley, Antarctica. *FEMS Microbiology Ecology, 92*, fiw051.

Van Goethem, M. W., Makhalanyane, T. P., Cowan, D. A., & Valverde, A. (2017). Cyanobacteria and Alphaproteobacteria May Facilitate Cooperative Interactions in Niche Communities. *Frontiers in Microbiology*. https://doi.org/10.3389/fmicb.2017.02099.

van Horn, D. J., et al. (2013). Factors controlling soil microbial biomass and bacterial diversity and community composition in a Cold Desert ecosystem: Role of geographic scale. *PLoS One, 8*, e66103. https://doi.org/10.1371/journal.pone.0066103.

Varin, T., Lovejoy, C., Jungblut, A. D., Vincent, W. F., & Corbeil, J. (2012). Metagenomic analysis of stress genes in microbial mat communities from Antarctica and the High Arctic. *Applied and Environmental Microbiology, 78*, 549–559.

Vishniac, H. (1993). The microbiology of Antarctic soils. *Antarctic Microbiology*, 297–341.

Wada, E., Shibata, R., & Torii, T. (1981). 15N abundance in Antarctica: Origin of soil nitrogen and ecological implications. *Nature, 292*, 327.

Walvoord, M. A., et al. (2003). A reservoir of nitrate beneath desert soils. *Science, 302*, 1021–1024.

Wei, S. T., et al. (2015a). Diverse metabolic and stress-tolerance pathways in chasmoendolithic and soil communities of Miers Valley, McMurdo Dry Valleys, Antarctica. *Polar Biology, 38*, 433–443.

Wei, S. T., Higgins, C. M., Adriaenssens, E. M., Cowan, D. A., & Pointing, S. B. (2015b). Genetic signatures indicate widespread antibiotic resistance and phage infection in microbial communities of the McMurdo Dry Valleys, East Antarctica. *Polar Biology, 38*, 919–925.

Wierzchos, J., de los Ríos, A., & Ascaso, C. (2013). Microorganisms in desert rocks: The edge of life on earth. *International Microbiology, 15*, 172–182.

Williams, T. J., et al. (2013). The role of planktonic Flavobacteria in processing algal organic matter in coastal East Antarctica revealed using metagenomics and metaproteomics. *Environmental Microbiology, 15*, 1302–1317.

Wong, F. K., et al. (2010). Hypolithic microbial community of quartz pavement in the high-altitude tundra of central Tibet. *Microbial Ecology, 60*, 730–739. https://doi.org/10.1007/s00248-010-9653-2.

Wood, S. A., Rueckert, A., Cowan, D. A., & Cary, S. C. (2008). Sources of edaphic cyanobacterial diversity in the Dry Valleys of Eastern Antarctica. *The ISME Journal, 2*, 308–320. https://doi.org/10.1038/ismej.2007.104.

Wynn-Williams, D. D. (1991). Aerobiology and colonization in Antarctica—the BIOTAS Programme. *Grana, 30*, 380–393.

Wynn-Williams, D. (1996). Antarctic microbial diversity: The basis of polar ecosystem processes. *Biodiversity and Conservation, 5*, 1271–1293.

Xie, M., et al. (2016). Metagenomic analysis reveals symbiotic relationship among bacteria in microcystis-dominated community. *Frontiers in Microbiology, 7*, 56.

Yergeau, E., et al. (2009). Environmental microarray analyses of Antarctic soil microbial communities. *The ISME Journal, 3*, 340.

Yung, C. C., et al. (2014). Characterization of Chasmoendolithic Community in Miers Valley, McMurdo Dry Valleys, Antarctica. *Microbial Ecology, 68*, 351–359. https://doi.org/10.1007/s00248-014-0412-7.

Zhang, L., et al. (2015). Cyanobacterial diversity in benthic mats of the McMurdo Dry Valley lakes, Antarctica. *Polar Biology, 38*, 1097–1110.

Chapter 2
Patterns of Microorganisms Inhabiting Antarctic Freshwater Lakes with Special Reference to Aquatic Moss Pillars

Ryosuke Nakai, Satoshi Imura, and Takeshi Naganuma

Abstract The Antarctic continent has ice-free areas with many freshwater lakes that support life. These lakes are generally ultra-oligotrophic and possess simplified food chains dominated by microorganisms with algal and cyanobacterial mats often occurring in the lake bottoms. In association with such mats, aquatic mosses sometimes form unique towerlike structures called "moss pillars." Previous microflora analysis revealed the presence of several key groups (e.g., *Leptolyngbya* and *Bradyrhizobium* species) and uncultivated novel lineages in the pillars and the fact that the bacterial communities differ among the pillar sections. A wide range of eukaryotic phylotypes associated with algae, ciliates, fungi, nematodes, rotifers, and tardigrades, as well as unclassified phylotypes, were detected in the pillars. Moss pillars colonizing the nutrient-limited lakes are likely formed by a synergistic association of diverse organisms including both primary producers and decomposers. Indeed, a potential functional zonation, possibly reflected by different redox conditions within the pillar structure, was identified during the analyses of functional genes (e.g., CO_2 fixation-coding genes). Interestingly, multiple sequences related to moss pillar-derived sequences were also observed in other Antarctic habitats. These findings provide clues toward solving a conundrum pertaining to Antarctic lake ecosystems: biomass-rich communities existing in the nutrient-poor conditions.

R. Nakai (✉)
Applied Molecular Microbiology Research Group, Bioproduction Research Institute, National Institute of Advanced Industrial Science and Technology (AIST), Sapporo, Japan
e-mail: nakai-ryosuke@aist.go.jp

S. Imura
National Institute of Polar Research (NIPR), Tachikawa, Tokyo, Japan

Department of Polar Science, SOKENDAI (The Graduate University for Advanced Studies), Tachikawa, Tokyo, Japan

T. Naganuma
Graduate School of Biosphere Science, Hiroshima University, Higashihiroshima, Japan

S. Castro-Sowinski (ed.), *The Ecological Role of Micro-organisms in the Antarctic Environment*, Springer Polar Sciences,
https://doi.org/10.1007/978-3-030-02786-5_2

Keywords Antarctic lakes · Freshwater · Microbial mats · Bacteria · Eukaryotes · Phylogenetic diversity

2.1 Introduction

Antarctica is the coldest, driest, and windiest continent on Earth. The Antarctic continent is covered by an ice sheet up to 4 km thick and is one of the most extreme environments for life anywhere in the world. On the other hand, there are ice-free areas in the coastal margins that possess lakes of various sizes, which were formed mainly by glacial processes (Laybourn-Parry and Wadham 2014). These lakes are called "polar oases" existing in polar deserts and are considered important places that support life. Beyond the ice-free areas, the basal regions of the thick ice sheet are known to harbor over 400 subglacial lakes (Wright and Siegert 2012; Siegert et al. 2016). Much scientific attention to the existence of microbial life inhabiting these lakes (e.g., Lake Vostok and Lake Whillans) has increased over the past two decades (Priscu et al. 1999; Petit et al. 2005; Christner et al. 2014).

Classic microbial studies relying on microscopy and culturing techniques revealed invisible microorganisms (e.g., bacteria, yeast, and fungi) thriving in Antarctic environments (Wynn-Williams 1990). Additionally, the development of molecular biological approaches using DNA extracted from environmental samples has shed light on the hidden diversity of Antarctic microorganisms. For instance, surveys of small subunit ribosomal RNA (rRNA) gene sequences (e.g., 16S rRNA genes for prokaryotes and 18S rRNA genes for eukaryotes) allow for the detection of almost all organisms without the need of cultivation (Creer et al. 2016). Such rRNA-based analysis has captured the species composition and phylogenetic novelty of the microorganisms within the targeted Antarctic ecosystems (Shivaji and Reddy 2010; Niederberger et al. 2012). Moreover, recent "omics" techniques (e.g., metagenomics, transcriptomics, and proteomics) allow the describing of the functional diversity of microorganisms in Antarctica (e.g., Casanueva et al. 2010; Lauro et al. 2010; Yau et al. 2013). Furthermore, these techniques are applied to investigations regarding virus diversity and dynamics, which have been being unraveled for a long time (López-Bueno et al. 2009). In short, molecular methodologies are revolutionizing our understanding of Antarctic ecosystems, which are dominated by microorganisms.

In this chapter, we examine general characteristics of Antarctic lakes (mainly freshwater lakes), as well as species diversity and functional traits of the biota associated with aquatic moss colonies studied using molecular biological methods. Through close inspection of these unique environs, we attempt to illustrate the critical factors for the development and maintenance of the ecosystems in lake environments.

2.2 General Characteristics of Antarctic Lakes

Antarctic lakes are mainly located on the coastal ice-free areas (e.g., Syowa Oasis and the Larsemann Hills in East Antarctica), the McMurdo Dry Valleys and the Maritime Antarctic. These lakes display various limnological features with salinity, redox (reduction/oxidation) conditions, and nutrient and organic carbon supply (Laybourn-Parry and Wadham 2014). Particularly, a remarkable diversity of lake salinity exists, ranging from freshwater to hyposaline, saline (similar to seawater), and hypersaline lakes. In coastal ice-free areas, entrapped seawater is "lyophilized" to become concentrated and yield high-saline water bodies. Under these conditions, several unique bacteria and archaea (e.g., *Halomonas* species and *Halorubrum*-related species) have been recorded (James et al. 1994; Bowman et al. 2000; Naganuma et al. 2005).

The coastal areas in East Antarctica carry a large number of freshwater lakes. For example, in Syowa Oasis along the Sôya Coast, over 40 freshwater lakes are listed with names, locations, and limnological features (Kudoh and Tanabe 2014). Of importance is that there are even more freshwater lakes that remain unnamed and unstudied. Similarly, there are 150 freshwater lakes (including ponds) in the Larsemann Hills (Gillieson et al. 1990). Most of these coastal lakes were formed since deglaciation after the last glacial maximum (Verleyen et al. 2011). In these areas, various types of ice-dependent lake are found: ice-proximal lakes (Fig. 2.1), ice-dammed lakes, ice-distal lakes, and ice-scoured lakes (Pienitz et al. 2008). Aside

Fig. 2.1 Tsubaki-Ike Lake, Skarvsnes, East Antarctica. This is an example of an ice-proximal freshwater lake directly fed by glacial meltwater. (Photo by R. Nakai)

from these lake types, supraglacial lakes formed on the top of glaciers are also common in Antarctica. Freshwater lakes provide valuable habitats for a wide variety of organisms under "edge of life" conditions in Antarctic terrestrial environments.

In Antarctic freshwater lakes, fish are not found; invisible prokaryotic microorganisms (archaea and bacteria) and eukaryotic microorganisms (mainly algae and protozoa) dominate. Cyanobacteria often account for a large portion of the biota in freshwater lakes with some benthic groups capable of forming mm-thick mats, films, and crusts over the lake bottoms (Vincent et al. 1993b). In association with benthic cyanobacteria, aquatic mosses, algae, and bacteria in some lakes form large structures (see below).

2.3 Freshwater Ecosystems

Antarctic freshwater lakes are generally classified as ultra-oligotrophic lakes based on the low levels of chlorophyll *a* as an indicator of phytoplankton biomass in the lake water with annual maximum concentrations <2.5 mg m^{-3} and annual mean concentration <1 mg m^{-3} (Lizotte 2008). Nevertheless, time-course monitoring of water column in oligotrophic lakes of East Antarctica indicates that seasonal increases in chlorophyll *a* concentration, so-called phytoplankton blooms, occur under dim-light conditions in the spring and autumn when the studied lakes are covered with ice (Tanabe et al. 2008). This monitoring also suggests that during the summer when the ice cover on the lakes is decreasing, photo-inhibition of phytoplankton occurs under strong light conditions. Similar phenomena have been suggested in other freshwater lakes (Goldman et al. 1963; Roos and Vincent 1998).

In contrast to the water column with a low phytoplankton biomass, the lake bottom is known to harbor high biomass, dominated by algae, cyanobacteria, and other microorganisms. Cyanobacteria often form luxuriant multilayered mat consortia (Vincent and Quesada 2012). On the other hand, cyanobacteria-based mats are exposed to ultraviolet (UV) radiation in shallow freshwater lakes, because oligotrophic lake water contains low levels of UV-absorbing material (Morris et al. 1995). To impede this challenge and stress, benthic mats produce photo-protective pigments and materials such as xanthophyll, scytonemin, and mycosporine-like amino acids (Vincent et al. 1993b; Hodgson et al. 2004; Tanabe et al. 2010). Indeed, a color zonation reflected by different pigment compositions is observed in several cyanobacteria-dominant mats from lakes in East Antarctica (Tanabe et al. 2010). Additionally, Hodgson et al. (2004) have recorded the pigment content and composition of microbial communities in 62 East Antarctic lakes.

Larger mat structures have also been found in some Antarctic freshwater lakes, stromatolite-like algal mats called "living stromatolites" in the McMurdo Dry Valley lakes (Parker et al. 1981), moss pillars in some lakes in Syowa Oasis (Imura et al. 1999), and dome-shaped conical stromatolites in Lake Untersee, Dronning Maud Land (Andersen et al. 2011). These large structures, as well as cyanobacteria-based mats, provide a biological mystery as to how large biomass-rich structures form in ultra-oligotrophic Antarctic lake environments. The following examines this conundrum using moss pillars as an example.

2.4 Biota of Aquatic Moss Pillars

2.4.1 Moss Pillars

The benthic tower-like vegetation "moss pillars" were first discovered in Hotoke-ike Lake, formerly known as B-4 Ike, around Syowa Station (69°00′S, 39°35′E), East Antarctica, by the 36th Japanese Antarctic Research Expedition (Fig. 2.2). The pillars rise from algal and cyanobacterial mats and grow up to about 40 cm in diameter and 60 cm high (Imura et al. 1999). The large colonies of pillars occur mainly at a 3–5 m depth. The main component is *Leptobryum* sp. (later described as *L. wilsonii*; Kato et al. 2013), sometimes in association with *Bryum pseudotriquetrum* (Imura et al. 1999). It is noteworthy that *Leptobryum* terrestrial taxon has never been observed in either the vicinity of Syowa Station or elsewhere in continental Antarctica. Molecular phylogenetic analysis suggests that *L. wilsonii* immigrated into Antarctic freshwater lakes during the Holocene epoch via long-distance dispersal, perhaps in the form of moss spores from South America since no genetic variation is detected between Antarctic samples and some Chilean

Fig. 2.2 Benthic moss pillars in Hotoke-ike Lake, Skarvsnes, East Antarctica. At the top of the pillar, a green-colored moss is easily recognized. The original figure was in Nakai et al. (2012a). (Photo by S. Imura)

samples (Kato et al. 2013). Such unique moss colonies have not been reported for any other lakes around the world and are found only in Syowa Oasis.

Since the first description of moss pillars (Imura et al. 1999), many studies have been performed on the growth rate (0.7 mm per year; Imura et al. 2000); distribution pattern (Imura et al. 2003); macrostructure and carbon, nitrogen, and chlorophyll *a* contents (Kudoh et al. 2003a); temperature and light environment of the habitat (Kudoh et al. 2003b; Tanabe et al. 2008; Kimura et al. 2010); photosynthetic activity (Kudoh et al. 2003c, 2009); and nutrients inside lakebeds (Tanabe et al. 2017). Among these, the following things are noteworthy from the point of microbiological aspects. The exterior section of the pillar is dominated by green moss shoots, while the interior is packed with decomposed black-colored moss tissues with a strong smell, likely derived from sulfur compounds (e.g., hydrogen sulfide) (Kudoh et al. 2003a). This suggests that oxygen and redox-potential gradients exist between the exterior and interior portions of the pillar structure. Moss pillars thus have distinct redox-affected sections with oxidative exteriors and reductive interiors.

Metabolisms of microorganisms are diverse, enabling aerobic, micro-aerobic, and anaerobic groups to coexist along the redox gradients. This redox factor may control the microbial zonation related to carbon, nitrogen, and sulfur cycling. For instance, some bacteria can "breathe" nitrate (NO_3^-) using it as a terminal electron acceptor in micro-aerobic environments (e.g., nitrate-reducing bacteria), while others can "breathe" sulfate (SO_4^{2-}) in anaerobic environments (e.g., sulfate-reducing bacteria). Consortial metabolic associations help mediate mutually effective cycling of potentially limiting nutrients and trace elements (Paerl and Pinckney 1996a; Paerl et al. 2000), and such associations are probably important in Antarctic environments that seem to be nutrient-limited. In fact, moss pillars harbor an unexpectedly high diversity of microorganisms, including both bacteria and eukaryotes (Figs. 2.3 and 2.4), but not archaea. These microorganisms possess various metabolisms, possibly that are related in synergistic associations (see below).

2.4.2 *Cyanobacteria*

Cyanobacteria are capable of fixing carbon dioxide (CO_2) through oxygenic photosynthesis. Some members can also fix atmospheric nitrogen (N_2), contributing to the nitrogen cycle. Fixation of N_2 by cyanobacteria has been recognized as an important nitrogen source in Antarctic aquatic ecosystems (Howard-Williams et al. 1989; Fernández-Valiente et al. 2001). Remarkably, such N_2 fixation is responsible for a large proportion of the total nitrogen input in ultra-oligotrophic Antarctic lakes. One source of this may be the moss pillars since their colonies are coated with cyanobacteria (Imura et al. 1999). Indeed, cyanobacteria-affiliated phylotypes are detected on the exteriors of moss pillars

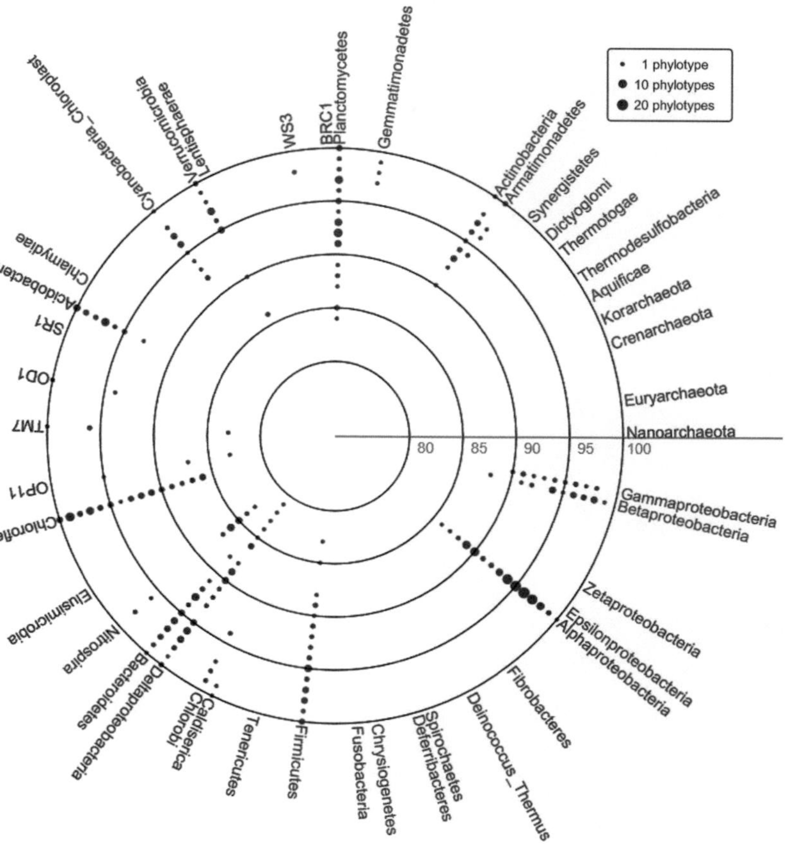

Fig. 2.3 Phylogenetic diversity and novelty of bacteria inhabiting moss pillars. Taxonomic affiliations of 564 phylotypes were determined and visualized using VITCOMIC ver. 2.0 (Mori et al. 2018; http://vitcomic.org/). Large concentric circles indicate sequence similarities corresponding to 80%, 85%, 90%, 95%, and 100% outward, respectively, for each phylotype sequence relative to the reference database consisting of genome-sequenced strains. (Sequence data from Nakai et al. 2012a)

(Fig. 2.3; Nakai et al. 2012a). Many of these phylotypes are closely related to previously described *Leptolyngbya*, *Nostoc*, *Phormidium*, and *Synechococcus* species.

The 16S rRNA gene sequence of the predominant *Leptolyngbya* phylotype, named MPB1-3, is completely identical to the sequence recovered from the benthic cyanobacterial mats in Lake Reid (69°23′S, 76°23′E), which is a hyposaline lake in the Larsemann Hills, East Antarctica (Taton et al. 2006a). This phylotype also has a high sequence similarity with that of an Antarctic isolate, *Leptolyngbya frigida*

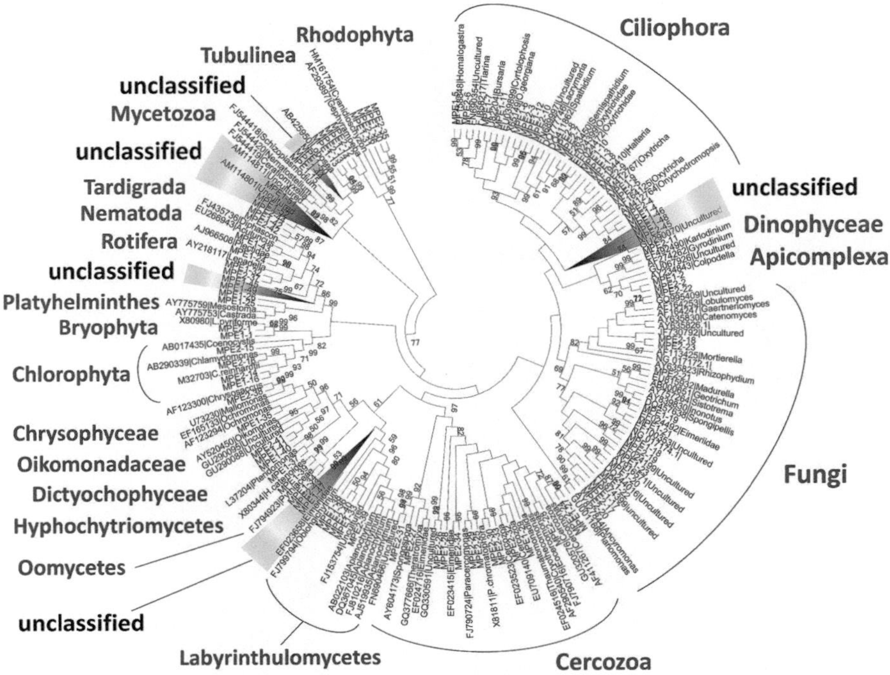

Fig. 2.4 Phylogenetic tree of eukaryotic phylotypes recovered from moss pillars. The evolutionary history of 96 phylotypes and related organisms was inferred using the maximum likelihood method based on the general time reversible model. Evolutionary analyses were conducted using MEGA7 (Kumar et al. 2016). Bootstrap values >50% based on 1000 replications are shown at the nodes. (Sequence data from Nakai et al. 2012b)

ANT.LH64B.1 (Taton et al. 2006b). *Leptolyngbya* species are often reported in various habitats of both the Arctic and Antarctic regions (Quesada and Vincent 2012). For example, *Leptolyngbya fritschiana* inhabiting Antarctic seepages are known to start to be dominant in early stages of mat formation and to then form fine mats (Komárek and Komárek 2010). The aforementioned *Phormidium* species are also known to form mats and are found in several Antarctic lakes (Wharton et al. 1983; Taton et al. 2006a).

Through analysis of genes encoding the CO_2-fixing enzyme ribulose-1,5-bisphosphate carboxylase/oxygenase (RuBisCO, E.C. 4.1.1.39), a *Leptolyngbya*-related RuBisCO genotype (named ORU5) is frequently detected (Nakai et al. 2012c), consistent with the results from 16S rRNA gene-based analysis (see above). On a phylogenetic tree, this genotype nearly aligns with a sequence recovered from Lake Bonney (77°43′S, 162° 23′E), a perennially ice-covered lake in the McMurdo Dry Valleys. A number of cyanobacterial species, including from the genera *Lyngbya*, *Nostoc*, *Oscillatoria*, and *Phormidium*, were detected in L. Bonney by microscopic observation (Vincent et al. 1993a; Wing and Priscu 1993; Paerl and Pinckney 1996b). Interestingly, although another genotype ORU13 is distantly related to known cyanobacterial

RuBisCO sequences, it has high sequence similarity with only the sequence recovered from L. Bonney. This suggests that cyanobacteria possessing a unique RuBisCO gene sequence exist in Antarctic freshwater lakes.

Culture-independent approaches also detect novel cyanobacteria in the moss pillars. Among these, the rare phylotype MPB2-6 is related to only Arctic soil-derived sequences in the public nucleotide database, implying a bipolar distribution or an anthropogenic dispersal of a unique cyanobacterial lineage (Lynch et al. 2012; Lynch and Neufeld 2015). Notably, these phylotypes recovered from polar habitats form a sister group with *Gloeobacter violaceus* PCC 7421 on a phylogenetic tree (Lynch et al. 2012). *G. violaceus* is a thylakoid-less cyanobacterium considered the most primitive among modern living cyanobacteria (Mareš et al. 2013). The finding of the sister group suggests that early cyanobacteria are more diverse than initially anticipated. Therefore, it appears that moss pillars harbor known cyanobacteria, as well as phylogenetically novel ones.

2.4.3 Proteobacteria *and Others*

The phylum *Proteobacteria* possesses the greatest number of cultivated isolates among the bacterial phyla and contains members with extremely diversified metabolisms relevant to carbon, nitrogen, and sulfur cycles (Kerters et al. 2006). Proteobacteria-affiliated phylotypes account for approximately 50% of the 16S rRNA gene clone libraries constructed from moss pillar specimens (Fig. 2.3; Nakai et al. 2012a). Although the predominant phylotype MPB1-33 is affiliated with the class *Alphaproteobacteria*, this phylotype does not show a high similarity to any other known genus. Its metabolism and ecological role remain unknown. Another subdominant phylotype MPB1-27 is related to *Bradyrhizobium* species and was recovered from both the oxidative exterior and reductive interior of moss pillars.

Bradyrhizobium species can grow heterotrophically in the presence of oxygen or can survive using denitrification as an alternative form of respiration under anaerobic conditions when nitrate is provided as an electron acceptor (van Berkum and Keyser 1985; Delgado et al. 2003). Furthermore, in micro-aerobic conditions, they can also grow utilizing light-independent chemoautotrophic CO_2 fixation by RuBisCO (Lepo et al. 1980). These versatile metabolisms may contribute to the wide distribution through the entire moss pillar structure. *Bradyrhizobium* species have been isolated from the aforementioned L. Bonney in the McMurdo Dry Valleys (Stingl et al. 2008). Generally, *Bradyrhizobium* species colonize soil environments, but it is possible that this group is also widely distributed in Antarctic lake bottoms as another niche.

During analysis of RuBisCO genotypes, it became clear that the frequency of detection of proteobacterial RuBisCO sequences gradually increased in proportion from the exterior of the pillar toward the deeper interior (Nakai et al. 2012c). As expected, *Bradyrhizobium*-related RuBisCO genotypes are detected. Surprisingly, another genotype ORU2 is closely related to endosymbiotic bacteria from the deep-sea mussel *Bathymodiolus azoricus*. Sunlight does not penetrate very deeply into the sea, so some deep-sea animals depend on microbial chemosynthesis as a primary source of organic matter (Dubilier et al. 2008). A similar situation appears to occur in the moss pillar consortia.

Considering the above discussion, there are two processes for CO_2 fixation in moss pillars. On the moss pillar surface, CO_2 fixation via photosynthesis by mosses and algae is believed to be a major carbon source, while interiorly, chemosynthesis may also contribute to carbon input. This functional zonation is possibly reflected by different microhabitat conditions (mainly redox conditions) within the two-layered structure of the pillar. A light-independent, chemosynthesis-based process may be important during the polar night or dark season. Although most research is restricted during the polar night due to logistical constraints, a pioneering study has detected RuBisCO diversity in chemoautotrophic bacteria inhabiting L. Bonney during this period (Kong et al. 2012).

Besides the frequently detected proteobacterial phylotypes, multiple phylotypes with high phylogenetic novelty have also been observed. These are classified in the following four phyla: *Bacteroidetes*, *Chloroflexi*, *Firmicutes*, and *Planctomycetes* (Fig. 2.3); the ecological role of each novel group is currently not known.

2.4.4 Uncultivated Novel Bacterial Lineages

In addition to phylotypes affiliated with known species, novel phylum-level bacterial lineages are observed in moss pillars that have no cultivated representatives and are known as candidate divisions OD1, OP11, TM7, and WS3 (Fig. 2.3; Nakai et al. 2012a). Candidate divisions OD1 and OP11 have recently been recognized as ultrasmall bacterial groups called candidate phyla radiation (CPR) (Brown et al. 2015). Inferred from genomic information and electron micrographs, several CPR bacteria may have symbiotic lifestyles associated with other organisms (Kantor et al. 2013; Luef et al. 2015; Nelson and Stegen 2015). Remarkably, sequences derived from OD1 and OP11 have been identified in other Antarctic lakes, Ace Lake (68°24′S, 78°11′E) in the Vestfold Hills (Lauro et al. 2010), and Lake Fryxell (77°37′S, 163°07′E) in the McMurdo Dry Valleys (Kwon et al. 2017).

Ultrasmall microorganisms are capable of passing through filters with a 0.2 μm pore-size, which are often used for collecting bacteria in water samples (Nakai et al. 2013; Luef et al. 2015). Because of this, it is possible that such filterable bacteria have been overlooked in microbial community surveys of Antarctic water samples. Isolation and physiological characterization of these overlooked agents are needed in future studies. One option for isolation of uncultivated microorganisms like the candidate divisions is to test and optimize culture conditions based on physiological characteristics and nutritional needs inferred from genomic/metagenomic information. Indeed, isolation of a key bacterium that dominates the Tammar Wallabies's gut microbiota was achieved by reconstructing the bacterium's metabolism from metagenomic data (Pope et al. 2011). It will continue to be essential to cultivate microorganisms living in Antarctic environments, even in the post-omics era.

2.4.5 Eukaryotes

Nutrient-rich algal mats are considered to be suitable habitats for archaea, bacteria, fungi, protists, and metazoans. Indeed, various eukaryotic taxa have been detected in mat communities from both the Antarctic and Arctic regions (Jungblut et al. 2012). Clone library analysis of 18S rRNA gene sequences show unexpectedly high diversity of eukaryotic phylotypes present in algal-coated moss pillars (Nakai et al. 2012b; Fig. 2.4). The phylotypes detected in the clone libraries generated from the pillars are dominated by novel fungi (approximately 27–75%), with the predominant fungal phylotype named MPE1-17. The closest related organism to MPE1-17 is *Entophlyctis helioformis* isolate AFTOL-ID 40, which is affiliated with the phylum *Chytridiomycota* (chytrids).

Chytrids are common fungi associated with aquatic habitats and moist soils and include parasitic species that infect hosts and saprophytic species that decompose organic material (Dewel et al. 1985; Gleason et al. 2008). Freeman et al. (2009) report chytrids as the dominant fungal communities in snowmelt-affected soils of the maritime Antarctic and high-elevation soils. Unique chytrid-dominated ecosystems have not been observed in other general environments. This implies that chytrids colonize and thrive in specific environments and play important ecological roles as parasites or decomposers. A *Chytridiomycota*-affiliated phylotype is also present in cyanobacterial mats of Fresh Pond (78°00′S, 165°32′E) on the McMurdo Ice Shelf (Jungblut et al. 2012).

Clone library surveys detect unique fungus-like protists in the pillars called Labyrinthulomycetes (also known as Labyrinthulea). These organisms prefer marine habitats such as estuarine, marine, and deep-sea waters (Raghukumar 2002; Nakai and Naganuma 2015). Freshwater species from this group remain largely unknown, except for a few lineages including species of the genera *Diplophrys* and *Fibrophrys* (Takahashi et al. 2014, 2016). While freshwater Labyrinthulomycetes appear to be rare, several Labyrinthulomycetes-affiliated phylotypes have been recovered from pillars inhabiting a freshwater lake (Nakai et al. 2012b). A preliminary microscopic study revealed *Fibrophrys*-like cells in the pillar specimens (Fig. 2.5).

By producing extracellular enzymes, members of Labyrinthulomycetes are able to degrade a wide variety of organic substrates, including relatively refractory substrates (Bremer 1995; Nagano et al. 2011). These protists also produce high amounts of long-chain polyunsaturated fatty acids (PUFAs) such as docosahexaenoic acid (DHA) and docosapentaenoic acid (DPA), which are essential fatty acids for other organisms (Yokoyama et al. 2007). As noted, in addition to some being saprophytic,

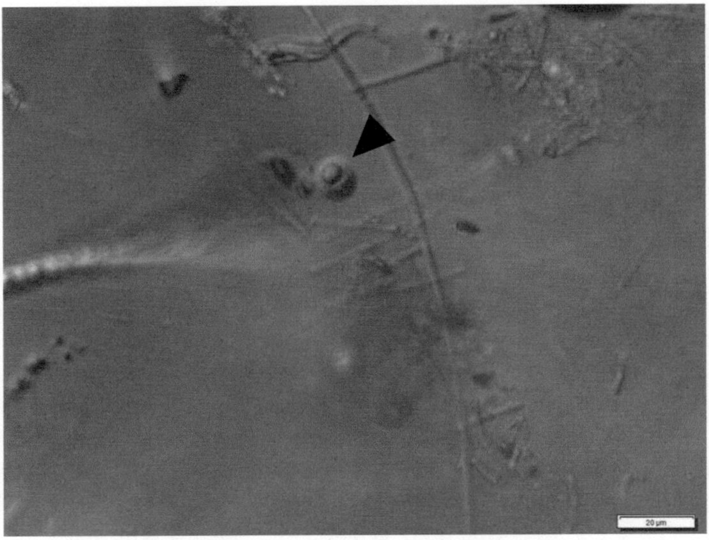

Fig. 2.5 Micrograph of labyrinthulomycete-like cell observed in the moss pillar specimen. This cell (arrowhead) contains an amber-colored lipid body and resembles *Fibrophrys* species (Amphifilidae, Labyrinthulomycetes; Takahashi et al. 2016). Scale bar, 20 μm. (Photo courtesy of Y. Takahashi & M. Yoshida)

others are known to be pathogenic parasites (Stokes et al. 2002). One Labyrinthulomycete phylotype (MPE2-39) in the pillars was unexpectedly related to a seaweed isolate ANT10.3 obtained from the southwestern Antarctic Peninsula (Mystikou et al. 2014). It was observed by microscopy that cells of isolate ANT10.3 were attached to algal filaments, suggesting a potential association between this protist and alga. Further analysis of the Labyrinthulomycetes inhabiting moss pillars and other Antarctic lakes may be needed due to their potentially distinctive characteristics.

Analyses of the phylotype composition reveal that the metazoan microfauna called "small animals" (nematodes, rotifers, and tardigrades) is patchily distributed in the exterior sections of the pillars (Nakai et al. 2012b). It is well-known that these metazoans exist in lakes and ponds in ice-free coastal areas, as well as in the ground nunataks (mountain peaks penetrating the ice sheet) of continental Antarctica (Sohlenius and Boström 2005). Clone library analysis demonstrated that moss pillars contain a previously described nematode (*Plectidae* sp.), tardigrades (*Acutuncus antarcticus*), and rotifer (*Lepadella* sp.). For tardigrades, a morphological study indicated that *Acutuncus antarcticus* is the most common and dominant species in the lake environments of the coastal regions around Syowa Station (Fig. 2.6; Tsujimoto et al. 2014).

In contrast to "small animals" inhabiting the exteriors of the pillars, certain eukaryotes are characteristically found in the pillar interiors (Nakai et al. 2012b). These phylotypes are related to cercomonads (Protozoa, Cercozoa). Cercomonads are gliding bacterivorous protozoa found globally in freshwater ecosystems and

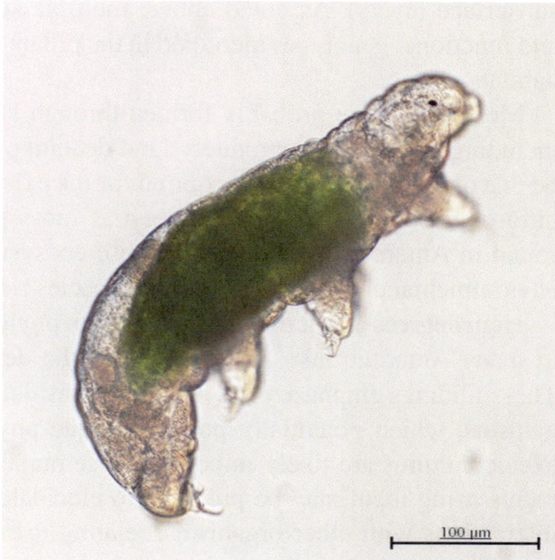

Fig. 2.6 An individual of the *Acutuncus antarcticus* strain isolated from Hamagiku-Ike Lake (formerly known as Abi-Ike Lake), Skarvsnes, East Antarctica. Scale bar, 100 μm. (Photo courtesy of M. Tsujimoto)

include species that can grow in anaerobic conditions (e.g., *Paracercomonas anaerobica*; Bass et al. 2009). Although the detected phylotypes are known distantly related species, Dumack et al. (2016) recently cultivated *Kraken carinae*, a novel lineage of Cercozoa whose sequences are highly similar (>99%) to one phylotype isolated from moss pillars. This species has the ability to prey on bacteria. Additionally, a phylotype related to ciliates (Protozoa, Ciliophora) that can be voracious grazers of bacteria was also detected in the pillar interiors. Ciliates have been well studied in multiple Antarctic freshwater lakes (Petz et al. 2007). The Ciliophora lineage contains species that have an anaerobic metabolism (e.g., nitrate respiration; Finlay et al. 1983; Kamp et al. 2015). The cercomonads and ciliates mentioned here may be major predators of bacteria in anaerobic habitats of Antarctic lakes.

As described above, several eukaryotic phylotypes detected in moss pillars are related to well-studied species. However, approximately 60% of all phylotypes have a low similarity (<95%) with known species or genera (Nakai et al. 2012b), suggesting novel taxa at the species, genus, or higher levels. Additionally, unclassified phylotypes are also identified on the phylogenetic tree (Fig. 2.4). Thus, the moss pillars clearly harbor unknown eukaryotes with high phylogenetic novelty.

2.5 Conclusions and Future Perspectives

Molecular ecological studies reveal bacterial and eukaryotic species inhabiting moss pillars that are more phylogenetically diverse than anticipated. It has also become clear that microbial zonation and functional diversity, reflected possibly by different redox conditions, exist within the pillar structure. Similar redox gradients seem to be occurred in other places in Antarctic lakes (e.g., between mat surface and subsurface layers). As noted above, multiple sequences related to the phylotypes and functional genotypes identified in the pillars are also observed in other Antarctic habitats.

Moss pillars are probably formed through an association of diverse organisms including both primary producers and decomposers. A synergistic association may be one of the crucial factors required for the existence of species-rich ecosystems in ultra-oligotrophic environments such as moss pillars and algal consortia that are found in Antarctic lake bottoms. These ecosystems are likely to be vulnerable to environmental changes that affect the species richness and biota composition.

High degrees of microbial diversity with phylogenetic novelty have been reported in many Antarctic lakes, in part due to the development of "omics" techniques. These findings emphasize the need for the isolation of as-yet-uncultivated microorganisms, which potentially possess unique physiological traits. In this situation, axenic cultures are likely important since numerous laboratory and in situ experiments using them may be prepared to elucidate physiological features, as well as interactions with other organisms, relating to competition, mutualism, and synergism. Furthermore, while our knowledge on virus diversity and activity in Antarctic lakes is limited, the virus-host relationship also seems to be important. A deeper understanding of microorganisms, including viruses, and their ways of life will stimulate additional discussion regarding how Antarctic ecosystems are shaped in extreme conditions.

Acknowledgment Photo in Fig. 2.1 was taken during the 56th Japanese Antarctic Research Expedition (JARE-56). We thank Drs. Y. Takahashi and M. Yoshida for providing unpublished data for Fig. 2.5 and Dr. M. Tsujimoto for providing the micrograph for Fig. 2.6. The work on novel labyrinthulomycete lineages in the moss pillars was supported by the Sasakawa Scientific Research Grant (no. 29-726) from The Japan Science Society.

References

Andersen, D. T., Sumner, D. Y., Hawes, I., Webster-Brown, J., & McKay, C. P. (2011). Discovery of large conical stromatolites in Lake Untersee, Antarctica. *Geobiology, 9*, 280–293.

Bass, D., Howe, A. T., Mylnikov, A. P., Vickerman, K., Chao, E. E., Edwards Smallbone, J., Snell, J., Cabral, C., & Cavalier-Smith, T. (2009). Phylogeny and classification of Cercomonadida (Protozoa, Cercozoa): *Cercomonas*, *Eocercomonas*, *Paracercomonas*, and *Cavernomonas* gen. nov. *Protist, 160*, 483–521.

Bowman, J. P., McCammon, S. A., Rea, S. M., & McMeekin, T. A. (2000). The microbial composition of three limnologically disparate hypersaline Antarctic lakes. *FEMS Microbiology Letters, 183*, 81–88.

Bremer, G. B. (1995). Lower marine fungi (labyrinthulomycetes) and the decay of mangrove leaf litter. *Hydrobiologia, 295*, 89–95.

Brown, C. T., Hug, L. A., Thomas, B. C., Sharon, I., Castelle, C. J., Singh, A., Wilkins, M. J., Wrighton, K. C., Williams, K. H., & Banfield, J. F. (2015). Unusual biology across a group comprising more than 15% of domain bacteria. *Nature, 523*, 208–211.

Casanueva, A., Tuffin, M., Cary, C., & Cowan, D. A. (2010). Molecular adaptations to psychrophily: The impact of 'omic' technologies. *Trends in Microbiology, 18*, 374–381.

Christner, B. C., Priscu, J. C., Achberger, A. M., Barbante, C., Carter, S. P., Christianson, K., Michaud, A. B., Mikucki, J. A., Mitchell, A. C., Skidmore, M. L., Vick-Majors, T. J., & Team, W. S. (2014). A microbial ecosystem beneath the West Antarctic ice sheet. *Nature, 512*, 310–313.

Creer, S., Deiner, K., Frey, S., Porazinska, D., Taberlet, P., Thomas, W. K., Potter, C., & Bik, H. M. (2016). The ecologist's field guide to sequence-based identification of biodiversity. *Methods in Ecology and Evolution, 7*, 1008–1018.

Delgado, M. J., Bonnard, N., Tresierra-Ayala, A., Bedmar, E. J., & Müller, P. (2003). The *Bradyrhizobium japonicum napEDABC* genes encoding the periplasmic nitrate reductase are essential for nitrate respiration. *Microbiology, 149*, 3395–3403.

Dewel, R. A., Joines, J. D., & Bond, J. J. (1985). A new chytridiomycete parasitizing the tardigrade *Milnesium tardigradum*. *Canadian Journal of Botany, 63*, 1525–1534.

Dubilier, N., Bergin, C., & Lott, C. (2008). Symbiotic diversity in marine animals: The art of harnessing chemosynthesis. *Nature Reviews Microbiology, 6*, 725–740.

Dumack, K., Schuster, J., Bass, D., & Bonkowski, M. (2016). A novel lineage of 'naked filose amoebae'; *Kraken carinae* gen. nov. sp. nov. (Cercozoa) with a remarkable locomotion by disassembly of its cell body. *Protist, 167*, 268–278.

Fernández-Valiente, E., Quesada, A., Howard-Williams, C., & Hawes, I. (2001). N_2-fixation in cyanobacterial Mats from ponds on the McMurdo ice shelf, Antarctica. *Microbial Ecology, 42*, 338–349.

Finlay, B. J., Span, A. S. W., & Harman, J. M. P. (1983). Nitrate respiration in primitive eukaryotes. *Nature, 303*, 333–336.

Freeman, K. R., Martin, A. P., Karki, D., Lynch, R. C., Mitter, M. S., Meyer, A. F., Longcore, J. E., Simmons, D. R., & Schmidt, S. K. (2009). Evidence that chytrids dominate fungal communities in high-elevation soils. *Proceedings of the National Academy of Sciences, 106*, 18315–18320.

Gillieson, D., Burgess, J., Spate, A., & Cochrane, A. (1990). *An atlas of the lakes of the Larsemann Hills, Princess Elizabeth Land, Antarctica*. Kingston: The Publications Office, Australian Antarctic Division.

Gleason, F. H., Kagami, M., Lefevre, E., & Sime-Ngando, T. (2008). The ecology of chytrids in aquatic ecosystems: Roles in food web dynamics. *Fungal Biology Reviews, 22*, 17–25.

Goldman, C. R., Mason, D. T., & Wood, B. J. B. (1963). Light injury and inhibition in Antarctic freshwater phytoplankton. *Limnology and Oceanography, 8*, 313–322.

Hodgson, D. A., Vyverman, W., Verleyen, E., Sabbe, K., Leavitt, P. R., Taton, A., Squier, A. H., & Keely, B. J. (2004). Environmental factors influencing the pigment composition of in situ benthic microbial communities in East Antarctic lakes. *Aquatic Microbial Ecology, 37*, 247–263.

Howard-Williams, C., Priscu, J. C., & Vincent, W. F. (1989). Nitrogen dynamics in two antarctic streams. *Hydrobiologia, 172*, 51–61.

Imura, S., Bando, T., Saito, S., Seto, K., & Kanda, H. (1999). Benthic moss pillars in Antarctic lakes. *Polar Biology, 22*, 137–140.

Imura, S., Takahashi, H., & Nakamura, T. (2000). Benthic moss pillars (Koke Bouzu) in Antarctic lakes–analysis of colonization and growth by 14C dating. *Summary of Research using AMS Nagoya University, XI*, 176–183 in Japanese with English abstract, table, and figures.

Imura, S., Bando, T., Seto, K., Ohtani, S., Kudoh, S., & Kanda, H. (2003). Distribution of aquatic mosses in the Sôya coast region, East Antarctica. *Polar Bioscience, 16*, 1–10.

James, S. R., Burton, H. R., McMeekin, T. A., & Mancuso, C. A. (1994). Seasonal abundance of *Halomonas meridiana*, *Halomonas subglaciescola*, *Flavobacterium gondwanense* and *Flavobacterium salegens* in four Antarctic lakes. *Antarctic Science, 6*, 325–332.

Jungblut, A. D., Vincent, W. F., & Lovejoy, C. (2012). Eukaryotes in Arctic and Antarctic cyanobacterial mats. *FEMS Microbiology Ecology, 82*, 416–428.

Kamp, A., Høgslund, S., Risgaard-Petersen, N., & Stief, P. (2015). Nitrate storage and dissimilatory nitrate reduction by eukaryotic microbes. *Frontiers in Microbiology, 6*, 1492.

Kantor, R. S., Wrighton, K. C., Handley, K. M., Sharon, I., Hug, L. A., Castelle, C. J., Thomas, B. C., & Banfield, J. F. (2013). Small genomes and sparse metabolisms of sediment-associated bacteria from four candidate phyla. *MBio, 4*, e00708–e00713.

Kato, K., Arikawa, T., Imura, S., & Kanda, H. (2013). Molecular identification and phylogeny of an aquatic moss species in Antarctic lakes. *Polar Biology, 36*, 1557–1568.

Kerters, K., De Vos, P., Gillis, M., Swings, J., Vandamme, P., & Stackebrandt, E. (2006). Introduction to the Proteobacteria. In M. Dworkin, S. Falkow, E. Rosenberg, K.-H. Schleifer, & E. Stackebrandt (Eds.), *The prokaryotes, a handbook on the biology of bacteria, Proteobacteria: Alpha and Beta Subclasses* (Vol. 5, pp. 3–37). New York: Springer.

Kimura, S., Ban, S., Imura, S., Kudoh, S., & Matsuzaki, M. (2010). Limnological characteristics of vertical structure in the lakes of Syowa Oasis, East Antarctica. *Polar Science, 3*, 262–271.

Komárek, O., & Komárek, J. (2010). Diversity and ecology of cyanobacterial microflora of Antarctic seepage habitats: Comparison of King George Island, Shetland Islands, and James Ross Island, NW Weddell Sea, Antarctica. In J. Seckbach & A. Oren (Eds.), *Microbial mats: Modern and ancient microorganisms in stratified systems* (pp. 515–539). Dordrecht: Springer.

Kong, W., Ream, D. C., Priscu, J. C., & Morgan-Kiss, R. M. (2012). Diversity and expression of RuBisCO genes in a perennially ice-covered Antarctic lake during the polar night transition. *Applied and Environmental Microbiology, 78*, 4358–4366.

Kudoh, S., & Tanabe, Y. (2014). Limnology and ecology of lakes along the Sôya Coast, East Antarctica. *Advances in Polar Science, 25*, 75–91.

Kudoh, S., Kashino, Y., & Imura, S. (2003a). Ecological studies of aquatic moss pillars in Antarctic lakes 3. Light response and chilling and heat sensitivity of photosynthesis. *Polar Bioscience, 16*, 33–42.

Kudoh, S., Tsuchiya, Y., Ayukawa, E., Imura, S., & Kanda, H. (2003b). Ecological studies of aquatic moss pillars in Antarctic lakes 1. Macro structure and carbon, nitrogen and chlorophyll a contents. *Polar Bioscience, 16*, 11–22.

Kudoh, S., Watanabe, K., & Imura, S. (2003c). Ecological studies of aquatic moss pillars in Antarctic lakes 2. Temperature and light environment at the moss habitat. *Polar Bioscience, 16*, 23–32.

Kudoh, S., Tanabe, Y., Matsuzaki, M., & Imura, S. (2009). In situ photochemical activity of the phytobenthic communities in two Antarctic lakes. *Polar Biology, 32*, 1617–1627.

Kumar, S., Stecher, G., & Tamura, K. (2016). MEGA7: Molecular evolutionary genetics analysis version 7.0 for bigger datasets. *Molecular Biology and Evolution, 33*, 1870–1874.

Kwon, M., Kim, M., Takacs-Vesbach, C., Lee, J., Hong, S. G., Kim, S. J., Priscu, J. C., & Kim, O.-S. (2017). Niche specialization of bacteria in permanently ice-covered lakes of the McMurdo Dry Valleys, Antarctica. *Environmental Microbiology, 19*, 2258–2271.

Lauro, F. M., DeMaere, M. Z., Yau, S., Brown, M. V., Ng, C., Wilkins, D., Raftery, M. J., Gibson, J. A. E., Andrews-Pfannkoch, C., Lewis, M., Hoffman, J. M., Thomas, T., & Cavicchioli, R. (2010). An integrative study of a meromictic lake ecosystem in Antarctica. *The ISME Journal, 5*, 879–895.

Laybourn-Parry, J., & Wadham, J. L. (2014). *Antarctic lakes*. Oxford: Oxford University Press.

Lepo, J. E., Hanus, F. J., & Evans, H. J. (1980). Chemoautotrophic growth of hydrogen-uptake-positive strains of *Rhizobium japonicum*. *Journal of Bacteriology, 141*, 664–670.

Lizotte, M. P. (2008). Phytoplankton and primary production. In W. F. Vincent & J. Laybourn-Parry (Eds.), *Polar lakes and rivers: Limnology of Arctic and Antarctic aquatic ecosystems* (pp. 157–178). London: University Press.

López-Bueno, A., Tamames, J., Velázquez, D., Moya, A., Quesada, A., & Alcamí, A. (2009). High diversity of the viral community from an Antarctic Lake. *Science, 326*, 858–861.

Luef, B., Frischkorn, K. R., Wrighton, K. C., Holman, H.-Y. N., Birarda, G., Thomas, B. C., Singh, A., Williams, K. H., Siegerist, C. E., Tringe, S. G., Downing, K. H., Comolli, L. R., & Banfield, J. F. (2015). Diverse uncultivated ultra-small bacterial cells in groundwater. *Nature Communications, 6*, 6372.

Lynch, M. D. J., & Neufeld, J. D. (2015). Ecology and exploration of the rare biosphere. *Nature Reviews Microbiology, 13*, 217–229.

Lynch, M. D. J., Bartram, A. K., & Neufeld, J. D. (2012). Targeted recovery of novel phylogenetic diversity from next-generation sequence data. *The ISME Journal, 6*, 2067–2077.

Mareš, J., Hrouzek, P., Kaňa, R., Ventura, S., Strunecký, O., & Komárek, J. (2013). The primitive thylakoid-less cyanobacterium *Gloeobacter* is a common rock-dwelling organism. *PLoS One, 8*, e66323.

Mori, H., Maruyama, T., Yano, M., Yamada, T., & Kurokawa, K. (2018). VITCOMIC2: Visualization tool for the phylogenetic composition of microbial communities based on 16S rRNA gene amplicons and metagenomic shotgun sequencing. *BMC Systems Biology, 12*, 30.

Morris, D. P., Zagarese, H., Williamson, C. E., Balseiro, E. G., Hargreaves, B. R., Modenutti, B., Moeller, R., & Queimalinos, C. (1995). The attenuation of solar UV radiation in lakes and the role of dissolved organic carbon. *Limnology and Oceanography, 40*, 1381–1391.

Mystikou, A., Peters, A. F., Asensi, A. O., Fletcher, K. I., Brickle, P., van West, P., Convey, P., & Küpper, F. C. (2014). Seaweed biodiversity in the South-Western Antarctic Peninsula: Surveying macroalgal community composition in the Adelaide Island/Marguerite Bay region over a 35-year time span. *Polar Biology, 37*, 1607–1619.

Nagano, N., Matsui, S., Kuramura, T., Taoka, Y., Honda, D., & Hayashi, M. (2011). The distribution of extracellular cellulase activity in marine eukaryotes, thraustochytrids. *Marine Biotechnology, 13*, 133–136.

Naganuma, T., Hua, P. N., Okamoto, T., Ban, S., Imura, S., & Kanda, H. (2005). Depth distribution of euryhaline halophilic bacteria in Suribati Ike, a meromictic lake in East Antarctica. *Polar Biology, 28*, 964–970.

Nakai, R., & Naganuma, T. (2015). Diversity and ecology of thraustochytrid protists in the marine environment. In S. Ohtsuka, T. Suzaki, T. Horiguchi, N. Suzuki, & F. Not (Eds.), *Marine protists: Diversity and dynamics* (pp. 331–346). Tokyo: Springer.

Nakai, R., Abe, T., Baba, T., Imura, S., Kagoshima, H., Kanda, H., Kanekiyo, A., Kohara, Y., Koi, A., Nakamura, K., Narita, T., Niki, H., Yanagihara, K., & Naganuma, T. (2012a). Microflorae of aquatic moss pillars in a freshwater lake, East Antarctica, based on fatty acid and 16S rRNA gene analyses. *Polar Biology, 35*, 425–433.

Nakai, R., Abe, T., Baba, T., Imura, S., Kagoshima, H., Kanda, H., Kohara, Y., Koi, A., Niki, H., Yanagihara, K., & Naganuma, T. (2012b). Eukaryotic phylotypes in aquatic moss pillars inhabiting a freshwater lake in East Antarctica, based on 18S rRNA gene analysis. *Polar Biology, 35*, 1495–1504.

Nakai, R., Abe, T., Baba, T., Imura, S., Kagoshima, H., Kanda, H., Kohara, Y., Koi, A., Niki, H., Yanagihara, K., & Naganuma, T. (2012c). Diversity of RuBisCO gene responsible for CO2 fixation in an Antarctic moss pillar. *Polar Biology, 35*, 1641–1650.

Nakai, R., Shibuya, E., Justel, A., Rico, E., Quesada, A., Kobayashi, F., Iwasaka, Y., Shi, G.-Y., Amano, Y., Iwatsuki, T., & Naganuma, T. (2013). Phylogeographic analysis of filterable bacteria with special reference to *Rhizobiales* strains that occur in cryospheric habitats. *Antarctic Science, 25*, 219–228.

Nelson, W., & Stegen, J. (2015). The reduced genomes of Parcubacteria (OD1) contain signatures of a symbiotic lifestyle. *Frontiers in Microbiology, 6*, 713.

Niederberger, T. D., McDonald, I. R., & Cary, S. C. (2012). Archaea. In R. V. Miller & L. G. Whyte (Eds.), *Polar microbiology: Life in a deep freeze* (pp. 32–61). Washington, DC: ASM Press.

Paerl, H. W., & Pinckney, J. L. (1996a). A mini-review of microbial consortia: Their roles in aquatic production and biogeochemical cycling. *Microbial Ecology, 31*, 225–247.

Paerl, H. W., & Pinckney, J. L. (1996b). Ice aggregates as a microbial habitat in Lake Bonney, Dry Valley lakes, Antarctica: Nutrient-rich microzones in an oligotrophic ecosystem. *Antarctic Journal of the United States, 31*, 220–222.

Paerl, H. W., Pinckney, J. L., & Steppe, T. F. (2000). Cyanobacterial–bacterial mat consortia: Examining the functional unit of microbial survival and growth in extreme environments. *Environmental Microbiology, 2*, 11–26.

Parker, B. C., Simmons, J. G. M., Love, F. G., Wharton, J. R. A., & Seaburg, K. G. (1981). Modern stromatolites in Antarctic Dry Valley lakes. *Bioscience, 31*, 656–661.

Petit, J. R., Alekhina, I., & Bulat, S. (2005). Lake Vostok, Antarctica: Exploring a subglacial lake and searching for life in an extreme environment. In M. Gargaud, B. Barbier, H. Martin, & J. Reisse (Eds.), *Lectures in astrobiology* (Vol. I, pp. 227–288). Berlin: Springer.

Petz, W., Valbonesi, A., Schiftner, U., Quesada, A., & Cynan Ellis-Evans, J. (2007). Ciliate biogeography in Antarctic and Arctic freshwater ecosystems: Endemism or global distribution of species? *FEMS Microbiology Ecology, 59*, 396–408.

Pienitz, R., Doran, P. T., & Lamoureux, S. F. (2008). Origin and geomorphology of lakes in the polar regions. In W. F. Vincent & J. Laybourn-Parry (Eds.), *Polar lakes and rivers: Limnology of Arctic and Antarctic aquatic ecosystems* (pp. 25–41). London: University Press.

Pope, P. B., Smith, W., Denman, S. E., Tringe, S. G., Barry, K., Hugenholtz, P., McSweeney, C. S., McHardy, A. C., & Morrison, M. (2011). Isolation of succinivibrionaceae implicated in low methane emissions from Tammar wallabies. *Science, 333*, 646–648.

Priscu, J. C., Adams, E. E., Lyons, W. B., Voytek, M. A., Mogk, D. W., Brown, R. L., McKay, C. P., Takacs, C. D., Welch, K. A., Wolf, C. F., Kirshtein, J. D., & Avci, R. (1999). Geomicrobiology of subglacial ice above Lake Vostok, Antarctica. *Science, 286*, 2141–2144.

Quesada, A., & Vincent, W. F. (2012). Cyanobacteria in the cryosphere: Snow, ice and extreme cold. In B. A. Whitton (Ed.), *Ecology of cyanobacteria II: Their diversity in space and time* (pp. 387–399). Dordrecht: Springer.

Raghukumar, S. (2002). Ecology of the marine protists, the Labyrinthulomycetes (Thraustochytrids and Labyrinthulids). *European Journal of Protistology, 38*, 127–145.

Roos, J. C., & Vincent, W. F. (1998). Temperature dependence of UV radiation effects on Antarctic cyanobacteria. *Journal of Phycology, 34*, 118–125.

Shivaji, S., & Reddy, G. S. N. (2010). Bacterial biodiversity of Antarctica: Conventional polyphasic and rRNA approaches. In A. K. Bej, J. Aislabie, & R. M. Atlas (Eds.), *Polar microbiology: The ecology, biodiversity and bioremediation potential of microorganisms in extremely cold environments* (pp. 61–93). Boca Raton: CRC Press.

Siegert, M. J., Ross, N., & Le Brocq, A. M. (2016). Recent advances in understanding Antarctic subglacial lakes and hydrology. *Philosophical Transactions of the Royal Society A, 374*, 20140306.

Sohlenius, B., & Boström, S. (2005). The geographic distribution of metazoan microfauna on East Antarctic nunataks. *Polar Biology, 28*, 439–448.

Stingl, U., Cho, J. C., Foo, W., Vergin, K. L., Lanoil, B., & Giovannoni, S. J. (2008). Dilution-to-extinction culturing of psychrotolerant planktonic bacteria from permanently ice-covered lakes in the McMurdo Dry Valleys, Antarctica. *Microbial Ecology, 55*, 395–405.

Stokes, N. A., Calvo, L. M. R., Reece, K. S., & Burreson, E. M. (2002). Molecular diagnostics, field validation, and phylogenetic analysis of quahog parasite unknown (QPX), a pathogen of the hard clam *Mercenaria mercenaria. Diseases of Aquatic Organisms, 52*, 233–247.

Takahashi, Y., Yoshida, M., Inouye, I., & Watanabe, M. M. (2014). *Diplophrys mutabilis* sp. nov., a new member of Labyrinthulomycetes from freshwater habitats. *Protist, 165*, 50–65.

Takahashi, Y., Yoshida, M., Inouye, I., & Watanabe, M. M. (2016). *Fibrophrys columna* gen. nov., sp. nov: A member of the family Amphifilidae. *European Journal of Protistology, 56*, 41–50.

Tanabe, Y., Kudoh, S., Imura, S., & Fukuchi, M. (2008). Phytoplankton blooms under dim and cold conditions in freshwater lakes of East Antarctica. *Polar Biology, 31*, 199–208.

Tanabe, Y., Ohtani, S., Kasamatsu, N., Fukuchi, M., & Kudoh, S. (2010). Photophysiological responses of phytobenthic communities to the strong light and UV in Antarctic shallow lakes. *Polar Biology, 33*, 85–100.

Tanabe, Y., Yasui, S., Osono, T., Uchida, M., Kudoh, S., & Yamamuro, M. (2017). Abundant deposits of nutrients inside lakebeds of Antarctic oligotrophic lakes. *Polar Biology, 40*, 603–613.

Taton, A., Grubisic, S., Balthasart, P., Hodgson, D. A., Laybourn-Parry, J., & Wilmotte, A. (2006a). Biogeographical distribution and ecological ranges of benthic cyanobacteria in East Antarctic lakes. *FEMS Microbiology Ecology, 57*, 272–289.

Taton, A., Grubisic, S., Ertz, D., Hodgson, D. A., Piccardi, R., Biondi, N., Tredici, M. R., Mainini, M., Losi, D., Marinelli, F., & Wilmotte, A. (2006b). Polyphasic study of Antarctic cyanobacterial strains. *Journal of Phycology, 42*, 1257–1270.

Tsujimoto, M., McInnes, S. J., Convey, P., & Imura, S. (2014). Preliminary description of tardigrade species diversity and distribution pattern around coastal Syowa Station and inland Sør Rondane Mountains, Dronning Maud Land, East Antarctica. *Polar Biology, 37*, 1361–1367.

van Berkum, P., & Keyser, H. H. (1985). Anaerobic growth and denitrification among different serogroups of soybean rhizobia. *Applied and Environmental Microbiology, 49*, 772–777.

Verleyen, E., Hodgson, D. A., Sabbe, K., Cremer, H., Emslie, S. D., Gibson, J., Hall, B., Imura, S., Kudoh, S., Marshall, G. J., McMinn, A., Melles, M., Newman, L., Roberts, D., Roberts, S. J., Singh, S. M., Sterken, M., Tavernier, I., Verkulich, S., de Vyver, E. V., Van Nieuwenhuyze, W., Wagner, B., & Vyverman, W. (2011). Post-glacial regional climate variability along the East Antarctic coastal margin—Evidence from shallow marine and coastal terrestrial records. *Earth-Science Reviews, 104*, 199–212.

Vincent, W. F., & Quesada, A. (2012). Cyanobacteria in high latitude lakes, rivers and seas. In B. A. Whitton (Ed.), *Ecology of cyanobacteria II: Their diversity in space and time* (pp. 371–385). Dordrecht: Springer.

Vincent, W. F., Castenholz, R. W., Downes, M. T., & Howard-Williams, C. (1993a). Antarctic cyanobacteria: Light, nutrients, and photosynthesis in the microbial mat environment. *Journal of Phycology, 29*, 745–755.

Vincent, W. F., Downes, M. T., Castenholz, R. W., & Howard-Williams, C. (1993b). Community structure and pigment organisation of cyanobacteria-dominated microbial mats in Antarctica. *European Journal of Phycology, 28*, 213–221.

Wharton, R. A., Parker, B. C., & Simmons, G. M. (1983). Distribution, species composition and morphology of algal mats in Antarctic dry valley lakes. *Phycologia, 22*, 355–365.

Wing, K. T., & Priscu, J. C. (1993). Microbial communities in the permanent ice cap of Lake Bonney, Antarctica: Relationships among chlorophyll-*a*, gravel and nutrients. *Antarctic JUS Review, 28*, 246–249.

Wright, A., & Siegert, M. (2012). A fourth inventory of Antarctic subglacial lakes. *Antarctic Science, 24*, 659–664.

Wynn-Williams, D. D. (1990). Ecological aspects of Antarctic microbiology. In K. C. Marshall (Ed.), *Advances in microbial ecology* (pp. 71–146). Boston: Springer.

Yau, S., Lauro, F. M., Williams, T. J., Demaere, M. Z., Brown, M. V., Rich, J., Gibson, J. A. E., & Cavicchioli, R. (2013). Metagenomic insights into strategies of carbon conservation and unusual sulfur biogeochemistry in a hypersaline Antarctic lake. *The ISME Journal, 7*, 1944–1961.

Yokoyama, R., Salleh, B., & Honda, D. (2007). Taxonomic rearrangement of the genus *Ulkenia* sensu lato based on morphology, chemotaxonomical characteristics, and 18S rRNA gene phylogeny (Thraustochytriaceae, Labyrinthulomycetes): Emendation for *Ulkenia* and erection of *Botryochytrium*, *Parietichytrium*, and *Sicyoidochytrium* gen. nov. *Mycoscience, 48*, 329–341.

Chapter 3
Yeast Activities Involved in Carbon and Nitrogen Cycles in Antarctica

Silvana Vero, Gabriela Garmendia, Adalgisa Martínez-Silveira, Ivana Cavello, and Michael Wisniewski

Abstract Antarctica and sub-Antarctic regions are characterized by extreme conditions for life such as low temperatures and nutrient availability, high solar irradiation, and dryness; however, microorganisms from the three domains of life have been found as common inhabitants of soils and waters from those zones. Despite bacteria being the most numerous microorganisms in those habitats, a great diversity of psychrotrophic and psychrophilic yeasts have been also isolated and described. Yeasts, as chemoheterotrophic microorganisms, are involved in the recycling and mineralization of organic matter, playing an important role in carbon cycle. The range of organic substrates that they can degrade is wide. Their ability to produce extracellular hydrolytic enzymes involved in the breakdown of natural organic polymers has been well documented. Moreover, they can also use other substrates as n-alkanes or polyphenolic compounds as a sole carbon and energy source, so they could play a role in bioremediation in human-impacted areas. Most yeast obtain their energy by aerobic respiration; however, in anaerobic conditions, some of them carry out fermentation or anaerobic respiration. The use of nitrate or nitrite as the final electron acceptor provides nitrous oxide (a greenhouse gas) as an end product. Thus, those yeasts can be considered as denitrifying microorganisms playing an important role in the nitrogen cycle.

Keywords Psychrotrophic and psychrophilic yeasts · Carbon biogeochemical cycling · Nitrogen biogeochemical cycling · Phosphorus biogeochemical cycling · Production of extracellular enzymes

---

S. Vero (✉) · G. Garmendia · A. Martínez-Silveira
Laboratory of Biotechnology, Microbiology Area, Department of Bioscience, Faculty of Chemistry, Universidad de la República, Montevideo, Uruguay
e-mail: svero@fq.edu.uy

I. Cavello
Research and Development Center for Industrial Fermentations, CINDEFI (CONICET-La Plata, UNLP), La Plata, Argentina

M. Wisniewski
USDA-ARS Appalachian Fruit Research Station, Kearneysville, WV, USA

S. Castro-Sowinski (ed.), *The Ecological Role of Micro-organisms in the Antarctic Environment*, Springer Polar Sciences,
https://doi.org/10.1007/978-3-030-02786-5_3

3.1 Yeasts in Antarctic and Sub-Antarctic Habitats

The extreme environmental conditions in Antarctica, such as low temperatures, high UV irradiation, oligotrophy, and low water availability in soils, represent a challenge for most organisms (Weinstein et al. 2000). Microbial life, however, has been found in every ice-free Antarctic and sub-Antarctic environment, ranging from the continental deserts of the McMurdo Dry Valleys to the maritime islands where conditions are milder (Lawley et al. 2004). In general, microbial abundance in Antarctic soils exhibits a positive relationship with vegetation, water content, the concentration of organic carbon, total nitrogen, and with increasing temperature (Yergeau et al. 2007b). Fungi (yeasts and molds) are organotrophic and heterotrophic microorganisms, requiring organic compounds as energy and carbon sources. Thus, their growth is associated with the presence of organic matter. Walton (1985) reported that fungi are the dominant decomposers in Antarctica, so their role in nutrient cycles is essential and needs to be considered. Yeasts appear to be better adapted than filamentous fungi to polar environments, and so they dominate in most locations (Zalar and Gunde-Cimerman 2014). For example, yeasts were the only fungi isolated from Taylor Valley, Mt. Discovery, Wright Valley, and two mountain peaks on South Victoria Land, located on continental Antarctica (Connell et al. 2008). In most cases, however, the abundance of yeasts recovered from Antarctic or sub-Antarctic environments has been low. Martinez et al. (2016), in their analysis of soil and water samples from King George island, found that yeasts could not be recovered directly from 64% of the samples without prior enrichment, indicating that the yeast concentration in those samples was lower than 10 cfu g^{-1} or 1 cfu ml^{-1} in solid and water samples, respectively. Additionally, culturable yeasts were less than 1 cfu in 10 g or 10 ml (dry and liquid samples, respectively) in 25% of the samples, since they could not be recovered even after enrichment. In fact, yeast counts were low in all of the examined samples, with 5×10^4 cfu g^{-1} being the highest concentration detected.

Both ascomycetous and basidiomycetous yeasts have been isolated from most ice-free Antarctic and sub-Antarctic habitats, although again not in high numbers. The majority of studies have indicated that basidiomycetous yeasts are better adapted to cold and dry environments than ascomycetous yeasts and so are more frequently recovered from Antarctic habitats (Turchetti et al. 2008; Connell et al. 2008; Moliné et al. 2011; Uetake et al. 2012; Carrasco et al. 2016). Among basidiomycetous yeasts, species of *Cryptococcus* and *Rhodotorula* have been most isolated, especially in arid soils (Connell et al. 2014; Buzzini et al. 2017). Studies of fungal diversity in soils from McMurdo Dry Valleys, which are acknowledged as the coldest, driest regions on Earth Wynn-Williams (1990), reported the presence of culturable yeasts identified as *Cryptococcus antarcticus*, *C. friedmannii*, *C. vishniacii* in samples collected from a variety of sites, including Lake Fryxell Basin, Mt. Fleming, and Allan Hills (Arenz et al. 2006). Martinez et al. (2016) reported that *Rhodotorula laryngis* (Fig. 3.1) was the predominant basidiomycetous yeast species found in water and soil samples from King George Island, in the sub-Antarctic

Fig. 3.1 *Rhodotorula laringys* F4A. (**a**) Sampling site; (**b**) growth on PDA at 25 °C during 48 h; (**c**) optical microscope observation (400×)

region, with *Cryptococcus victoriae* being the second most abundant species. The ability of these taxa to survive in dry, oligotrophic, and cold environmental conditions has been associated with the presence of certain traits. These are (1) a characteristic polysaccharide capsule that confers desiccation resistance (Connell et al. 2008); (2) high-affinity and low-specificity transporters, which facilitate nutrient uptake from oligotrophic environments; and (3) a membrane with a much higher content of unsaturated fatty acids, which could maintain fluidity at low temperatures (Buzzini and Margesin 2014). Additionally, the production of photoprotective compounds such as carotenoids, which have been extensively reported in *Rhodotorula* species, is also a strategy to survive in places with high UV radiation such as Antarctica.

Among ascomycetous yeasts, *Candida* and *Debaryomyces* species have been commonly isolated from cold habitats (Duarte et al. 2013). In particular *D. hansenii* (Fig. 3.2), a psychro-, xero-, and halotolerant yeast species, has been isolated from sea, freshwater, and sub-Antarctic soils, as well as from maritime and inland Antarctica soils (Martinez et al. 2016; Buzzini et al. 2017). It was also the only yeast species isolated from oligotrophic soil samples collected from the extreme inland site of McKelvey Valley, where total organic carbon was only 0.11 g/100 g of soil (Rao et al. 2012). *D. hansenii* was also the predominant ascomycetous yeast species found by Martinez et al. (2016) in samples from King George Island.

Other yeasts and yeast-like fungi have been found as part of cryptoendolithic communities in Antarctic rocks. These yeasts are commonly known as black yeasts

Fig. 3.2 *Debaryomyces hansenii* F18A. (**a**) Sampling site; (**b**) growth on PDA at 25 °C during 48 h; (**c**) optical microscope observation (400×)

due to their melanized cell walls. Black yeasts exhibit high resistance to solar irradiation, desiccation, and oligotrophic conditions (Selbmann et al. 2008; Zalar et al. 2008). *Cryomyces antarcticus* is one example of a black yeast that can colonize extreme environments. Since the habitat *C. antarcticus* occupies is similar to habitats found on Mars, it has been used as a eukaryotic model in exobiological studies (Zalar and Gunde-Cimerman 2014).

3.2 The Role of Yeasts in the Carbon Cycle

The supply of elements that constitute biomass, such as carbon, nitrogen, sulfur, phosphorous, hydrogen, and oxygen, is finite, and so their recycling through biogeochemical cycles is necessary. Biogeochemical cycles can be defined as natural pathways that allow essential elements of living matter to circulate through the biotic and abiotic parts of an ecosystem. Microbial activity is an essential part of these cycles. In fact, biogeochemical cycles in Antarctic terrestrial habitats are almost exclusively driven by microbial activity (Yergeau et al. 2007a). The presence and growth of psychrophilic and cold-adapted yeasts in most ice-free Antarctic habitats are indicative of their active role in the recycling of numerous elements (Thomas-Hall et al. 2009), even in oligotrophic environments.

The carbon cycle involves various microbial activities including CO_2 fixation, organic matter decomposition, mineralization to CO_2, oxidation of methane (methanotrophy), and production of methane (methanogenesis). The latter is an exclusively anaerobic process, while the other activities may occur under both aerobic and anaerobic conditions (Fig. 3.3) (Cui et al. 2015; Madigan et al. 2015).

Production of methane from organic matter is a multi-step process, in which methanogenesis, carried out by microorganisms from the Archaea domain, is the final step. Oxidation of methane is a bacterial process, while carbon dioxide fixation can be carried out by chemolithotrophic bacteria and by oxygenic and anoxygenic photosynthetic organisms, which does not include yeasts. As chemoheterotrophic microorganisms, the specific role of yeasts in the carbon cycle involves the recycling and mineralization of organic matter in both aerobic and anaerobic environments (Fig. 3.3). Yeasts, like other fungi, use only organic compounds as energy sources and release CO_2 in that process, mainly as a product of respiration. Most yeasts are able to obtain energy through aerobic respiration (*Rhodotorula* spp., *Cryptococcus* spp.); however, in anaerobic environments, many yeasts can also carry out (i) anaerobic respiration with nitrite or nitrate as the final electron acceptor or (ii) fermentation (Tsuruta et al. 1998; Merico et al. 2007). In either case, CO_2 (a greenhouse gas) is released to the atmosphere, but to a varying extent. When an organic compound is used as an energy source in aerobic or anaerobic respiration, it is completely oxidize to CO_2. During fermentation, however, CO_2 production is lower, since a portion of the energy source remains as fermentation products of which ethanol is the most abundant. Since redox-cofactor balancing constrains product yields in fermentation processes, other organic products, such as glycerol,

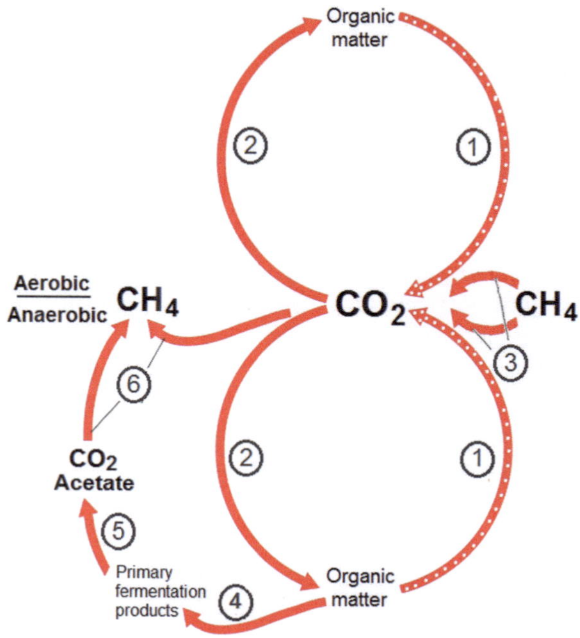

Fig. 3.3 Microbial activities involved in carbon cycle. Dotted and full arrows show the steps performed and not performed by fungi. Numbers are as follows: (*1*) organic matter decomposition and mineralization to CO_2, (*2*) CO_2 fixation, (*3*) oxidation of methane (methanotrophy), (*4*) fermentation, (*5*) fermentation involving syntrophic bacteria, (*6*) production of methane (methanogenesis)

may also be obtained (Guadalupe-Medina et al. 2013). For example, *Candida sake* H14Cs, an Antarctic isolate, produces glycerol during ethanolic fermentation (Ballester-Tomás et al. 2017). This capacity leads to a lower CO_2 production and the emergence of a different organic molecule that could be used as carbon source by other microorganisms. In case of facultative fermentative yeasts, which exhibit both respiratory and fermentative metabolism, fermentation normally occurs in anaerobic conditions. The inhibition of fermentative metabolism by oxygen has been described as the Pasteur effect (Merico et al. 2007). Some yeasts, referred to as Crabtree-positive yeasts, however, have evolved the ability to ferment even in presence of O_2, although the production of energy is less efficient (Pfeiffer and Morley 2014). In this case, the flux distribution between respiration and fermentation depends on the environmental conditions and in particular on the concentration of the energy source (Pfeiffer and Morley 2014). In fact, high glucose levels induce the production of fermentation products. Crabtree-positive yeasts include species within the genera *Saccharomyces*, *Zygosaccharomyces*, *Dekkera*, and *Schizosaccharomyces* (Kregiel 2008). Thus, the appearance of yeast fermentation products in Antarctic aerobic environments would not be expected since these genera have not been frequently found in Antarctic habitats.

Yeasts utilize only organic compounds as a carbon source for structural components and organic metabolites such as phenylethanol, a volatile compound produced by the Antarctic isolate *Candida sake* 41E (Arrarte et al. 2017). In most cases, the carbon source is the same compound used to provide energy. Carbohydrates are generally preferred as carbon and energy sources; however, many yeasts can also grow on other organic compounds, including lower aliphatic alcohols, sugar alcohols, organic acids, fatty acids, amino acids, and even alkanes or aromatic compounds (Kavanagh 2005). Most monomeric or polymeric carbohydrates in Antarctic areas are provided by vegetation through plant exudates, root turnover, or dead plant material (Melick et al. 1994). Therefore, the type and magnitude of vegetation cover determine the diversity and abundance of the soil microbiota. Soluble carbohydrates like sucrose, fructose, and glucose that leached mainly from Antarctic bryophytes and arabitol that leached from Antarctic lichens have been found in various Antarctic ecosystems (Melick and Seppelt 1992; Melick and Seppelt 1994). Many types of yeasts can readily absorb and metabolize these soluble carbohydrate compounds thus enabling yeast to be found near living plants. Some yeast species can also grow using biopolymers such as starch, cellulose, hemicellulose, proteins, or even lignin as carbon source and can also serve as decomposers of dead plant material. In this case, yeasts must breakdown the polymers using extracellular enzymes, such as cellulases, pectinases, hemicellulases, etc., prior to absorbing the resulting lower molecular weight molecules. Yeast and other microorganisms can use these oligomeric and monomeric molecules as carbon or energy source.

3.2.1 Production of Enzymes Involved in Biopolymer Degradation

The range of biopolymers that yeasts can hydrolyze is considerable and well documented (Brizzio et al. 2007; De García et al. 2007; Kurtzman et al. 2011). Antarctic yeasts are capable of producing extracellular enzymes, such as proteases, glycosidases, esterases, and lipases, that play a role in the breakdown of natural organic polymers (Buzzini et al. 2012; Carrasco et al. 2012; Martinez et al. 2016; Carrasco et al. 2016) (Fig. 3.4). Among other reports, it has been shown the occurrence of cold-active hydrolyzing yeast associated with the Antarctic worm *Grania* sp., an invertebrate that presents a microbial population that may contribute to the macroalgae decomposition and nutrient cycling in the Antarctic ecosystem (Herrera et al. 2017).

Pectinases, xylanases, and cellulases play an important role in degradation of cellulose-based materials, such as plant cell walls, and are examples of enzymes involved in the natural cycling of carbon. Carrasco et al. (2012) reported that 46% of the yeasts isolated from soil samples in the sub-Antarctic region of King George Island had the ability to produce cellulases, and Vaz et al. (2011) reported that cellulases were among the most common enzymes produced by yeasts from that

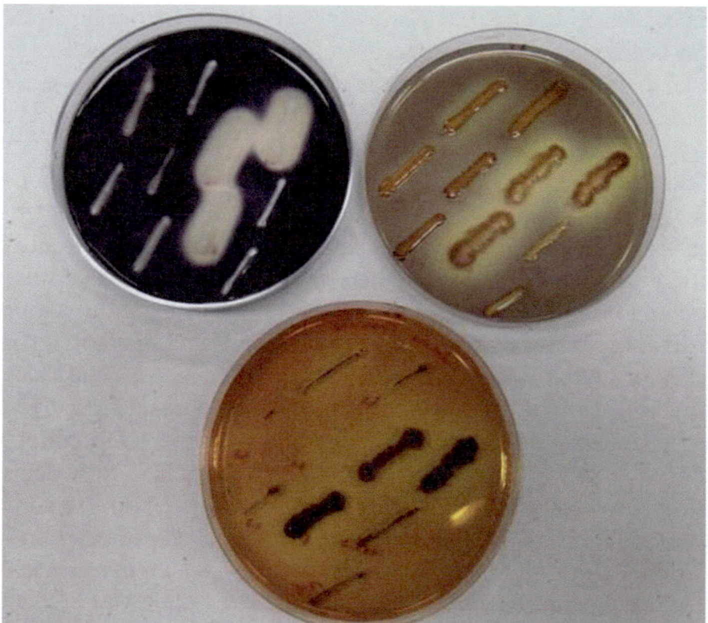

Fig. 3.4 Screening for amylases, pectinases, and inulinases cold-adapted-producing yeast in solid medium containing starch, pectin, or inuline as carbon source, respectively. Activities were detected after the addition of Lugol's solution. A clear halo around the colonies indicates the production of amylases, pectinases, and inulinases

region. Duarte et al. (2013) documented the production of xylanases by Antarctic yeast and found that 37% of 97 yeast isolates recovered from the Antarctic continent had the ability to produce xylanase. Martinez et al. (2016) and Carrasco et al. (2012), however, reported lower percentages of xylanolytic yeasts (15% and 4%, respectively) from King George Island. The reported percentage of yeast strains capable of producing pectinases has also been variable, ranging from 8% to 46% in different surveys (Vaz et al. 2011; Carrasco et al. 2012; Martinez et al. 2016).

The ability of Antarctic yeasts to produce ligninolytic enzymes has also been demonstrated. Rovati et al. (2013) reported that 38% of the yeasts isolated from King George Island exhibited significant ligninolytic activity and that 25% of the examined yeasts were able to produce laccases, which are known to catalyze the depolymerization of lignin and a broad range of other phenolic substrates (Mayer and Staples 2002). The presence of microorganisms producing such enzymes ensures that the initial step of degradation of lignocellulosic material at specific sites can occur. The turnover of plant-based polymeric materials plays an important role in the global carbon cycle (Kasana and Gulati 2011), especially since cellulose is the most abundant biopolymer on Earth (McNamara et al. 2015).

Chitin is the most abundant renewable polymer in marine environments where it serves as an important source of carbon and nitrogen (Souza et al. 2011). Thus, the initial degradation of chitin by chitinases is a key step in the cycling of organic matter in marine habitats. The production of chitinases by Antarctic yeasts has been demonstrated (Ramli et al. 2011; Carrasco et al. 2012). Extracellular chitinolytic activity has been reported in yeasts recovered from soil samples collected on King George Island, and these microorganisms were identified as *Mrakia psychrophila*, *Sporidiobolus salmonicolor*, *Metschnikowia* sp., *Leuconeurospora* sp., and *Dioszegia fristingensis* (Carrasco et al. 2012).

Other common hydrolytic enzymes associated with Antarctic yeast are esterases (Carrasco et al. 2012; Martinez et al. 2016). Esterases, enzymes that catalyze the hydrolysis of glycerol esters of short-chain fatty acids, have been frequently found in Antarctic yeasts. Carrasco et al. (2012) reported that hydrolysis of tributyrin (an ester composed of butyric acid and glycerol) was one of the most common enzymatic activities observed in yeasts from King George Island. Moreover, Białkowska et al. (2017) found that all of the Antarctic yeasts examined in their study had the ability to hydrolyze tributyrin. None of them, however, produced true lipases which according to Szczesna Antczak et al. (2009) are esterases with the ability to hydrolyze water-insoluble triacylglycerols composed of long-chain fatty acids. Duarte et al. (2013), however, demonstrated that nearly 47% of 97 yeast isolates obtained from marine and terrestrial samples collected in Antarctica exhibited true lipolytic activity. Lipases from Antarctic isolates of *Leucosporidium antarcticum* (Rashid et al. 2010) and *Pseudozyma antarctica* (Domínguez De María et al. 2005) have been already identified and characterized. In fact, lipases from the yeast *Pseudozyma antarctica* have such unique biochemical properties, that the authors protected their results with a patent and the enzymes are currently commercially produced (Shivaji and Prasad 2009). The production of lipases by Antarctic yeasts enables them to hydrolyze fats and oils to produce glycerol and fatty acids, which can then be used as a carbon or energy source by themselves and other microorganisms.

Turkiewicz et al. (2003) has stated that the secretion of proteases by yeasts is not common; however, proteolytic activity by Antarctic yeasts has been reported. Duarte et al. (2013) found that 14 out of 97 examined Antarctic yeast isolates were able to produce proteases when tested on skim milk, with a strain of *Rhodotorula mucilaginosa* exhibiting the highest activity. Martinez et al. (2016) found that 21% of the yeasts isolated from King George Island were able to produce proteases when cultured at 8 °C, but among them only 7% had the ability to produce proteases at 20 °C. Matsui et al. (2016) isolated 71 microorganisms from three Antarctic freshwater lakes and found that only eight were protease-producing yeasts, all identified as *Glaciozyma antarctica*. Although protease production in yeasts is not common, these enzymes could play an important role in hydrolyzing proteins, which are one of the major high-molecular-weight compounds produced by living organisms in Antarctica (Matsui et al. 2016), further contributing to the circulation of carbon in the ecosystem.

3.2.2 Degradation of Organic Pollutants

Although Antarctica is one of the least polluted and most stringently protected places on Earth, the presence of organic and inorganic contaminants has been reported in soil and water samples (Corsolini 2009; Lo Giudice et al. 2013). The biodegradation of pollutants in those habitats requires the activity of cold-adapted microorganisms. The ability of bacteria to degrade organic pollutants has been well documented (Aislabie et al. 1998; de Jesus and Peixoto 2015); however, the role of fungi, in particular of yeasts, in this process is currently being explored. Fernández et al. (2017) demonstrated that phenol, methanol, and hexadecane molecules could be used as carbon source by many Antarctic yeasts. Among pollutants, hexadecane was the most frequently used carbon source (78% of the tested isolates belonging to both *Ascomycota* and *Basidiomycota*), while only approximately 6% (eight isolates) could also utilize phenol and methanol. Rovati et al. (2013) studied the degradation of polyphenolic substrates and dyes by cold-adapted yeasts from King George Island. They detected that isolates from both ascomycetous and basidiomycetous genera have the ability to degrade such compounds. These studies collectively provide information about the role of cold-adapted yeasts in the carbon cycle and demonstrate the potential use of these microorganisms in the bioremediation of the Antarctic environment.

3.2.3 Fate of Assimilated Organic Compounds

After being absorbed, carbon sources are metabolized and become the building blocks of macromolecules that form yeast cells and their collective biomass. When carbon is in excess and nitrogen is limited, two major metabolites, triacylglycerol (TAG) or glycogen, are accumulated within yeast cells as a strategy to deal with extended periods of starvation as reported by Bhutada et al. (2017). Nitrogen has been reported as a limiting factor in Antarctic ecosystems (Yergeau et al. 2007a), so the intracellular accumulation of TAG and glycogen polymers by yeasts most likely occurs in these environments. Indeed, oleaginous yeasts, such as *Candida glaebosa*, *Leucosporidiella fragariae*, and *Leucosporidium scottii*, have been found in Antarctic habitats (Pereyra et al. 2014; Martinez et al. 2016). Oleaginous yeasts are defined as yeasts that are able to accumulate more than 20% of their dry mass as lipids (Ratledge 2004), which are stored in intracellular lipid droplets. These organelles vary greatly in size, ranging from 1 μm to 100 μm in diameter, and contain a neutral lipid core composed mainly of sterol esters or TAG surrounded by a phospholipid monolayer and associated proteins (Farese Jr and Walther 2009). Stored lipids are hydrolyzed into fatty acids during periods of carbon deprivation to obtain energy. This ability provides a competitive evolutionary advantage to oleaginous yeasts, increasing their chances to survive in environments where energy supplies fluctuate, as they do in Antarctic habitats.

3.3 Role of Yeasts in the Nitrogen Cycle

The nitrogen cycle is one of the most important nutrient cycles in terrestrial ecosystems and involves four basic microbiological processes: nitrogen fixation, nitrification, denitrification, and mineralization (Hayatsu et al. 2008). A fifth process, however, where soluble forms of nitrogen are incorporated in microbial biomass, will also be discussed. Fungi are capable of performing all of the mentioned processes, excluding nitrogen fixation, which is exclusively a prokaryotic process (Kneip et al. 2007) (Fig. 3.5).

Nitrification is defined as the biological oxidation of ammonium to nitrite and nitrate or more generally as the biological transformation of reduced to oxidized forms of nitrogen (Simek 2000). Both autotrophic and heterotrophic nitrification have been described in microorganisms, but fungi only perform heterotrophic nitrification (Bernhard 2010). The latter process constitutes the oxidation of inorganic and organic reduced forms of nitrogen to nitrate, and unlike autotrophic nitrification, it is not necessarily coupled to energy conservation (Hayatsu et al. 2008).This process occurs in a wide range of heterotrophic microorganisms including bacteria and fungi.

Both filamentous fungi such as *Aspergillus flavus* and yeasts such as *Barnettozyma californica* and *Candida rugosa* have been characterized as heterotrophic nitrifiers

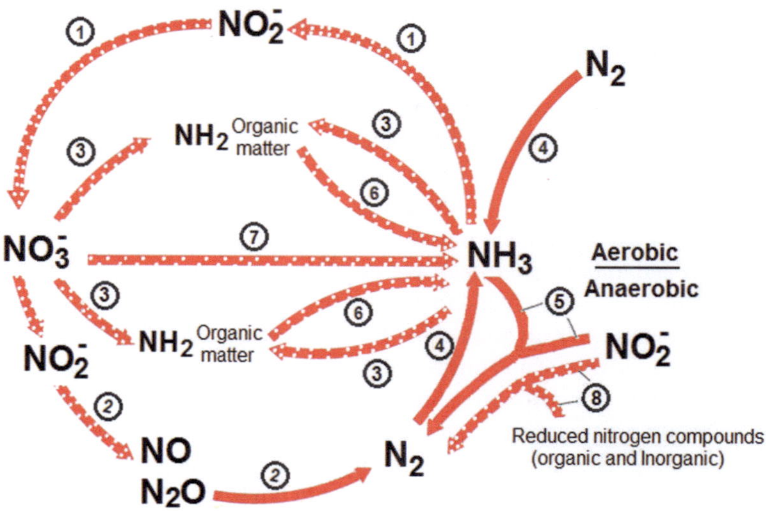

Fig. 3.5 Microbial activities involved in nitrogen cycle. Dotted and full arrows show the steps performed and not performed by fungi, respectively. Numbers are as follows: (*1*) autotrophic and heterotrophic nitrification (only heterotrophic nitrification may be produced by fungi), (*2*) denitrification processes, (*3*) incorporation of nitrogen into biomass, (*4*) nitrogen fixation, (*5*) anaerobic ammonium oxidation (Anammox), (*6*) mineralization, (*7*) dissimilatory nitrate reduction to ammonia (DNRA) processes (fungi are involved only in ammonia fermentation process), (*8*) co-denitrification

(Kurtzman et al. 2011; Stein 2011). Heterotrophic nitrification is favored over autotrophic nitrification in acidic soils where the latter process is inhibited (Ward 2008). It has been demonstrated that nitrifying fungal activity predominates over autotrophic nitrification in the acidic soils of coniferous forests (Jordan et al. 2005; Zhang et al. 2011). To date, studies on the role of molds and yeasts in heterotrophic nitrification in Antarctic soils have not been conducted; thus, the contribution of yeasts in this kind of process would be a matter of future research.

Another process within the nitrogen cycle that includes fungal activity is nitrogen mineralization or ammonification. This is the process by which inorganic nitrogen (ammonia) is obtained by the degradation of organic nitrogenous compounds. The action of enzymes such as urease or proteases plays an essential role in this process. Many fungi, including yeasts commonly found in Antarctic habitats, have the ability to produce such enzymes. *Rhodotorula muscorum*, *Rhodotorula mucilaginosa*, *Cryptococcus aerius*, and *Cryptococcus albidus* are examples of cold-adapted yeasts found in Antarctic environments (Buzzini et al. 2012) that are able to produce urease (Kurtzman et al. 2011). Ammonia can also be produced by fungal ammonia fermentation, a dissimilatory metabolic pathway that supplies energy under anoxic conditions (Takasaki et al. 2004). It comprises the reduction of nitrate to ammonium, coupled with substrate-level phosphorylation, and the catabolic oxidation of ethanol, to acetate (Greben et al. 2007). This process was first described in *Fusarium oxysporum* but, since then, has been detected in various filamentous fungi belonging to other species in the phylum *Ascomycota* (Zhou et al. 2002). However, such kind of activity has not been reported in yeasts.

Yeasts or other microorganisms, in assimilatory processes, can use ammonia as a nitrogen source for the formation of microbial biomass. Fungi, in fact, can assimilate many inorganic and organic nitrogenous compounds, and so are considered to play an important role in soils, as a nitrogen storage (Gorfer et al. 2014). In this regard, many molds and yeasts, including some isolated from Antarctic habitats, have the ability to assimilate nitrate. For example, species from the genera *Cryptococcus* and *Rhodotorula*, such as *Cryptococcus antarcticus* and *Rhodotorula muscorum*, which have been found in Antarctica (Buzzini et al. 2012), have the ability to assimilate nitrate (Kurtzman et al. 2011). The existence of a transport system to uptake nitrate and nitrate and nitrite reductases to convert nitrate to ammonium is essential for a microorganism to assimilate nitrate. The genes coding for the necessary proteins are commonly clustered together on a chromosome. Some evidence suggests that this cluster was probably obtained by horizontal gene transfer from an oomycete donor to a common ancestor of different phyla within the fungal kingdom (Slot et al. 2007; Slot and Hibbett 2007; Gorfer et al. 2011). In some species, such as those in the genus *Saccharomyces*, the cluster was partially or completely lost. In other cases, such as in *Debaryomyces* species, which are commonly found in Antarctic habitats (Vishniac 2006), nitrate reductase activity and the corresponding genes have not been detected. In those yeasts only nitrite reductase is expressed, thus providing those yeasts the ability to assimilate nitrite (Dujon et al. 2004).

Although nitrate is one of the most abundant nitrogen sources in nature, most yeast utilize ammonium more readily. Nitrate must be reduced to the ammonium oxidation level in order to be incorporated into biomass, so the energetic cost of nitrate assimilation is higher compared with ammonium assimilation (Siverio 2002). Thus, assimilation of nitrate must be carefully regulated in order to save energy, and so in many cases, the presence of ammonium represses the uptake and assimilation of nitrate.

Nitrate and nitrite can be also used by many fungi, including yeasts, in dissimilatory pathways such as denitrification. In that process, both anions are transformed in gases like NO or N_2O which are released to the atmosphere, causing the depletion of fixed nitrogen forms. N_2O is a greenhouse gas with a global warming potential 265- to 298-fold higher than CO_2, as reported over a 100-year timescale (IPCC 2007) and so has a much greater impact. Denitrification is an anaerobic respiration process in which nitrate or nitrite constitute the external electron acceptor. Denitrification in fungi was first described for *Fusarium oxysporum* (Shouns and Tanimoto 1991). Since then, the process has been associated to many other fungal species, including yeasts (Mothapo et al. 2015). It has been demonstrated, in different habitats, that 50% or more of total soil N_2O emissions could be attributed to fungal activity (Crenshaw et al. 2008; Chen et al. 2014). Tsuruta et al. (1998) concluded that the ability for denitrification is widely distributed among yeasts. The authors found that 50% of the tested yeasts, selected among a wide range of genera, had the ability to produce NO or N_2O. Among denitrifiers, they found an Antarctic isolate identified as *Candida* sp. and a mesophilic isolate of *Trichosporon cutaneum*, species which has been found in Antarctic soils (Vishniac 2006).

It is important to outline that N_2O and NO, but not N_2, have been detected as result of fungal denitrification. Genes encoding nitrous oxide reductase, which reduces N_2O to N_2, are not found in fungi (Shoun et al. 2012). In fact, complete denitrification with N_2 production is considered a prokaryotic process (Madigan et al. 2015). In some cases, however, N_2 can be formed through a process called co-denitrification, which can be mediated by fungi. In such process, hybrid dinitrogen (N_2) molecules are formed by combining nitrogen atoms from nitrite, and other more reduced nitrogen compounds (e.g., amines) under denitrifying conditions. Co-denitrification is widely viewed as an anaerobic, enzymatically mediated nitrosation process, but the specific pathway is still unknown (Phillips et al. 2016). Three fungal species (order *Hypocreales*) have been experimentally demonstrated to perform co-denitrification: *Fusarium oxysporum*, *Fusarium solani*, and *Cylindrocarpon tonkinense* (Spott et al. 2011). Co-denitrification has an important role when organic nitrogen is present in a soil. In fact, fungal co-denitrification contributes with up to 92% of the N_2 produced in grassland soils (Laughlin and Stevens 2002). However, co-denitrification by yeast and its role in cold habitats, such as Antarctic soils, have not been evaluated yet.

3.4 Role of Yeasts in Phosphorous Cycle

Phosphorous is an essential element for animal and plants. Its biogeochemical cycle differs from nitrogen and carbon cycle in that it does not include a gas phase (Madigan et al. 2015). Very little phosphorus circulates in the atmosphere because at Earth's normal temperatures and pressures, phosphorus and its various compounds are not found in gas state. The largest reservoir of phosphorus is found in sedimentary rock. The phosphorous cycle starts with rain and weathering which cause rocks to release phosphate ions and other minerals. Inorganic phosphate is then distributed in soils and water where it is taken up by living organisms to form biomass. When organisms die and decay, the organic phosphate is turned back to the soil. Within the soil, organic forms of phosphate can be hydrolyzed releasing phosphate to the environment in a process known as mineralization. Microbial phosphatases play an important role in this process. In soils, organic phosphorus is mainly found as phytates (inositol hexa- and penta-phosphates) from which phosphate is released by the action of specific phosphatases called phytases (Singh and Satyanarayana 2011). There are many reports about phytase production by bacteria and filamentous fungi, but data about yeasts is scarce. The occurrence of phytase-producing yeast isolates from King George Island, in the sub-Antarctic region of Antarctica, was recently reported (Białkowska et al. 2017). The authors suggested that phytase-producing yeast are widely distributed in this location, since almost all analyzed strains showed the ability to utilize phytate.

In natural environments, however, the formation of insoluble phytates with cations such as Ca^{+2}, Al^{+3}, or Fe^{+2} limits the utilization of phytates by yeasts or other microorganisms. The secretion of organic acids could help to facilitate phytate availability by chelating cations (Singh and Satyanarayana 2011). Yeasts expressing both activities, phytase and organic acid productions, would play an important role in mineralization processes within the phosphorous cycle.

Table 3.1 Examples of Antarctic yeasts involved in C, N, and P cycles

Cycle	Process	Substrate	End-product	Antarctic yeast species	References
C	Hydrolysis of organic compounds	Proteins	Peptides and amino acids	*Glaciozyma antarctica*	Matsui et al. (2016)
C	Hydrolysis of organic compounds	Triglycerides	Long chain fatty acids and glycerol	*Pseudozyma antarctica*	Domínguez De María et al. (2005)
C	Hydrolysis of organic compounds	Tributyrin	Butyric acid and glycerol	*Cryptococcus victoriae*	Martínez et al. (2016)
C	Hydrolysis of organic compounds	Pectin	Oligo-galacturonates	*Candida davisiana*	Martínez et al. (2016)
C	CO_2 production (fermentation)	Monosaccharides	CO_2	*Candida sake*	Ballester-Tomás et al. (2017)
C	CO_2 production (aerobic respiration)	Monosaccharides	CO_2	*Rhodotorula laringys*	Kurtzman et al. (2011)
N	Denitrification (anaerobic respiration)	Nitrite	N_2O	*Trichosporum cutaneum*	Mothapo et al. (2015) and Tsuruta et al. (1998)
N	Nitrite assimilation	Nitrite	Nitrogen compounds in biomass	*Debaryomyces* spp.	Dujon et al. (2004)
N	Nitrate assimilation	Nitrate	Nitrogen compounds in biomass	*Cryptococcus antarcticus*	Kurtzman et al. (2011)
N	Heterotrophic nitrification	Ammonium	Nitrate	*Candida rugosa*	Stein (2011)
N	Hydrolysis of organic compounds	Urea	Ammonium	*Cryptococcus aerius*	Kurtzman et al. (2011)
P	Hydrolysis of organic compounds	Phytate	Phosphate and inositol	*Debaryomyces hansenii*	Białkowska et al. (2017)

3.5 Concluding Remarks

Yeasts are common inhabitants of water and soils of Antarctic and sub-Antarctic ice-free areas. Their role as chemoorganotrophic and heterotrophic microorganisms in the cycling of organic matter is fundamental. The extracellular production of many cold-adapted hydrolytic enzymes has been demonstrated in many Antarctic yeast, and these degrading activities ensure that Antarctic yeast plays an active role as organic matter decomposers (Table 3.1).

Most yeast obtain energy by aerobic respiration, and therefore they release carbon dioxide, a greenhouse gas, into the atmosphere. In addition, by anaerobic respiration, some yeast also produce N_2O, another greenhouse gas with a global warming potential greater than that of CO_2. Yeasts are involved in assimilatory and dissimilatory pathways of nitrogen and carbon cycles, and they also have an important role in the release of phosphate from organic compounds such as phytates (Table 3.1).

References

Aislabie, J., McLeod, M., & Fraser, R. (1998). Potential for biodegradation of hydrocarbons in soil from the Ross Dependency, Antarctica. *Applied Microbiology and Biotechnology, 49*, 210–214. https://doi.org/10.1007/s002530051160.

Arenz, B. E., Held, B. W., Jurgens, J. A., Farrell, R. L., & Blanchette, R. A. (2006). Fungal diversity in soils and historic wood from the Ross Sea Region of Antarctica. *Soil Biology and Biochemistry, 38*, 3057–3064. https://doi.org/10.1016/j.soilbio.2006.01.016.

Arrarte, E., Garmendia, G., Rossini, C., Wisniewski, M., & Vero, S. (2017). Volatile organic compounds produced by Antarctic strains of Candida sake play a role in the control of postharvest pathogens of apples. *Biological Control, 109*, 14–20.

Ballester-Tomás, L., Prieto, J. A., Gil, J. V., Baeza, M., & Randez-Gil, F. (2017). The Antarctic yeast Candida sake: Understanding cold metabolism impact on wine. *International Journal of Food Microbiology, 245*, 59–65. https://doi.org/10.1016/j.ijfoodmicro.2017.01.009.

Bernhard, A. (2010). The nitrogen cycle: Processes. *Nat Educ Knowl, 2*, 1–8.

Bhutada, G., Kavšček, M., Ledesma-Amaro, R., Thomas, S., Rechberger, G. N., Nicaud, J. M., & Natter, K. (2017). Sugar versus fat: Elimination of glycogen storage improves lipid accumulation in Yarrowia lipolytica. *FEMS Yeast Research, 17*, fox020. https://doi.org/10.1093/femsyr/fox020.

Białkowska, A. M., Szulczewska, K. M., Krysiak, J., Florczak, T., Gromek, E., Kassassir, H., Kur, J., & Turkiewicz, M. (2017). Genetic and biochemical characterization of yeasts isolated from Antarctic soil samples. *Polar Biology, 40*, 1787–1803.

Brizzio, S., Turchetti, B., De Garcia, V., Libkind, D., Buzzini, P., & Van Broock, M. (2007). Extracellular enzymatic activities of basidiomycetous yeasts isolated from glacial and subglacial waters of Northwest Patagonia (Argentina). *Canadian Journal of Microbiology, 53*, 519–525.

Buzzini, P., & Margesin, R. (2014). Cold-adapted yeasts: A lesson from the cold and a challenge for the XXI century. In *Cold-adapted yeasts: Biodiversity, adaptation strategies and biotechnological significance* (pp. 3–22). Berlin: Springer.

Buzzini, P., Branda, E., Goretti, M., & Turchetti, B. (2012). Psychrophilic yeasts from worldwide glacial habitats: Diversity, adaptation strategies and biotechnological potential. *FEMS Microbiology Ecology, 82*, 217–241.

Buzzini, P., Turk, M., Perini, L., Turchetti, B., & Gunde-Cimerman, N. (2017). Yeasts in polar and subpolar habitats. In *Yeasts in natural ecosystems: Diversity* (pp. 331–365). Berlin: Springer.

Carrasco, M., Rozas, J. M., Barahona, S., Alcaíno, J., Cifuentes, V., & Baeza, M. (2012). Diversity and extracellular enzymatic activities of yeasts isolated from King George Island, the sub-Antarctic region. *BMC Microbiology, 12*, 251. https://doi.org/10.1186/1471-2180-12-251.

Carrasco, M., Villarreal, P., Barahona, S., Alcaíno, J., Cifuentes, V., & Baeza, M. (2016). Screening and characterization of amylase and cellulase activities in psychrotolerant yeasts. *BMC Microbiology, 16*, 21. https://doi.org/10.1186/s12866-016-0640-8.

Chen, J., Vandelle, E., Bellin, D., & Delledonne, M. (2014). Nitric oxide detection and function of nitric oxide during the hypersensitive response in Arabidopsis thaliana: Where there's a will there's a way. *Nitric Oxide, 43*, 81–88. https://doi.org/10.1016/j.niox.2014.06.008.

Connell, L., Redman, R., Craig, S., Scorzetti, G., Iszard, M., & Rodriguez, R. (2008). Diversity of soil yeasts isolated from South Victoria Land, Antarctica. *Microbial Ecology, 56*, 448–459. https://doi.org/10.1007/s00248-008-9363-1.

Connell, L. B., Rodriguez, R. R., Redman, R. S., & Dalluge, J. J. (2014). Cold-adapted yeasts in Antarctic deserts. In *Cold-adapted yeasts* (pp. 75–98). Berlin: Springer.

Corsolini, S. (2009). Industrial contaminants in Antarctic biota. *Journal of Chromatography. A, 1216*, 598–612.

Crenshaw, C. L., Lauber, C., Sinsabaugh, R. L., & Stavely, L. K. (2008). Fungal control of nitrous oxide production in semiarid grassland. *Biogeochemistry, 87*, 17–27. https://doi.org/10.1007/s10533-007-9165-4.

Cui, M., Ma, A., Qi, H., Zhuang, X., & Zhuang, G. (2015). Anaerobic oxidation of methane: An "active" microbial process. *Microbiologyopen, 4*, 1–11.

De García, V., Brizzio, S., Libkind, D., Buzzini, P., & Van Broock, M. (2007). Biodiversity of cold-adapted yeasts from glacial meltwater rivers in Patagonia, Argentina. *FEMS Microbiology Ecology, 59*, 331–341.

de Jesus H. E., & Peixoto, R. S. (2015) Bioremediation in Antarctic soils. *Journal of Petroleum and Environmental Biotechnology, 6*. https://doi.org/10.4172/2157-7463.1000248.

Domínguez De María, P., Carboni-Oerlemans, C., Tuin, B., Bargeman, G., Van Der Meer, A., & Van Gemert, R. (2005). Biotechnological applications of Candida antarctica lipase A: State-of-the-art. *Journal of Molecular Catalysis B: Enzymatic, 37*, 36–46. https://doi.org/10.1016/j.molcatb.2005.09.001.

Duarte, A. W. F., Dayo-Owoyemi, I., Nobre, F. S., Pagnocca, F. C., Chaud, L. C. S., Pessoa, A., Felipe, M. G. A., & Sette, L. D. (2013). Taxonomic assessment and enzymes production by yeasts isolated from marine and terrestrial Antarctic samples. *Extremophiles, 17*, 1023–1035. https://doi.org/10.1007/s00792-013-0584-y.

Dujon, B., Sherman, D., Fischer, G., Durrens, P., Casaregola, S., Lafontaine, I., De Montigny, J., Marck, C., Neuvéglise, C., & Talla, E. (2004). Genome evolution in yeasts. *Nature, 430*, 35.

Farese, R. V., Jr., & Walther, T. C. (2009). Lipid droplets finally get a little RESPECT. *Cell, 139*, 855–860.

Fernández, P. M., Martorell, M. M., Blaser, M. G., Ruberto, L. A. M., de Figueroa, L. I. C., & Mac Cormack, W. P. (2017). Phenol degradation and heavy metal tolerance of Antarctic yeasts. *Extremophiles, 21*, 445–457. https://doi.org/10.1007/s00792-017-0915-5.

Gorfer, M., Blumhoff, M., Klaubauf, S., Urban, A., Inselsbacher, E., Bandian, D., Mitter, B., Sessitsch, A., Wanek, W., & Strauss, J. (2011). Community profiling and gene expression of fungal assimilatory nitrate reductases in agricultural soil. *The ISME Journal, 5*, 1771–1783. https://doi.org/10.1038/ismej.2011.53.

Gorfer, M., Klaubauf, S., Berger, H., & Strauss, J. (2014). The fungal contribution to the nitrogen cycle in agricultural soils. In *Metagenomics microbial nitrogen cycle: Theory, methods and applications* (pp. 209–225). Norfolk: Caister Academic Press.

Greben, H. A., Joubert, L. M., Tjatji, M. P., Whites, H. E., & Botha, A. (2007). Biological nitrate removal from synthetic wastewater using a fungal consortium in one stage bioreactors. *Water SA, 33*, 285–290. https://doi.org/10.4314/wsa.v33i2.49094.

Guadalupe-Medina, V., Wisselink, H. W., Luttik, M. A., De Hulster, E., Daran, J. M., Pronk, J. T., & Van Maris, A. J. A. (2013). Carbon dioxide fixation by Calvin-Cycle enzymes improves ethanol yield in yeast. *Biotechnology for Biofuels, 6*, 125. https://doi.org/10.1186/1754-6834-6-125.

Hayatsu, M., Tago, K., & Saito, M. (2008). Various players in the nitrogen cycle: Diversity and functions of the microorganisms involved in nitrification and denitrification. *Soil Science & Plant Nutrition, 54*, 33–45.

Herrera, L. M., García-Laviña, C. X., Marizcurrena, J. J., Volonterio, O., de León, R. P., & Castro-Sowinski, S. (2017). Hydrolytic enzyme-producing microbes in the Antarctic oligochaete Grania sp.(Annelida). *Polar Biology, 40*(4), 947–953.

IPCC. (2007). *Intergovernmental panel on climate change*. Summary for policy makers.

Jordan, F. L., Cantera, J. J. L., Fenn, M. E., & Stein, L. Y. (2005). Autotrophic ammonia-oxidizing bacteria contribute minimally to nitrification in a nitrogen-impacted forested ecosystem. *Applied and Environmental Microbiology, 71*, 197–206. https://doi.org/10.1128/AEM.71.1.197-206.2005.

Kasana, R. C., & Gulati, A. (2011). Cellulases from psychrophilic microorganisms: A review. *Journal of Basic Microbiology, 51*, 572–579.

Kavanagh, K. (2005). Fungal fermentation systems and products. In *Fungi: Biology and applications*. Hoboken: Wiley.

Kneip, C., Lockhart, P., Voß, C., & Maier, U. G. (2007). Nitrogen fixation in eukaryotes – New models for symbiosis. *BMC Evolutionary Biology, 7*, 55.

Kregiel, D. (2008). Physiology and metabolism of Crabtree-negative yeast Debaryomyces occidentalis. *Food Chem Biotechnol, 72*, 35–44.

Kurtzman, C. P., Fell, J. W., & Boekhout, T. (2011). *The yeasts: A taxonomic study*. New York: Elsevier.

Laughlin, R. J., & Stevens, R. J. (2002). Evidence for fungal dominance of denitrification and codenitrification in a grassland soil. *Soil Science Society of America Journal, 66*, 1540. https://doi.org/10.2136/sssaj2002.1540.

Lawley, B., Ripley, S., Bridge, P., & Convey, P. (2004). Molecular analysis of geographic patterns of eukaryotic diversity in Antarctic soils molecular analysis of geographic patterns of eukaryotic diversity in Antarctic soils. *American Society for Microbiology, 70*, 5963–5972. https://doi.org/10.1128/AEM.70.10.5963.

Lo Giudice, A., Casella, P., Bruni, V., & Michaud, L. (2013). Response of bacterial isolates from Antarctic shallow sediments towards heavy metals, antibiotics and polychlorinated biphenyls. *Ecotoxicology, 22*, 240–250. https://doi.org/10.1007/s10646-012-1020-2.

Madigan, M. T., Martinko, J. M., Stahl, D., & Clark, D. P. (2015). *Brock biology of microorganisms*. New York: Pearson.

Martinez, A., Cavello, I., Garmendia, G., Rufo, C., Cavalitto, S., & Vero, S. (2016). Yeasts from sub-Antarctic region: Biodiversity, enzymatic activities and their potential as oleaginous microorganisms. *Extremophiles, 20*, 759–769. https://doi.org/10.1007/s00792-016-0865-3.

Matsui, M., Kawamata, A., Kosugi, M., Imura, S., & Kurosawa, N. (2016). Diversity of proteolytic microbes isolated from Antarctic freshwater lakes and characteristics of their cold-active proteases. *Polar Science, 13*, 82–90.

Mayer, A. M., & Staples, R. C. (2002). Laccase: New functions for an old enzyme. *Phytochemistry, 60*, 551–565.

McNamara, J. T., Morgan, J. L. W., & Zimmer, J. (2015). A molecular description of cellulose biosynthesis. *Annual Review of Biochemistry, 84*, 895–921. https://doi.org/10.1146/annurev-biochem-060614-033930.

Melick, D. R., & Seppelt, R. D. (1992). Loss of soluble carbohydrates and changes in freezing point of Antarctic bryophytes after leaching and repeated freeze-thaw cycles. *Antarctic Science, 4*, 399–404. https://doi.org/10.1017/S0954102092000592.

Melick, D. R., & Seppelt, R. D. (1994). The effect of hydration on carbohydrate levels, pigment content and freezing point of Umbilicaria decussata at a continental Antarctic locality. *Cryptogamic Botany, 4*, 212–217.

Melick, D. R., Hovenden, M. J., & Seppelt, R. D. (1994). Phytogeography of bryophyte and lichen vegetation in the Windmill Islands, Wilkes Land, continental Antarctica. *Vegetation, 111*, 71–87. https://doi.org/10.1007/BF00045578.

Merico, A., Sulo, P., Piškur, J., & Compagno, C. (2007). Fermentative lifestyle in yeasts belonging to the Saccharomyces complex. *The FEBS Journal, 274*, 976–989. https://doi.org/10.1111/j.1742-4658.2007.05645.x.

Moliné, M., Libkind, D., Van Broock, M., & Rosa, C. A. (2011). The diversity, extracellular enzymatic activities and photoprotective compounds of yeasts isolated in Antarctic. *Brazilian Journal of Microbiology, 42*, 937–947.

Mothapo, N., Chen, H., Cubeta, M. A., Grossman, J. M., Fuller, F., & Shi, W. (2015). Phylogenetic, taxonomic and functional diversity of fungal denitrifiers and associated N2O production efficacy. *Soil Biology and Biochemistry, 83*, 160–175.

Pereyra, V., Martinez, A., Rufo, C., & Vero, S. (2014). Oleaginous yeasts form Uruguay and Antarctica as renewable raw material for diodiesel production. *American Journal of Bioscience, 2*, 251. https://doi.org/10.11648/j.ajbio.20140206.20.

Pfeiffer, T., & Morley, A. (2014). An evolutionary perspective on the Crabtree effect. *Frontiers in Molecular Biosciences, 1*, 1–6. https://doi.org/10.3389/fmolb.2014.00017.

Phillips, R. L., Song, B., McMillan, A. M. S., Grelet, G., Weir, B. S., Palmada, T., & Tobias, C. (2016). Chemical formation of hybrid di-nitrogen calls fungal codenitrification into question. *Scientific Reports, 6*, 39077. https://doi.org/10.1038/srep39077.

Ramli, A. N. M., Mahadi, N. M., Rabu, A., Murad, A. M. A., Bakar, F. D. A., & Illias, R. M. (2011). Molecular cloning, expression and biochemical characterisation of a cold-adapted novel recombinant chitinase from Glaciozyma antarctica PI12. *Microbial Cell Factories, 10*, 94. https://doi.org/10.1186/1475-2859-10-94.

Rao, S., Chan, Y., Lacap, D. C., Hyde, K. D., Pointing, S. B., & Farrell, R. L. (2012). Low-diversity fungal assemblage in an Antarctic dry valleys soil. *Polar Biology, 35*, 567–574. https://doi.org/10.1007/s00300-011-1102-2.

Rashid, F. A. A., Rahim, R. A., & Ibrahim, D. (2010). Identification of lipase-producing psychrophilic yeast, Leucosporidium sp. *Internet Journal of Microbiology, 9*.

Ratledge, C. (2004). Fatty acid biosynthesis in microorganisms being used for single cell oil production. *Biochimie, 86*, 807–815.

Rovati, J. I., Pajot, H. F., Ruberto, L., Mac Cormack, W., & Figueroa, L. I. C. (2013). Polyphenolic substrates and dyes degradation by yeasts from 25 de Mayo/King George Island (Antarctica). *Yeast, 30*, 459–470. https://doi.org/10.1002/yea.2982.

Selbmann, L., de Hoog, G. S., Zucconi, L., Isola, D., Ruisi, S., Gerrits van den Ende, A. H. G., Ruibal, C., De Leo, F., Urzì, C., & Onofri, S. (2008). Drought meets acid: Three new genera in a dothidealean clade of extremotolerant fungi. *Studies in Mycology, 61*, 1–20. https://doi.org/10.3114/sim.2008.61.01.

Shivaji, S., & Prasad, G. S. (2009). Antarctic yeasts: Biodiversity and potential applications. In *Yeast biotechnology: Diversity and applications* (pp. 3–18). Dordrecht: Springer.

Shoun, H., Fushinobu, S., Jiang, L., Kim, S.-W., & Wakagi, T. (2012). Fungal denitrification and nitric oxide reductase cytochrome P450nor. *Philosophical Transactions of the Royal Society B Biological Sciences, 367*, 1186–1194. https://doi.org/10.1098/rstb.2011.0335.

Shouns, H., & Tanimoto, T. (1991). Denitrification by the fungus Fusarium oxysporum and involvement nitrite reduction * of cytochrome P-450 in the respiratory. *Biochemistry, 266*, 11078–11082.

Simek, M. (2000). Nitrification in soil – Terminology and methodology (review). *Rostl Vyroba, 46*, 385–395.

Singh, B., & Satyanarayana, T. (2011). Microbial phytases in phosphorus acquisition and plant growth promotion. *Physiology and Molecular Biology of Plants, 17*, 93–103.

Siverio, J. M. (2002). Assimilation of nitrate by yeasts. *FEMS Microbiology Reviews, 26*, 277–284.

Slot, J. C., & Hibbett, D. S. (2007). Horizontal transfer of a nitrate assimilation gene cluster and ecological transitions in fungi: A phylogenetic study. *PLoS One, 2*, e1097. https://doi.org/10.1371/journal.pone.0001097.

Slot, J. C., Hallstrom, K. N., Matheny, P. B., & Hibbett, D. S. (2007). Diversification of NRT2 and the origin of its fungal homolog. *Molecular Biology and Evolution, 24*, 1731–1743. https://doi.org/10.1093/molbev/msm098.

Souza, C. P., Almeida, B. C., Colwell, R. R., & Rivera, I. N. G. (2011). The importance of chitin in the marine environment. *Marine Biotechnology, 13*, 823–830.

Spott, O., Russow, R., & Stange, C. F. (2011). Formation of hybrid N2O and hybrid N2due to codenitrification: First review of a barely considered process of microbially mediated N-nitrosation. *Soil Biology and Biochemistry, 43*, 1995–2011.

Stein, L. Y. (2011). Heterotrophic nitrification and nitrifier denitrification. In *Nitrification* (pp. 95–114). Washington, DC: ASM Press.

Szczesna Antczak, M., Kubiak, A., Antczak, T., & Bielecki, S. (2009). Enzymatic biodiesel synthesis – Key factors affecting efficiency of the process. *Renewable Energy, 34*, 1185–1194. https://doi.org/10.1016/j.renene.2008.11.013.

Takasaki, K., Shoun, H., Yamaguchi, M., Takeo, K., Nakamura, A., Hoshino, T., & Takaya, N. (2004). Fungal ammonia fermentation, a novel metabolic mechanism that couples the dissimilatory and assimilatory pathways of both nitrate and ethanol. *ASBMB, 279*, 12414–12420. https://doi.org/10.1074/jbc.M313761200.

Thomas-Hall, S. R., Turchetti, B., Buzzini, P., Branda, E., Boekhout, T., Theelen, B., & Watson, K. (2009). Cold-adapted yeasts from Antarctica and the Italian Alps-description of three novel species: Mrakia robertii sp. nov., Mrakia blollopis sp. nov. and Mrakiella niccombsii sp. nov. *Extremophiles, 14*, 47–59. https://doi.org/10.1007/s00792-009-0286-7.

Tsuruta, S., Takaya, N., Zhang, L., Shoun, H., Kimura, K., Hamamoto, M., & Nakase, T. (1998). Denitrification by yeasts and occurrence of cytochrome P450nor in Trichosporon cutaneum. *FEMS Microbiology Letters, 168*, 105–110.

Turchetti, B., Buzzini, P., Goretti, M., Branda, E., Diolaiuti, G., D'Agata, C., Smiraglia, C., & Vaughan-Martini, A. (2008). Psychrophilic yeasts in glacial environments of alpine glaciers. *FEMS Microbiology Ecology, 63*, 73–83. https://doi.org/10.1111/j.1574-6941.2007.00409.x.

Turkiewicz, M., Pazgier, M., Kalinowska, H., & Bielecki, S. (2003). A cold-adapted extracellular serine proteinase of the yeast Leucosporidium antarcticum. *Extremophiles, 7*, 435–442. https://doi.org/10.1007/s00792-003-0340-9.

Uetake, J., Yoshimura, Y., Nagatsuka, N., & Kanda, H. (2012). Isolation of oligotrophic yeasts from supraglacial environments of different altitude on the Gulkana Glacier (Alaska). *FEMS Microbiology Ecology, 82*, 279–286. https://doi.org/10.1111/j.1574-6941.2012.01323.x.

Vaz, A. B. M., Rosa, L. H., Vieira, M. L. A., de Garcia, V., Brandão, L. R., Teixeira, L. C. R., Moliné, M., Libkind, D., Van Broock, M., & Rosa, C. A. (2011). The diversity, extracellular enzymatic activities and photoprotective compounds of yeasts isolated in Antarctica. *Brazilian Journal of Microbiology, 42*, 937–947.

Vishniac, H. S. (2006). Yeast biodiversity in the Antarctic. In *Biodiversity and ecophysiology* (pp. 419–440). London: Springer. https://doi.org/10.1007/3-540-30985-3_16.

Walton, D. W. H. (1985). *Cellulose decomposition and its relationship to nutrient cycling at South Georgia* (pp. 192–199). New York: Antarctica Nutrient Cycles and Food Webs. https://doi.org/10.1007/978-3-642-82275-9_27.

Ward, B. B. (2008). Nitrification in marine systems. In *Nitrogen in the marine environment* (pp. 199–261). Amsterdam: Elsevier.

Weinstein, R. N., Montiel, P. O., & Johnstone, K. (2000). Influence of growth temperature on lipid and soluble carbohydrate synthesis by fungi isolated from fellfield soil in the maritime Antarctic. *Mycologia, 92*, 222–229. https://doi.org/10.2307/3761554.

Wynn-Williams, D. D. (1990). Ecological aspects of Antarctic microbiology. *Advances in Microbial Ecology, 11*, 71–146. https://doi.org/10.1007/978-1-4684-7612-5_3.

Yergeau, E., Kang, S., He, Z., Zhou, J., & Kowalchuk, G. A. (2007a). Functional microarray analysis of nitrogen and carbon cycling genes across an Antarctic latitudinal transect126. *IsmeJ, 1*, 163–179.

Yergeau, E., Newsham, K. K., Pearce, D. A., & Kowalchuk, G. A. (2007b). Patterns of bacterial diversity across a range of Antarctic terrestrial habitats. *Environmental Microbiology, 9*, 2670–2682. https://doi.org/10.1111/j.1462-2920.2007.01379.x.

Zalar, P., & Gunde-Cimerman, N. (2014). Cold-adapted yeasts in arctic habitats. In *Cold-adapted yeasts: Biodiversity, adaptation strategies and biotechnological significance* (pp. 49–74). Berlin: Springer.

Zalar, P., Gostinčar, C., de Hoog, G. S., Uršič, V., Sudhadham, M., & Gunde-Cimerman, N. (2008). Redefinition of Aureobasidium pullulans and its varieties. *Studies in Mycology, 61*, 21–38. https://doi.org/10.3114/sim.2008.61.02.

Zhang, J., Müller, C., Zhu, T., Cheng, Y., & Cai, Z. (2011). Heterotrophic nitrification is the predominant NO3- production mechanism in coniferous but not broad-leaf acid forest soil in subtropical China. *Biology and Fertility of Soils, 47*, 533–542. https://doi.org/10.1007/s00374-011-0567-z.

Zhou, Z., Takaya, N., Nakamura, A., Yamaguchi, M., Takeo, K., & Shoun, H. (2002). Ammonia fermentation, a novel anoxic metabolism of nitrate by fungi. *The Journal of Biological Chemistry, 277*, 1892–1896. https://doi.org/10.1074/jbc.M109096200.

Chapter 4
Living with Pigments: The Colour Palette of Antarctic Life

Juan José Marizcurrena, María Fernanda Cerdá, Diego Alem, and Susana Castro-Sowinski

Abstract The production of pigments is a common feature that may help microorganisms to cope with the harsh conditions found in Antarctica. They have functions such as protection against UV irradiation and superoxide and nitrogen reactive species (antioxidant activity) and modulation of membrane fluidity under cold stress. In addition, they act as antibiotics, modulating the microbial communities in their natural environments, and harvest light for increasing the efficiency of photosynthesis, thus influencing the biogeochemical cycles. This chapter deals with the chemistry and the biological role of microbial pigments (except chlorophylls) in the Antarctic environment and also includes a brief overview of the potential biotechnological use of pigments.

Keywords Microbial production of pigments · UV-resistance · Antioxidant activity · Membrane fluidity · Photosynthesis · Antimicrobial activity

---

J. J. Marizcurrena · S. Castro-Sowinski (✉)
Biochemistry and Molecular Biology, Faculty of Sciences,
Universidad de la República, Montevideo, Uruguay
e-mail: scs@fcien.edu.uy

M. F. Cerdá
Laboratory of Biomaterials, Faculty of Sciences, Universidad de la República,
Montevideo, Uruguay

D. Alem
Biochemistry and Molecular Biology, Faculty of Sciences,
Universidad de la República, Montevideo, Uruguay

Epigenetics and Genomics Instability Laboratory, Instituto de Investigaciones Biológicas
Clemente Estable, Montevideo, Uruguay

S. Castro-Sowinski (ed.), *The Ecological Role of Micro-organisms in the Antarctic Environment*, Springer Polar Sciences,
https://doi.org/10.1007/978-3-030-02786-5_4

4.1 Introduction

The ultraviolet (UV) solar radiation is divided into UVC (170–280 nm), UVB (280–315 nm) and UVA (315–400 nm). UVC, as well as most UVB radiation, does not reach the Earth's surface due to the filtering effect of the ozone layer located at the stratosphere. This UV irradiation is damaging for the life on Earth, but many life forms produce pigments as protecting molecules among other adaptations to protect themselves. Biological pigments are substances whose presence in tissues or cells colours them. They are found in different sources, as animals and higher plants, but they are also produced by bacteria, algae, lichen or moss. Dyes act as light-harvesting molecules, protecting organism from radiation, and are also used for synthetic purposes.

During the early stage of life, the selective pressure of solar irradiation may have slanted the microbial evolution (Garcia-Pichel 1998; Cockell and Horneck 2001). The new life forms should have to develop different mechanisms to cope with the harmful effect of solar irradiation, and probably, by natural selection, they produced different pigments as cellular products that served for multiple purposes. Pigments conferred unique features, such as photon harvesting for energy income, and protection of vital molecules like DNA and proteins (Wynn-Williams et al. 2002).Thus, constant UV irradiation would have been shielded by these new pigments, avoiding oxidative stress and direct DNA damage, protecting cells from the UV mutagenic potential (Mulkidjanian and Junge 1997).

Currently, as a consequence of the depletion of the ozone layer, organisms from Earth are exposed to increased levels of UV irradiation. Therefore, various organisms have developed mechanisms for protection against the damage caused by UV irradiation. They can produce and accumulate absorbing molecules to decrease the transmission of incident light. Alternatively, they can produce UV screening compounds such as phenolic compounds, able to absorb and re-emit the UV radiation as fluorescence or thermal radiation. A molecule is able to absorb light from the HOMO to the LUMO orbital, and then several pathways can dissipate the energy. The energy can return to the basal state dissipating the excess of energy by producing heat or by radioactive processes (fluorescence or phosphorescence). Instead, transfer to other molecules, inducing structural modifications via reversible or irreversible pathways, can dissipate the energy. Finally, the last protective strategy is the production of repairing macromolecules such as the production of photolyases, enzymes that repair the DNA damage induced by UV irradiation (Sancar et al. 2004). Among others, Antarctic bacteria are a good source of highly efficient photolyases (Marizcurrena et al. 2017).

This chapter deals with the chemistry and the biological role of pigments produced by microorganisms, with the focus in Antarctic microbes. We will describe a set of pigments, but the well-known photosynthetic pigment chlorophyll has not been included in this chapter.

4.2 Living with Carotenoids

Carotenoids are organic liposoluble pigments from the group of isoprenoids, found naturally in plants and other photosynthetic organisms, and are responsible for most of the yellow, orange or red colours in nature. They are divided into two groups: (1) non-oxygenated molecules as β-carotene and (2) oxygenated molecules as xanthophylls (Fig. 4.1). The maximum absorptions of carotenoids are between 440 and 520 nm, with a strong molar absorption coefficient (ca. 10^5 L mol^{-1} cm^{-1}). They protect cells from photo-oxidative damage by (i) preventing the formation of reactive oxygen species through the thermal dissipation of the excess of energy; and (ii) quenching of the excited states of chlorophyll and singlet oxygen.

During their emergence in the early life, carotenoids may have function as light-harvesting antenna, acting as accessory pigments for chlorophylls. As carotenoids collect energy from a wider spectrum of visible light compared with chlorophylls, it may increase the photosynthetic efficiency (Lichtenthaler 1987), thus

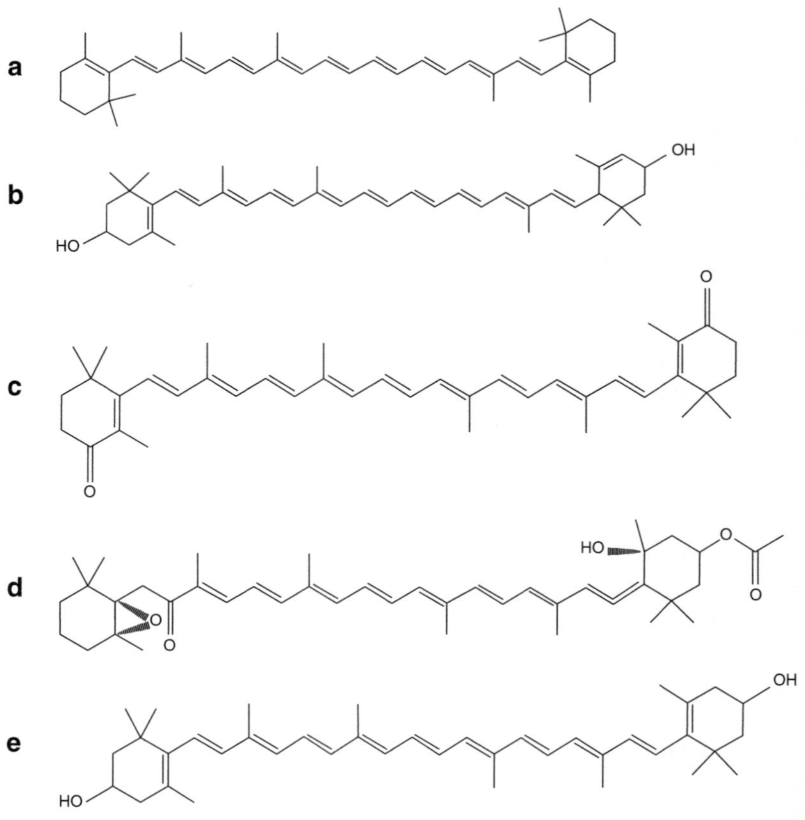

Fig. 4.1 Structures of (**a**) β-carotene (**b**) and lutein (**c**) cathaxanthin (**d**) fucoxanthin and (**e**) zeaxanthin

influencing the biogeochemical cycling of carbon. On a geological time scales, the evidence support that there is an important impact of photosynthesis in biogeochemical cycles (Falkowski 1994). Pigments also provide photoprotection against UV irradiation, absorbing light and preserving DNA and proteins from photodamage and oxidizing agents that produce reactive oxygen species (ROS; antioxidant properties). Alternatively, carotenoids provide protection against other physiological stresses such as low temperature, another common feature in Antarctica (Tian and Hua 2010).

Carotenoids are also involved in cell differentiation and cell cycle regulation. They also act as growth factor regulators, as immune systems inducers and as intracellular signalling molecules, among others functions (Fiedor and Burda 2014). Chemically, they have hydrophobic chains that attach them into the lipid bilayer of cell membranes; thus, changes in the amount of membrane carotenoids influence the thickness, fluidity and rigidity of the membranes, an important feature in a roasting early habitat and also in the cold Antarctic environment (Wisniewska and Subczynski 1998, 2006). The advantage of counting on a wide range of carotenoids (β-carotenes and xanthophylls) not only influences the stability of the cell membrane, it also improves the resistance to ROS. Carotenoids like β-carotene absorb UVA and UVC light, both inducing oxidative stress and DNA photodamage, respectively.

In Antarctica, microorganisms have been detected in all habitats such as lakes, ponds, rivers, streams, rocks and soil. Among pigments, carotenoids are the main pigments found in microbes. They have been reported as cryo- or solar radiation protectants as well as light harvesters in photovoltaic cells (Montagni et al. 2018). Carotenoids such as nostoxanthin and zeaxanthin have been found in *Pseudomonas*, whereas canthaxanthin or 2′-hydroxyflexixanthin has been found in *Hymenobacter* isolates (Klassen and Foght 2008). Also the phytoplankton is a rich source of pigments including chlorophylls, xanthins and α- and β-carotenes; these photosynthetic and photoprotective pigments may allow the adaptation of life under low light conditions (Ferreira et al. 2017).

Among other carotenoid-producing Antarctic microorganisms, it has been shown that the bacterium *Sphingobacterium antarcticus* produces at least three carotenoids (zeaxanthin, β-cryptoxanthin and β-carotene). Another example of an Antarctic carotenoid-producing bacterium is *Arthrobacter agilis*, isolated from a sea ice sample, that increases the content of carotenoids (C-50 bacterioruberin-type carotenoid and its glycosylated derivatives) at low temperatures of growth; interestingly, the production of carotenoid decreased when growing at high-salinity conditions (Fong et al. 2001). In addition to cold adaptation, the production of carotenoids by Antarctic heterotrophic bacteria is a strategy to withstand other environmental stresses, such as freeze-thaw cycles and solar irradiation (Dieser et al. 2010). The production of carotenoid-like pigments by Arctic microorganisms is also involved in physiological plasticity, a rapid response to a gradual decrease in temperature and freeze-thaw (Singh et al. 2017). Finally, the very well-known radio-resistant bacterium *Deinococcus radiodurans* produces the carotenoid deinoxanthin, a pigment that scavenges superoxide anions and effectively quenches free radicals containing nitrogen (reactive nitrogen species). Interestingly, proteins might be the principal target of free radicals, and the level of oxidative protein damage determines the degree of cell resistance indeed (Dong et al. 2015).

In summary, microbial carotenoids may play an important role as reactive species scavenger but also in the adaptation to cold stress, solar irradiation and freeze-thaw cycles.

4.3 Living with Phycobiliproteins

Phycobiliproteins (phycocyanins, phycoerythrins and allophycocyanin; blue, red and green, respectively) are proteins formed by α and β subunits that covalently bond phycobilins (tetrapyrroles that act as chromophores) by thioether bonds to cysteine residues (Ficner and Huber 1993). There are four major phycobilins in photosynthetic organisms, but the main ones are phycocyanobilin and phycoerythrobilin (Fig. 4.2) (chromophores present in phycoerythrin and phycocyanin or allophycocyanin, respectively). These pigmented proteins mainly act as accessory photosynthetic pigments and have antioxidant activities (Eriksen 2008; Patel et al. 2018). Interestingly, red algae phycobiliproteins contain a third type of

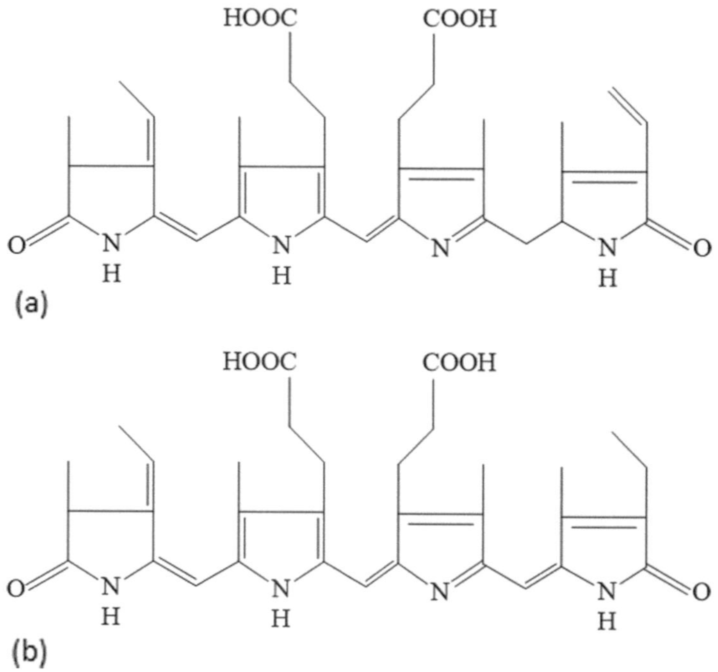

Fig. 4.2 General structure of the major phycobilins: (**a**) phycoerythrobilin and (**b**) phycocyanobilin

chromophore-bearing protein subunit, the γ one; thus, red algae phycobiliproteins are $(\alpha\beta)_3$ trimers and $(\alpha\beta)_6\gamma$ hexamers that covalently bond phycobilins.

Phycocyanin is a protein-pigment with an accessory function to chlorophylls, acting as a photosynthetic light-harvesting molecule (Wang et al. 2001; Enciso and Cerdá 2016). This pigment is produced by cyanobacteria (blue-green algae), found in damp sand and gravel around lakes, pools along melt water streams or in low-lying areas and semi-permanent to permanent snowy or ice-covered polar regions. However, the experiments carried out by Quesada and Vincent (Quesada and Vincent 1997), using two Antarctic cyanobacterial strains (*Phormidium murrayi* and *Oscillatoria priestleyi*, isolated from the McMurdo Ice Shelf), showed that the cellular concentrations of phycobiliproteins and to a lesser extent chlorophyll diminished when cells were exposed to UV radiation; but carotenoid increased to a threshold after UVB irradiation. In summary, the information suggests that the exposure to UV may lead to photoinhibition, phycobiliprotein degradation and chlorophyll bleaching; thus pigments like carotenoids, scytonemin and mycosporine-like amino acids take relevance because they are produced for protection against UV radiation as reported by Seckbach and Oren (Seckbach and Oren 2010). Antarctic mats have a seasonally ice-free 'moat' zone (that use light inefficiently) and two under-ice zones. The upper under-ice community contains both cyanobacteria and diatoms but also a high amount of phycoerythrin. At the lower under-ice zone, the concentration of the pigment declines with depth. Therefore, this pigment probably has a function of increasing the efficiency of the incident light utilization in light-limiting conditions (Vopel and Hawes 2006; Hawes and Schwarz 1999). Probably, phycobilins act as light-harvesting chromophores rather than reactive species scavengers or photoprotective pigments.

4.4 Living with Scytonemin

Scytonemin is a small hydrophobic alkaloid secondary metabolite responsible for yellowish-brown colour (Fig. 4.3a). This pigment is mainly produced by cyanobacteria when exposed to UVA-blue wavelengths (Garcia-Pichel and Castenholz 1991; Fleming and Castenholz 2007). Scytonemin-synthesizing cyanobacteria often inhabit highly solar-irradiated terrestrial, freshwater and coastal environments, acting as a highly efficient protective biomolecule that filters the damaging UV radiation, but also collecting energy for photosynthesis. Scytonemins and their methylated and methoxylated derivatives are frequently found in the upper layers of microbial mats, probably as a central protective biosignature of extreme environments, a

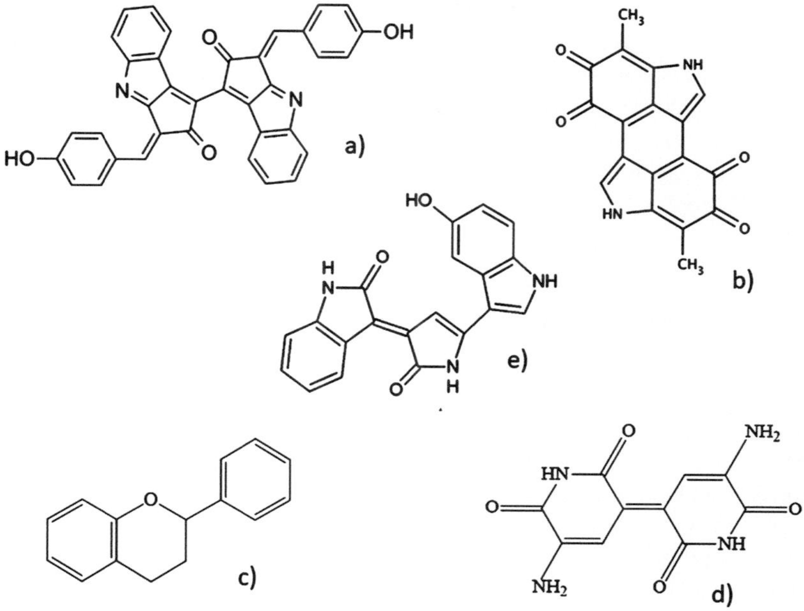

Fig. 4.3 General structure of other pigments. (**a**) Scytonemin, (**b**) melanin, (**c**) flavonoids, (**d**) indigoidine, (**e**) violacein

characteristic that points out this pigment as a potential biomolecular marker for the evaluation of putative exobiology habitats.

UV irradiation induces the accumulation of cyanobacterial carotenoids, detoxifying enzymes or radical quenchers, antioxidants, UV-absorbing/UV-screening substances such as mycosporine-like amino acids (MAAs), and scytonemin (Sinha et al. 1998). Thus, a set of biomolecules are produced to shield and protect cyanobacteria from the germicide radiation, molecules that could have a relevant role in the Antarctic environment during the early life on Earth (Dillon and Castenholz 1999). A palaeolimnological study performed in 62 east Antarctic lakes showed that in deeper lakes the pigment composition was dominated by chlorophylls, in intermediate depth lakes by chlorophylls and carotenoids and in shallow lakes by scytonemins. This differential pigment content is probably due to the influence of the light climate. As cyanobacteria can regulate their pigment content, among other properties, this difference may be a consequence of the ability of pigments to harvest the energy from light at different lake depths (Sabbe et al. 2004).

Among Antarctic scytonemin-producing cyanobacteria, the following genera have been reported: *Aphanocapsa* (Garcia-Pichel and Castenholz 1991; Vinocur and Pizarro 1995), *Calothrix* (Wynn-Williams et al. 1999), *Chlorogloeopsis* (Wynn-Williams et al. 1999), *Chroococcus* (Garcia-Pichel and Castenholz 1991),

Chroococcidiopsis (Wynn-Williams and Edwards 2000), *Diplocolon* (Garcia-Pichel and Castenholz 1991), *Entophysalis* (Garcia-Pichel and Castenholz 1991), *Gloeocapsa* (Vincent et al. 1993), *Haplosiphon* (Mushir et al. 2014), *Lyngbya* (Edwards et al. 2004), *Microcoleus* (Taton et al. 2003), *Nostoc* (Vincent et al. 1993), *Pleurocapsa* (Vincent et al. 1993), *Rivularia* (Pentecost and Edwards 2002), *Scytonema* (Wynn-Williams et al. 1999), *Stigonema* (Fernández-Carazo et al. 2012) and *Synechococcus* (Zakhia et al. 2008), among others.

4.5 Living with Melanin

Melanin is a dark biological high-molecular-weight pigment (dark green to brown, or totally black colour) found in the skin, hair, feathers, scales, eyes and some internal membranes. Chemically, it is the product of the polymerization of phenolic compounds and indolic rings (Nguyen et al. 2013), being the amino acid tyrosin the main precursor (Fig. 4.3b). However, when other precursors are used, the polymerization may render chemically different structures as the brown-black eumelanin, the yellow-red phaeomelanin and a heterogeneous group of allomelanins. According to the variety in the structure and occurrence of melanin, its biogenesis is not a single and universal process (Solano 2014). The production of melanin has been linked to UV- and visible light-irradiation resistance, protection against oxidizing and reducing agents, resistance to attack by cell-wall enzymes, antiviral activity and enhanced survival and competitive abilities under environmental stresses (Solano 2014; Castro-Sowinski et al. 2007). Interestingly, reduced melanin from *Shewanella algae* acts as an electron conduit to Fe(III) minerals, producing Fe(III)-reducing compound that may function as final electron acceptor (Turick et al. 2002). Melanin is generally insoluble in both aqueous and organic solvents, but the Antarctic bacterium *Lysobacter oligotrophicus* produces a water-soluble heteropolymer (Lo-melanin) characterized by a covalent bonding to a polysaccharide that showed to be related to its solubility (Kimura et al. 2015). Lo-melanin has the function of UV protection like other melanin pigments and also can scavenge radicals and ROS.

Melanin strongly absorbs UVB light and provides a similar protective function to microbes and the human skin against the harmful effects of UV irradiation. In addition, as melanized microorganisms have been commonly found in high-radiation environments, space stations, Antarctic mountains or at reactor cooling water combined with phenomenon of 'radiotropism', it has been suggested that melanins have functions analogous to other energy harvesting pigments such as chlorophylls (Dadachova and Casadevall 2008). This pigment is also commonly found in polar dark septate hyphae, and it has been proposed that melanin protects hyphae from extreme temperatures, playing a significant role in their persistence from year to year (Robinson 2001). Among Antarctic microfungi, *Friedmanniomyces endolithicus* produces very thick-walled highly melanized cells involved in UV resistance and also produces exopolysaccharides that protect cells from desiccation and freeze and thawing cycles (Onofri et al. 2004). Rock-inhabiting fungi (RIF) of Antarctic rocky deserts (considered as a closely related Martian habitat) is an example of adaptation to the extreme environment. RIF produces melanin-like pigments that protect them from excessive heat or cold, extreme pH or osmotic conditions, polychromatic UV radiation and tolerance against metals (Selbmann et al. 2015).

In summary, melanin-producing microbes may have a wider number of functions in the Antarctic environment including UV protection, tolerance to low temperature, desiccation, freezing and thawing cycles and scavenging of ROS, among others. Interestingly, these microorganisms can be used as microbial models for astrobiological/exobiological studies.

4.6 Living with Flavonoids

Flavonoids (yellow in nature) are a class of secondary metabolites with a general structure of a 15-carbon skeleton (two phenolic rings and a heterocyclic ring covalently joined; C6-C3-C6) that are chemically classified into the following subclasses: chalcone, flavone, isoflavone, flavonol, flavanone and isoflavonoid compounds (Fig. 4.3c). They have been recognized as important signalling molecules in microbe-plant interaction events (Morel and Castro-Sowinski 2013); thus, they probably contribute in the microbe-plant communication in the Antarctic environment. Only two native Antarctic vascular plants, *Deschampsia antarctica* and *Colobanthus quitensis*, are found (Benavent-González et al. 2018). Both have a diverse community of rhizospheric microbes dominated by Actinobacteria and Firmicutes (Teixeira et al. 2013), among others. When exposed to UV radiation, *D. antarctica* induces alterations in the plant chemistry, increasing the exudation of carbohydrates, carboxylic acids and flavonoids influencing the composition of the rhizosphere microbial community (Avery et al. 2003). It has been hypothesized that the UV irradiation may influence the quality or quantity of root exudates, being flavonoids among the most important molecules involved in the microbe-plant communication (Morel and Castro-Sowinski 2013).

4.7 Living with Indigoidine

Indigoidine (Fig. 4.3d) is a water-soluble brilliant blue pigment produced by a few microbes (Sutthiwong et al. 2014), including *Vogesella indigofera* (Kuhn et al. 1965; Day et al. 2017), *Erwinia chrysanthemi* (Reverchon et al. 2002), *Phaeobacter* sp. (Cude et al. 2012), *Streptomyces chromofuscus* (Yu et al. 2013) and isolates from the Antarctic genus *Arthrobacter* (Sutthiwong et al. 2014).

Isolates from the genus *Arthrobacter* (phylum Actinobacteria, family Micrococcaceae) are commonly found in different Antarctic environments, including soils and sediments (Dsouza et al. 2015; Reddy et al. 2003; Ganzert et al. 2011; Chen et al. 2005), among others). Bacteria from this genus produce different pigments such as yellow carotenoids and yellow riboflavins, blue indigoidine and blue indochrome and red porphyrins (Sutthiwong et al. 2014). The physiological and/or ecological role of indigoidine is still unknown, but it has been recently reported that indigoidine may provide resistance to oxidative stress (protection against the reactive oxygen species produced during the plant defence response) (Reverchon et al. 2002), as well as protection by its antimicrobial activity as reported in *Leisingera* (formerly *Phaeobacter*) isolates (Gromek et al. 2016). Thus, indigoidine-producing microorganisms may have a competitive advantage in their natural environments, mainly due to the antioxidant and antibiotic properties; moreover its functions as intracellular signalling molecules associated with motility have also been reported (Reverchon et al. 2002; Cude et al. 2012). In addition, indigoidine may have a role in the microbial adaptation to iron-rich environments as shown by when working with *Vogesella* sp. strain EB, isolated from Andean Patagonia (Day et al. 2017).

4.8 Living with Violacein

Chemically, violacein is bis-indole purple-pigment (Fig. 4.3e) with a few interesting biological activities (Beckstead et al. 2017). This pigment is produced by *Chromobacterium violaceum*, but also by *Alteromonas*, *Janthinobacterium*, *Pseudoalteromonas*, *Duganella* and *Collimonas* spp. (Durán et al. 2016; Smith et al. 2016; Shivaji et al. 1991), in a quorum-sensing-dependent manner (Mcclean et al. 1997). Violacein has a peak of absorbance in the UV range ($\lambda = 260$ nm), suggesting its potential role in protection against visible and ultraviolet radiation. In addition to UV resistance, it has been shown that violacein-enriched extracts from Antarctic *Janthinobacterium* isolates have antimicrobial activity against Gram-negative and Gram-positive bacteria, and fungus (Mojib et al. 2010; Asencio et al. 2014), and inhibit protozoan feeding, thus providing grazing protection against protozoans (Matz et al. 2004).

4.9 Microbial Mats: A Pool of Pigments

Microbial mats are benthic, multilayered sheet of microorganisms and self-sustaining communities that develop in various environments, mainly composed by cyanobacteria (Fernández-Valiente et al. 2007; Vincent 2000).The microbial species found within the mat and its sediment composition, combined with the environmental conditions, determine the morphological structure and the striking colours of the microbial mats (Stal 2012). The typical vertical colour pattern of the mats, that shows the stratification of the microbial communities, is caused by the different pigments produced by the different phototrophic microorganisms within each layer (Vincent 2000).

Antarctica is characterized by a seasonality that causes substantial variation in the light photoperiod (Mueller et al. 2005); thus, microbial mats are indeed exposed to high UV irradiation and low temperatures, leaving them vulnerable to photodamage, among other stresses (Roos and Vincent 1998). In this scenario, mats produce molecules involved in photoprotection that efficiently repair the photodamage, including pigments. Mats are well known to protect the microbial community from the high UV irradiation, fighting against free radicals and also filtering radiation (Ehling-Schulz and Scherer 1999; Cockell and Knowland 1999).

4.10 The Biotechnological Use of Pigments

In addition to physiological and ecological functions, pigments have many potential biotechnological applications. From 2007 to 2011, the global sales of natural pigments have arisen a 29%, reaching an estimated USD 600 million (Tuli et al. 2015). This economic growth and trade of pigments justify the search for new pigments and potential biotechnological applications. Currently, pigments from natural sources are used in food, fabric, cosmetic and pharmaceutical industries, among others.

The use of colour additives or pigments is under the Federal Food, Drug, and Cosmetic Act (Chapter VII, section 721) approval before they can be used in food, drugs or cosmetics or in medical devices for people or animals for a significant period of time (https://www.fda.gov/). Some pigments have been approved and are commonly used in foods, such as the yellow riboflavin (flavin, from *Bacillus subtilis*), orange-yellow β-carotene (carotenoid, from *Blakeslea trispora*, *Dunaliella salina*), yellow to red lycopene (carotenoid, from *Blakeslea trispora*), yellow to red astaxanthin (carotenoid, from *Haematococcus pluvialis*) and orange-red canthaxanthin (carotenoid, from *Haematococcus lacustris*) (Nigam and Luke 2016). However, microbial pigments are also used in fabrics, contributing to the development of the textile industry, such as the red pigment prodigiosin from *Vibrio* strains (commonly used to dye wool, nylon, silk and acrylic fibres) (Narsing Rao et al. 2017) and violacein from *Chromobacterium violaceum* (used for dying silk, cotton, acrylic and polyester) (Venil et al. 2013). Interestingly, violacein also has potential applications in the pharmaceutical industry, due its antiproliferative, antimicrobial, antiparasitic, antifungal and antiviral activities (Durán et al. 2012). The antitumor activity of violacein was demonstrated at least on cell lines derived from melanoma, colon cancer and breast cancer lines (Melo et al. 2003; de Carvalho et al. 2006; Bromberg et al. 2010; Alshatwi et al. 2016) and also sensitizes cells from colorectal cancer (Kodach et al. 2006). Antarctic pigment-producing microorganisms can also be considered as potential electron sources, so used for manufacturing dye-sensitized solar cells (DSSC), a promising alternative to conventional photovoltaic-silicon cells based. Thus, DSSC could represent an interesting alternative that partially may solve the need of energy at Antarctica. For example, the orange-xanthophyll pigment from the UVC-resistant *Hymenobacter* sp. UV11 isolate (Marizcurrena et al. 2017) was used in manufacturing DSSC (Montagni et al. 2018; Enciso and Cerdá 2016; Woronowicz et al. 2012; Calogero et al. 2015; Zhao et al. 2014; Órdenes-Aenishanslins et al. 2016).

4.11 Concluding Remarks

Pigment production is a common feature in microbes, and Antarctic microbes are not the exception (Fig. 4.4). The identification of Antarctic pigment-producing microorganisms has been far reported, and they can be considered an impressive

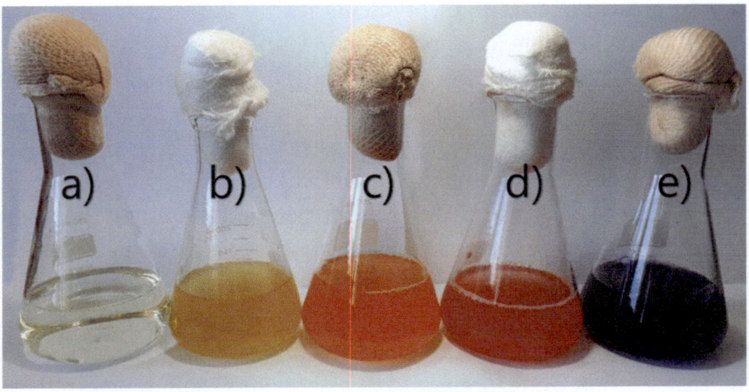

Fig. 4.4 Pigment-producing Antarctic bacteria. (**a**) Cell-free culture medium; (**b**–**e**) growth cultures from the Antarctic isolates *Pseudomonas* sp. AU10, *Sphingomonas* sp. UV9, *Hymenobacter* sp. UV11 and *Janthinobacterium* sp. UV13, respectively (Marizcurrena et al. 2017; Martínez-Rosales and Castro-Sowinski 2011)

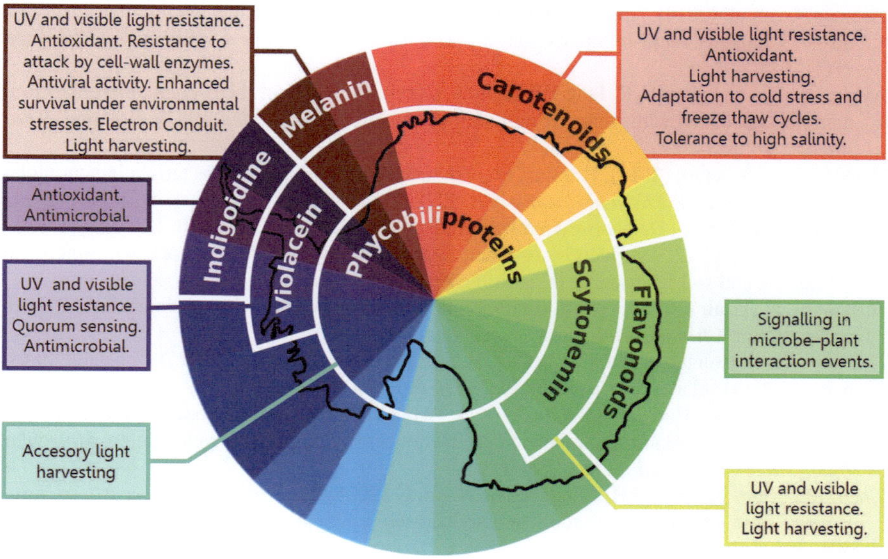

Fig. 4.5 Microbial pigments: an overview of their role

colour palette painting the white continent canvas. These pigments have many functions such as protection against UV irradiation and superoxide and nitrogen reactive species (antioxidant activity), modulation of membrane fluidity under cold stress, harvesting light for increasing the efficiency of photosynthesis and acting as antibiotics (Fig. 4.5). All these functions may help microorganisms to cope with the harsh conditions found in Antarctica (high UV irradiation, low temperatures, oxidative stress, freeze-thaw and variation in the light photoperiod cycles). Unfortunately, the knowledge about the ecological role of pigments from Antarctic microbes has not appropriately been faced yet.

Acknowledgements The authors thank the Uruguayan Antarctic Institute for the logistic support during the stay in the Antarctic Base Artigas. S. Castro-Sowinski, J.J. Marizcurrena, M.F. Cerdá and D. Alem are members of the National Research System (SNI, Sistema Nacional de Investigadores).

This work was partially supported by PEDECIBA (Programa de Desarrollo de las Ciencias Básicas), CSIC (Comisión Sectorial de Investigación Científica; Project C667) and ANII (Agencia Nacional de Investigación e Innovación, Project FMV_3_2016_1_1226654). The work of JJM and DA was supported by ANII and CAP (Comisión Académica de Posgrado, UdelaR).

References

Alshatwi, A. A., Subash-Babu, P., & Antonisamy, P. (2016). Violacein induces apoptosis in human breast cancer cells through up regulation of BAX, p53 and down regulation of MDM2. *Experimental and Toxicologic Pathology, 68*(1), 89–97.

Asencio, G., et al. (2014). Antibacterial activity of the Antarctic bacterium *Janthinobacterium* sp. SMN 33.6 against multi-resistant Gram-negative bacteria. *Electronic Journal of Biotechnology, 17*(1), 1–5.

Avery, L. M., Lewis Smith, R. I., & West, H. M. (2003). Response of rhizosphere microbial communities associated with Antarctic hairgrass (*Deschampsia antarctica*) to UV radiation. *Polar Biology, 26*, 525–529.

Beckstead, A. A., et al. (2017). Ultrafast excited-state deactivation of the bacterial pigment violacein. *The Journal of Physical Chemistry B, 121*, 7855–7861.

Benavent-González, A., Delgado-Baquerizo, M., Fernández-Brun, L., Singh, B. K., Maestre, F. T., & Sancho, L. G. (2018). Identity of plant, lichen and moss species connects with microbial abundance and soil functioning in maritime Antarctica. *Plant and Soil, 429*, 35–52.

Bromberg, N., et al. (2010). Growth inhibition and pro-apoptotic activity of violacein in Ehrlich ascites tumor. *Chemico-Biological Interactions, 186*(1), 43–52.

Calogero, G., Bartolotta, A., Di Marco, G., Di Carlo, A., & Bonaccorso, F. (2015). Vegetable-based dye-sensitized solar cells. *Chemical Society Reviews, 44*, 3244–3294.

Castro-Sowinski, S., Matan, O., Bonafede, P., & Okon, Y. (2007). A thioredoxin of Sinorhizobium meliloti CE52G is required for melanin production and symbiotic nitrogen fixation. *Molecular Plant-Microbe Interactions, 20*, 986–993.

Chen, M., Xiao, X., Wang, P., Zeng, X., & Wang, F. (2005). *Arthrobacter ardleyensis* sp. nov., isolated from Antarctic lake sediment and deep-sea sediment. *Archives of Microbiology, 183*, 301–305.

Cockell, C. S., & Horneck, G. (2001). The history of the UV radiation climate of the earth – theoretical and space-based observations. *Photochemistry and Photobiology, 73*, 447–451.

Cockell, C. S., & Knowland, J. (1999). Ultraviolet radiation screening compounds. *Biological Reviews, 74*, 311–345.

Cude, W. N., et al. (2012). Production of the antimicrobial secondary metabolite indigoidine contributes to competitive surface colonization by the marine roseobacter *Phaeobacter* sp. strain Y4I. *Applied and Environmental Microbiology, 78*, 4771–4780.

Dadachova, E., & Casadevall, A. (2008). Ionizing radiation: How fungi cope, adapt, and exploit with the help of melanin. *Current Opinion in Microbiology, 11*, 525–531.

Day, P. A., Villalba, M. S., Herrero, O. M., Arancibia, L. A., & Alvarez, H. M. (2017). Formation of indigoidine derived-pigments contributes to the adaptation of Vogesella sp. strain EB to cold aquatic iron-oxidizing environments. *Antonie van Leeuwenhoek, International Journal of General and Molecular Microbiology, 110*(3), 415–428.

de Carvalho, D. D., Costa, F. T. M., Duran, N., & Haun, M. (2006). Cytotoxic activity of violacein in human colon cancer cells. *Toxicology in Vitro, 20*(8), 1514–1521.

Dieser, M., Greenwood, M., & Foreman, C. M. (2010). Carotenoid pigmentation in antarctic heterotrophic bacteria as a strategy to withstand environmental stresses. *Arctic, Antarctic, and Alpine Research, 42*(4), 396–405.

Dillon, J. G., & Castenholz, R. W. (1999). Scytonemin, a cyanobacterial sheath pigment, protects against uvc radiation: Implications for early photosynthetic life. *Journal of Phycology, 35*, 673–681.

Dong, N., Li, H.-R., Yuan, M., Zhang, X.-H., & Yu, Y. (2015). *Deinococcus antarcticus* sp. nov., isolated from soil. *International Journal of Systematic and Evolutionary Microbiology, 65*(Pt 2), 331–335.

Dsouza, M., Taylor, M. W., Turner, S. J., & Aislabie, J. (2015). Genomic and phenotypic insights into the ecology of *Arthrobacter* from Antarctic soils. *BMC Genomics, 16*, 36.

Durán, M., Ponezi, A. N., Faljoni-Alario, A., Teixeira, M. F. S., Justo, G. Z., & Durán, N. (2012). Potential applications of violacein: A microbial pigment. *Medicinal Chemistry Research, 21*(7), 1524–1532.

Durán, N., et al. (2016). Advances in *Chromobacterium violaceum* and properties of violacein-its main secondary metabolite: A review. *Biotechnology Advances, 34*, 1030–1045.

Edwards, H. G. M., de Oliveira, L. F. C., Cockell, C. S., Ellis-Evans, J. C., & Wynn-Williams, D. D. (2004). Raman spectroscopy of senescing snow algae: Pigmentation changes in an Antarctic cold desert extremophile. *International Journal of Astrobiology, 3*, 125–129.

Ehling-Schulz, M., & Scherer, S. (1999). Uv protection in cyanobacteria. *European Journal of Phycology, 34*, 329–338.

Enciso, P., & Cerdá, M. F. (2016). Solar cells based on the use of photosensitizers obtained from Antarctic red algae. *Cold Regions Science and Technology, 126*, 51–54.

Eriksen, N. T. (2008). Production of phycocyanin – a pigment with applications in biology, biotechnology, foods and medicine. *Applied Microbiology and Biotechnology, 80*, 1–14.

Falkowski, P. G. (1994). The role of phytoplankton photosynthesis in global biogeochemical cycles. *Photosynthesis Research, 39*, 235–258.

Fernández-Carazo, R., Namsaraev, Z., Mano, M. J., Ertz, D., & Wilmotte, A. (2012). Cyanobacterial diversity for an anthropogenic impact assessment in the Sør Rondane Mountains area, Antarctica. *Antarctic Science, 24*, 229–242.

Fernández-Valiente, E., Camacho, A., Rochera, C., Rico, E., Vincent, W. F., & Quesada, A. (2007). Community structure and physiological characterization of microbial mats in byers peninsula, Livingston Island (South Shetland Islands, Antarctica). *FEMS Microbiology Ecology, 59*, 377–385.

Ferreira, A., Ciotti, Á. M., Mendes, C. R. B., Uitz, J., & Bricaud, A. (2017). Phytoplankton light absorption and the package effect in relation to photosynthetic and photoprotective pigments in the northern tip of Antarctic Peninsula. *Journal of Geophysical Research, Oceans, 122*, 7344–7363.

Ficner, R., & Huber, R. (1993). Refined crystal structure of phycoerythrin from *Porphyridium cruentum* at 0.23-nm resolution and localization of the γ subunit. *European Journal of Biochemistry, 218*, 103–106.

Fiedor, J., & Burda, K. (2014). Potential role of carotenoids as antioxidants in human health and disease. *Nutrients, 6*, 466–488.

Fleming, E. D., & Castenholz, R. W. (2007). Effects of periodic desiccation on the synthesis of the UV-screening compound, scytonemin, in cyanobacteria. *Environmental Microbiology, 9*, 1448–1455.

Fong, N. J. C., Burgess, M. L., Barrow, K. D., & Glenn, D. R. (2001). Carotenoid accumulation in the psychrotrophic bacterium Arthrobacter agilis in response to thermal and salt stress. *Applied Microbiology and Biotechnology, 56*, 750–756.

Ganzert, L., Bajerski, F., Mangelsdorf, K., Lipski, A., & Wagner, D. (2011). *Arthrobacter livingstonensis* sp. nov. and arthrobacter cryotolerans sp. nov., salt-tolerant and psychrotolerant species from antarctic soil. *International Journal of Systematic and Evolutionary Microbiology, 61*, 979–984.

Garcia-Pichel, F. (1998). Solar ultraviolet and the evolutionary history of cyanobacteria. *Origins of Life and Evolution of the Biosphere, 28*, 321–347.

Garcia-Pichel, F., & Castenholz, R. W. (1991). Characterization and biological implications of scytonemin, a cyanobacterial sheath pigment. *Journal of Phycology, 27*, 395–409.

Gromek, S. M., et al. (2016). *Leisingera sp.* JC1, a bacterial isolate from hawaiian bobtail squid eggs, produces indigoidine and differentially inhibits vibrios. *Frontiers in Microbiology, 7*, 1342.

Hawes, I., & Schwarz, A.-M. (1999). Photosynthesis in an extreme shade environment: Benthic microbial mats from lake hoare, a permanently ice-covered antarctic lake. *Journal of Phycology, 35*, 448–459.

Kimura, T., Fukuda, W., Sanada, T., & Imanaka, T. (2015). Characterization of water-soluble dark-brown pigment from Antarctic bacterium, Lysobacter oligotrophicus. *Journal of Bioscience and Bioengineering, 120*(1), 58–61.

Klassen, J. L., & Foght, J. M. (2008). Differences in carotenoid composition among Hymenobacter and related strains support a tree-like model of carotenoid evolution. *Applied and Environmental Microbiology, 74*, 2016–2022.

Kodach, L. L., Bos, C. L., Durán, N., Peppelenbosch, M. P., Ferreira, C. V., & Hardwick, J. C. H. (2006). Violacein synergistically increases 5-fluorouracil cytotoxicity, induces apoptosis and inhibits Akt-mediated signal transduction in human colorectal cancer cells. *Carcinogenesis, 27*(3), 508–516.

Kuhn, R., Starr, M. P., Kuhn, D. A., Bauer, H., & Knackmuss, H. J. (1965). Indigoidine and other bacterial pigments related to 3,3′-bipyridyl. *Archives of Microbiology, 51*, 71–84.

Lichtenthaler, H. K. (1987). Chlorophylls and carotenoids: Pigments of photosynthetic biomembranes. *Methods in Enzymology, 148*, 350–382.

Marizcurrena, J. J., Morel, M. A., Braña, V., Morales, D., Martinez-López, W., & Castro-Sowinski, S. (2017). Searching for novel photolyases in UVC-resistant Antarctic bacteria. *Extremophiles, 21*(2), 409–418.

Martínez-Rosales, C., & Castro-Sowinski, S. (2011). Antarctic bacterial isolates that produce cold-active extracellular proteases at low temperature but are active and stable at high temperature. *Polar Research, 30*, 7123–7130.

Matz, C., et al. (2004). Impact of violacein-producing bacteria on survival and feeding of bacterivorous nanoflagellates. *Applied and Environmental Microbiology, 70*(3), 1593–1599.

Mcclean, K. H., et al. (1997). Quorum sensing and Chrornobacteriurn violaceurn: Exploitation of violacein production and inhibition for the detection of N-acyl homoserine lactones. *Microbiology, 143*, 3703–3711.

Melo, P. S., Justo, G. Z., De Azevedo, M. B. M., Durán, N., & Haun, M. (2003). Violacein and its β-cyclodextrin complexes induce apoptosis and differentiation in HL60 cells. *Toxicology, 186*(3), 217–225.

Mojib, N., Philpott, R., Huang, J. P., Niederweis, M., & Bej, A. K. (2010). Antimycobacterial activity in vitro of pigments isolated from Antarctic bacteria. *Antonie van Leeuwenhoek, International Journal of General and Molecular Microbiology, 98*(4), 531–540.

Montagni, T., et al. (2018). Dye sensitized solar cells based on Antarctic Hymenobacter sp. UV11 dyes. *Environmental Sustainability, 1*, 89–97.

Morel, M. A., & Castro-Sowinski, S. (2013). The complex molecular signaling network in microbe–plant interaction. In *Plant microbe symbiosis: Fundamentals and advances* (pp. 169–199). New Delhi: Springer.

Mueller, D. R., Vincent, W. F., Bonilla, S., & Laurion, I. (2005). Extremotrophs, extremophiles and broadband pigmentation strategies in a high arctic ice shelf ecosystem. *FEMS Microbiology Ecology, 53*(1), 73–87.

Mulkidjanian, A. Y., & Junge, W. (1997). On the origin of photosynthesis as inferred from sequence analysis. A primordial UV-protector as common ancestor of reaction centers and antenna proteins. *Photosynthesis Research, 51*, 27–42.

Mushir, S., Deep, S., & Fatma, T. (2014). Screening of cyanobacterial strains for UV screening compound scytonemin – environmental perspectives. *The International Journal of Innovative Research in Science, Engineering and Technology, 3*, 12–20.

Narsing Rao, M. P., Xiao, M., & Li, W. J. (2017). Fungal and bacterial pigments: Secondary metabolites with wide applications. *Frontiers in Microbiology, 8.*, no. JUN, 1–13.

Nguyen, K.-H., Chollet-Krugler, M., Gouault, N., & Tomasi, S. (2013). UV-protectant metabolites from lichens and their symbiotic partners. *Natural Product Reports, 30*, 1490–1508.

Nigam, P. S., & Luke, J. S. (2016, June). *2016 pigments in food additives ScienceDirect Food additives: Production of microbial pigments and their antioxidant properties.*

Onofri, S., Selbmann, L., Zucconi, L., & Pagano, S. (2004). Antarctic microfungi as models for exobiology. *Planetary and Space Science, 52*, 229–237.

Órdenes-Aenishanslins, N., Anziani-Ostuni, G., Vargas-Reyes, M., Alarcón, J., Tello, A., & Pérez-Donoso, J. M. (2016). Pigments from UV-resistant Antarctic bacteria as photosensitizers in dye sensitized solar cells. *Journal of Photochemistry and Photobiology B: Biology, 162*, 707–714.

Patel, S. N., et al. (2018). Antioxidant activity and associated structural attributes of Halomicronema phycoerythrin. *International Journal of Biological Macromolecules, 111*, 359–369.

Pentecost, A., & Edwards, H. G. M. (2002). Raman spectroscopy and light microscopy of a modern and sub-fossil microstromatolite: *Rivularia haematites* (cyanobacteria, Nostocales). *International Journal of Astrobiology, 1*, 357–363.

Quesada, A., & Vincent, W. F. (1997). Strategies of adaptation by antarctic cyanobacteria to ultraviolet radiation. *European Journal of Phycology, 32*, 335–342.

Reddy, G. S. N., Prakash, J. S. S., Prabahar, V., Matsumoto, G. I., Stackebrandt, E., & Shivaji, S. (2003). Kocuria polaris sp. nov., an orange-pigmented psychrophilic bacterium isolated from an Antarctic cyanobacterial mat sample. *International Journal of Systematic and Evolutionary Microbiology, 53*, 977–984.

Reverchon, S., Rouanet, C., Expert, D., & Nasser, W. (2002). Characterization of indigoidine biosynthetic genes in Erwinia chrysanthemi and role of this blue pigment in pathogenicity. *Journal of Bacteriology, 184*, 654–665.

Robinson, C. H. (2001). Cold adaptation in Arctic and Antarctic fungi. *New Phytologist, 151*, 341–353.

Roos, J. C., & Vincent, W. F. (1998). Temperature dependence of UV radiation effects on Antarctic cyanobacteria. *Journal of Phycology, 34*, 118–125.

Sabbe, K., et al. (2004). Salinity, depth and the structure and composition of microbial mats in continental Antarctic lakes. *Freshwater Biology, 49*(3), 296–319.

Sancar, A., Lindsey-Boltz, L. A., Ünsal-Kaçmaz, K., & Linn, S. (2004). Molecular mechanisms of mammalian DNA repair and the DNA damage checkpoints. *Annual Review of Biochemistry, 73*, 39–85.

Seckbach, J., & Oren, A. (2010). *Microbial Mats*.

Selbmann, L., Zucconi, L., Isola, D., & Onofri, S. (2015). Rock black fungi: Excellence in the extremes, from the Antarctic to space. *Current Genetics, 61*, 335–345.

Shivaji, S., Ray, M. K., Kumar, G. S., Reddy, G. S. N., Saisree, L., & Wynn-Williams, D. D. (1991). Identification of *Janthinobacterium lividum* from the soils of the islands of Scotia Ridge and from Antarctic peninsula. *Polar Biology, 11*, 267–271.

Singh, A., Krishnan, K. P., Prabaharan, D., & Sinha, R. K. (2017). Lipid membrane modulation and pigmentation: A cryoprotection mechanism in Arctic pigmented bacteria. *Journal of Basic Microbiology, 57*(9), 770–780.

Sinha, R. P., Klisch, M., Gröniger, A., & Häder, D. P. (1998). Ultraviolet-absorbing/screening substances in cyanobacteria, phytoplankton and macroalgae. *Journal of Photochemistry and Photobiology B: Biology, 47*, 83–94.

Smith, H. J., Foreman, C. M., Akiyama, T., Franklin, M. J., Devitt, N. P., & Ramaraj, T. (2016). Genome sequence of *Janthinobacterium* sp. CG23_2, a Violacein-producing isolate from an Antarctic supraglacial stream. *Genome Announcements, 4*, e01468–e01415.

Solano, F. (2014). Melanins: Skin pigments and much more—Types, structural models, biological functions, and formation routes. *New Journal of Science, 498276*, 1–28.

Stal, L. J. (2012). Cyanobacterial mats and stromatolites. In *Ecology of cyanobacteria II: Their diversity in space and time*. Dordrecht: Springer.

Sutthiwong, N., Fouillaud, M., Valla, A., Caro, Y., & Dufossé, L. (2014). Bacteria belonging to the extremely versatile genus Arthrobacter as novel source of natural pigments with extended hue range. *Food Research International, 65*, 56–162.

Taton, A., Grubisic, S., Brambilla, E., De Wit, R., & Wilmotte, A. (2003). Cyanobacterial diversity in natural and artificial microbial mats of Lake Fryxell (McMurdo Dry Valleys, Antarctica): A morphological and molecular approach. *Applied and Environmental Microbiology, 69*, 5157–5169.

Teixeira, L. C. R. S., Peixoto, R. S., & Rosado, A. S. (2013). Bacterial diversity in rhizosphere soil from Antarctic vascular plants of Admiralty Bay in Maritime Antarctica. In *Molecular microbial ecology of the rhizosphere*. Hoboken: Wiley-Blackwell.

Tian, B., & Hua, Y. (2010). Carotenoid biosynthesis in extremophilic Deinococcus-Thermus bacteria. *Trends in Microbiology, 18*, 512–520.

Tuli, H. S., Chaudhary, P., Beniwal, V., & Sharma, A. K. (2015). Microbial pigments as natural color sources: Current trends and future perspectives. *Journal of Food Science and Technology, 52*(8), 4669–4678.

Turick, C. E., Tisa, L. S., & Caccavo, F. (2002). Melanin production and use as a soluble electron shuttle for Fe(III) oxide reduction and as a terminal electron acceptor by Shewanella algae BrY. *Applied and Environmental Microbiology, 68*, 2436–2444.

Venil, C. K., Zakaria, Z. A., & Ahmad, W. A. (2013). Bacterial pigments and their applications. *Process Biochemistry, 48*, 1065–1079.

Vincent, W. F. (2000). Cyanobacterial dominance in the polar regions. In *The ecology of cyanobacteria*. Dordrecht: Kluwer Academic Publishers.

Vincent, W. F., Downes, M. T., Castenholz, R. W., & Howard-Williams, C. (1993). Community structure and pigment organisation of cyanobacteria-dominated microbial mats in Antarctica. *European Journal of Phycology, 28*(4), 213–221.

Vinocur, A., & Pizarro, H. (1995). Periphyton flora of some lotic and lentic environments of Hope Bay (Antarctic Peninsula). *Polar Biology, 15*, 401–414.

Vopel, K., & Hawes, I. (2006). Photosynthetic performance of benthic microbial mats in Lake Hoare, Antarctica. *Limnology and Oceanography, 51*, 1801–1812.

Wang, X. Q., et al. (2001). Structure of C-phycocyanin from Spirulina platensis at 2.2 Å resolution: A novel monoclinic crystal form for phycobiliproteins in phycobilisomes. *Acta Crystallographica, Section D: Biological Crystallography, 57*, 784–792.

Wisniewska, A., & Subczynski, W. K. (1998). Effects of polar carotenoids on the shape of the hydrophobic barrier of phospholipid bilayers. *Biochimica et Biophysica Acta, Biomembranes, 1368*, 235–246.

Wisniewska, A., & Subczynski, W. K. (2006). Accumulation of macular xanthophylls in unsaturated membrane domains. *Free Radical Biology & Medicine, 40*, 1820–1826.

Woronowicz, K., et al. (2012). Near-IR absorbing solar cell sensitized with bacterial photosynthetic membranes. *Photochemistry and Photobiology, 88*, 1467–1472.

Wynn-Williams, D. D., & Edwards, H. G. M. (2000). Proximal analysis of regolith habitats and protective biomolecules in situ by laser Raman spectroscopy: Overview of terrestrial Antarctic habitats and Mars analogs. *Icarus, 144*, 486–503.

Wynn-Williams, D. D., Edwards, H. G. M., & Garcia-Pichel, F. (1999). Functional biomolecules of antarctic stromatolitic and endolithic cyanobacterial communities. *European Journal of Phycology, 34*, 381–391.

Wynn-Williams, D. D., Edwards, H. G. M., Newton, E. M., & Holder, J. M. (2002). Pigmentation as a survival strategy for ancient and modern photosynthetic microbes under high ultraviolet stress on planetary surfaces. *International Journal of Astrobiology, 1*(1), 39–49.

Yu, D., Xu, F., Valiente, J., Wang, S., & Zhan, J. (2013). An indigoidine biosynthetic gene cluster from Streptomyces chromofuscus ATCC 49982 contains an unusual IndB homologue. *Journal of Industrial Microbiology & Biotechnology, 40*, 159–168.

Zakhia, F., Jungblut, A. D., Taton, A., Vincent, W. F., & Wilmotte, A. (2008). Cyanobacteria in cold ecosystems. In *Psychrophiles: From biodiversity to biotechnology*. Cham: Springer.

Zhao, Y. L., Song, D. M., Qiang, Y. H., Gu, X. Q., Zhu, L., & Song, C. B. (2014). Dye-sensitized solar cells based on TiO2 hollow spheres/TiO2 nanotube array composite films. *Applied Surface Science, 309*, 85.

Part II
Invasive Colonization and Human Perturbations in Antarctic Microbial Ecosystems

Chapter 5
Horizontal Gene Transfer Elements: Plasmids in Antarctic Microorganisms

Matías Giménez, Gastón Azziz, Paul R. Gill, and Silvia Batista

Abstract Plasmids play an important role in the evolution of microbial communities. These mobile genetic elements can improve host survival and may also be involved in *h*orizontal *g*ene *t*ransfer (HGT) events between individuals. Diverse culture-dependent and culture-independent approaches have been used to characterize these mobile elements. Culture-dependent methods are usually associated with classical microbiological techniques. In the second approach, development of specific protocols for analysis of metagenomes involves many challenges, including assembly of sequences and availability of a reliable database, which are crucial. In addition, alternative strategies have been developed for the characterization of plasmid DNA in a sample, generically referred to as plasmidome.

The Antarctic continent has environments with diverse characteristics, including some with very low temperatures, humidity levels, and nutrients. The presence of microorganisms and genetic elements capable of being transferred horizontally has been confirmed in these environments, and it is generally accepted that some of these elements, such as plasmids, actively participate in adaptation mechanisms of host microorganisms.

Information related to structure and function of HGT elements in Antarctic bacteria is very limited compared to what is known about HGT in bacteria from temperate/tropical environments. Some studies are done with biotechnological objectives. The search for mobile elements, such as plasmids, may be related to improve the expression of heterologous genes in host organisms growing at very low temperatures. More recently, however, additional studies have been done to detect plasmids in isolates, associated or not with specific phenotypes such as drug resistance. Although various Antarctic metagenomes are available in public databases, corre-

M. Giménez · G. Azziz · S. Batista (✉)
Unidad Microbiología Molecular, Instituto de Investigaciones Biológicas Clemente Estable, Montevideo, Uruguay
e-mail: sbatista@iibce.edu.uy

P. R. Gill
Nevada City Biolabs, Nevada City, CA, USA

S. Castro-Sowinski (ed.), *The Ecological Role of Micro-organisms in the Antarctic Environment*, Springer Polar Sciences,
https://doi.org/10.1007/978-3-030-02786-5_5

sponding studies of plasmidomes are needed. The difficulties usually associated with the study of metagenomes are increased in these cases by the limited number of sequences in functionally characterized databases.

Keywords Horizontal gene transfer · Plastmidome · Global climate change · Animal and human influence · Antibiotic resistance genes

5.1 Definition of Biogeographic Regions

Classically, the Antarctic Polar Frontal Zone is classified into three regions with distinct climates and accompanying biota: Continental Antarctica, Maritime Antarctica, and Sub-Antarctica (Convey 2010). The well-studied McMurdo Dry Valleys of Continental Antarctica that have been ice-free for many years are considered to be a polar desert. Maritime Antarctica, which includes northwestern parts of the Antarctic Peninsula and surrounding Islands, has a cool moist climate and is less extreme in terms of high irradiance and low temperature compared with Continental Antarctica. The sub-Antarctic region includes various islands located between 35 and 60°S. These islands are breeding sites for seabird and mammal species and exhibit contrasting higher biodiversity compared with both the other regions.

The incorporation of new data and criteria for classification has allowed the identification of 16 Antarctic Conservation Biogeographic Regions (ACBRs) (Terauds and Lee 2016). These bioregions, located in ice-free areas (comprising 0.4% of the continent's surface), are internationally recognized and studied for different purposes including biodiversity analysis (Fig. 5.1).

5.2 Characteristic of Antarctic Soil in ACBRs

Soil properties of different ACBR sites are diverse, depending on their location, topography, local climate, and associated biota. Some soils have low carbon (C) and nitrogen (N) content, whereas those exposed to birds, like penguins, have relatively high C and N levels (Cowan 2014). Salt content is variable, depending on distance to a coastline. Some of these soils are especially dry, like those in the Dry Valleys, while others have comparatively higher humidity. Human activities, including those associated with scientific bases and tourism, are concentrated in these regions with their inevitable effects. Global climate change is also affecting these bioregions, in which retreating glaciers are exposing new surfaces with their particular soils.

5.3 Effect of Global Climate Change

Effects of global climate change have been analyzed on the continent. Recently, measurements of changes in Antarctic ice sheets under the surface of the Southern ocean was determined using novel strategies, including satellite altimeter

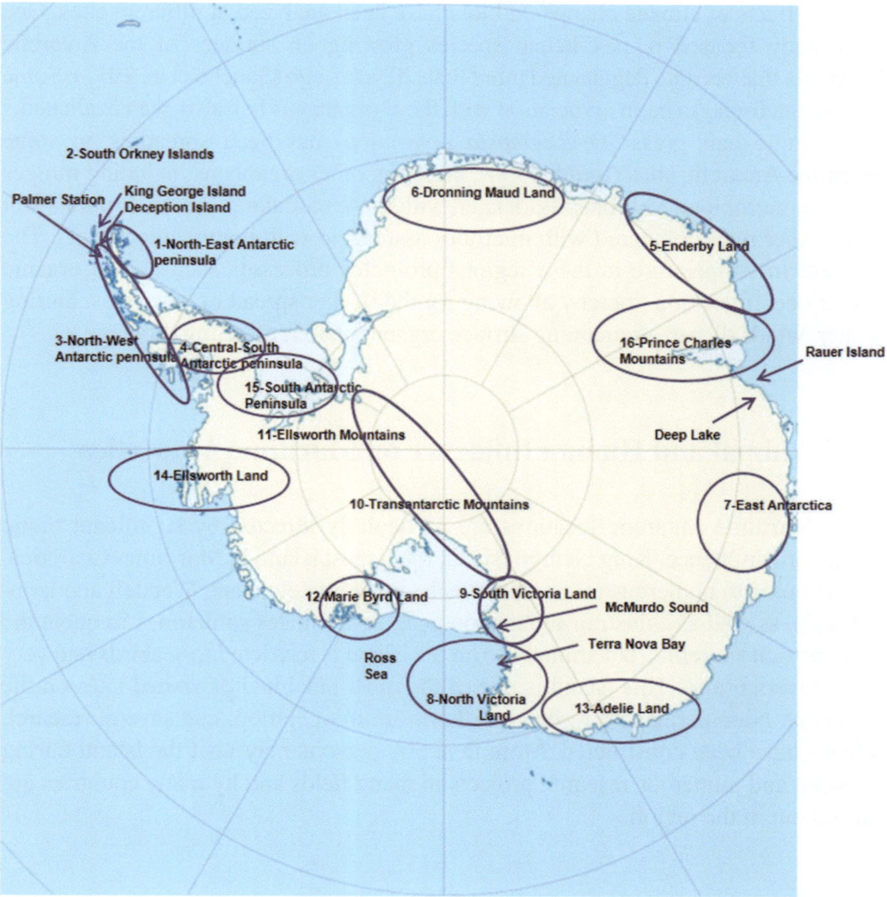

Fig. 5.1 Location of biogeographic regions in Antarctica, numbered from 1 to 16. The arrows show approximately the sampling sites mentioned in the studies included in the chapter

determinations of ice-elevation changes and ice geometry measurements (Konrad et al. 2018). They showed that 1463 km^2 of ice melted between 2010 and 2016. In about 10.7% of oceanfront glaciers, the "grounding line," where ocean, ice, and bedrock meet, glaciers are melting at a significant rate, probably because of the influence of warmer ocean water. In contrast, only 1.9% of these glaciers are actually still expanding.

Some Antarctic regions are especially vulnerable to global warming. Between 1979 and 1997, mean annual temperature in the Antarctic Peninsula increased about 0.32 °C per decade but dropped −0.47 °C per decade, from 1998 to 2014 (Sancho et al. 2017). These changes have further influenced public opinion especially with the occurrence of dramatic events, like the collapse of relatively large ice shelves (Shepherd et al. 2012).

The effects of climate change on biota have been analyzed at different sites. One such study focused on six lichen species growing on surfaces of the Antarctic Peninsula that became deglaciated more than 20 years ago (Sancho et al. 2017). Some microbiota found to be in association with these organisms has also been evaluated.

Antarctic hair grass (*Deschampsia antarctica*) has been spreading in some Maritime Antarctic sites (Smith 1994), affecting other organisms including mosses and soil microbiota. Microbes associated with these vascular plants incorporate soil N at higher rates compared with microbes associated with nonvascular plants. The increase in temperature in these regions promotes processes such as soil organic matter decomposition, thereby allowing for the further spread of hair grass. During winter, winds disperse remaining surface organic matter, affecting more areas.

5.4 Animal and Human Influence on Maritime Antarctica

Some Maritime Antarctic locations are particularly affected by significant biotic and abiotic influence. King George Island, the largest island in Maritime Antarctica, e.g., is home to numerous marine mammals, including elephant, Weddell and leopard seals, as well as chinstrap and gentoo penguins. Whales sometimes frequent the nutrient-rich shoreline, but animal life on the island is restricted to seabirds and very small invertebrates. This island is one of the most inhabited or visited sites on the continent. Nine permanent research stations are in operation, and several research refuges have been constructed. More than 500 persons may visit the island during summer, and numerous research projects in many fields and by many countries are carried out in the region.

5.5 Strategies for the Study of Microbial Communities

Microorganisms are essential for maintenance of diverse processes including soil structure formation, decomposition of xenobiotics, metabolism of organic matter, and support of biogeochemical cycles of specific elements such as C, N, and phosphorous (P). Also, the study of microorganisms and their interactions with biotic and abiotic factors could be used for the development of bioremediation strategies, biotechnological processes, evaluation of soil management, etc.

The first microbiological studies in Antarctica were initiated during the twentieth century (Darling and Siple 1941; Straka and Stokes 1960). These studies involved culture and isolation of microorganisms, and results obtained supported the idea of there being relatively low abundance and diversity in microbial assemblages.

Microscopy has also been used to evaluate abundance, succession, and distribution of microorganisms in Antarctic soil, microbial mats, etc. Ramsay (1983) studied bacterial presence in ornithogenic soils by fluorescence microscopy using acridine orange, a dye that emits green fluorescence when bound to dsDNA and red fluorescence when bound to ssDNA or RNA. Samples were collected at Cape Bird (Ross Island), and differences in numbers and forms of bacteria were found, depending on sampling location and particularly near rookeries. In that study, the author mentions problems that some investigators had when attempting to recover cultured bacteria from these types of sampling locations. Akiyama et al. (1986) concluded that abundance and diversity of bacteria and fungi in rookery soils were lower compared with other Antarctic soils and that this was due to the presence of antimicrobial compounds including acrylic acid, produced by common marine phytoplankton (Akiyama et al. 1986).

Other notable microscopic studies were developed by Komârek and Komârek (1999, 2003). Freshwater Antarctic algae and cyanobacteria were initially studied by microscopy. Later, developing a polyphasic strategy, other methods were incorporated for taxonomic purposes. In one of these studies, a comparison was made of biotopes in seepages from Maritime Antarctica (King George Island) and James Ross Island in the northwest of the Weddell Sea. These studies allowed for establishing that these microbial communities were dominated by cyanobacteria that formed mats with a characteristic stratified structure and composition. These communities support some of the highest levels of biomass and productivity on the continent.

The structure of mats from meltwater ponds on the McMurdo Ice Shelf was also analyzed using different techniques, including optical and fluorescence microscopy, confocal scanning laser microscopy, scanning electron microscopy with backscattered electron-imaging mode, low-temperature scanning electron microscopy, and microanalytical X-ray energy dispersive spectroscopy (de los Rios et al. 2004).

The incorporation of molecular methods allowed for detection of higher microbial richness in Antarctic microbial communities than anticipated by traditional microscopy methods (Cowan 2014). As such, use of culture-independent techniques was focused on description of community composition by 16S rRNA gene analysis and the potential physiological processes they may perform, evaluating for the presence of functional genes, e.g., nitrogenase, denitrification genes, etc. (Jungblut and Neilan 2010; Callejas et al. 2011, 2018; Alcántara-Hernández et al. 2014). Several related molecular studies were developed oriented to describe microbial community structure exposed to different environments.

These methods showed that Antarctic bacterial soil communities were relatively diverse, but some phyla were usually present in each community, including *Actinobacteria*, *Proteobacteria*, *Bacteroidetes*, *Acidobacteria*, *Gemmatimonadetes*, *Deinococcus-Thermus*, and *Cyanobacteria* (Cowan 2014). Also, these communities contained a high proportion of microorganisms that had not been described previously.

Jungblut and Neilan (2010) determined the presence of N_2-fixing microorganisms in microbial mats established in meltwater of Orange Pond (McMurdo Ice Shelf) using molecular methods. They analyzed *nifH* gene sequences from clone libraries and *nifH* gene transcripts, also complementing the study with measurement of nitrogenase enzyme activities using an acetylene reduction assay (ARA). In their study, 18 phylotypes of cyanobacteria were identified, as well as Firmicutes, beta-, gamma-, and delta-Proteobacteria, Spirochaetes, and Verrumicrobia. The *nifH* transcripts belonged to *Nostoc* spp., indicating the importance of these organisms in N_2 fixation at these sites.

High-throughput sequence analysis was used to describe the drastic changes that Antarctic microbial mats have suffered in recent years. The presence of lysis plaques having clear aspects within various microbial mats on Livingston Island (Antarctic Peninsula) was first seen in 2011 and subsequently analyzed. These blighting circles were studied using diverse methods, including high-throughput sequencing of 16S and 18S SSU rRNA genes, viral metagenome analysis, microscopy, isotope label uptake for C and N and ARA. According to results, the authors (Velázquez et al. 2016) proposed that these mats were undergoing decomposition by some organisms present in the community. The original cause for this process was not established, since they did not find any infective agent. They proposed that warmer temperatures could have improved dispersion of some fungi competing for resources in these systems.

Today, the culture of Antarctic microorganisms is being pursued as an important tool for understanding physiological processes, for biotechnological applications and identification of new organisms for taxonomic purposes, etc. Peeters et al. (2012), made a collection of more than 2000 heterotrophic bacteria isolated from aquatic microbial mat samples, collected from different sites in Continental Antarctica and the Antarctic Peninsula. 16S rRNA sequence analysis allowed them to identify five phyla, *Actinobacteria*, *Bacteroidetes*, *Proteobacteria*, *Firmicutes*, and *Deinococcus-Thermus*. A proportion of these sequences (36.9%) were similar to others previously identified on the Antarctic continent.

5.6 Psychrophilic and Psychrotolerant Bacteria

Psychrophilic bacteria were defined for the first time in 1902, although some confusion was generated since assignment as such was based only on minimal growth temperature (Helmke and Weyland 2004). To alleviate confusion, Morita (1975) defined psychrophilic microorganisms as those whose minimum growth temperature is 0 °C or below this and ca. 15 °C and 20 °C for optimum and maximum growth temperatures, respectively. Those microorganisms with higher optimal and maximum growth temperatures were termed as psychrotrophic. Today, the term psychrotrophic has been replaced with psychrotolerant (De Maayer et al. 2014). This definition should be reevaluated again, considering the growth temperatures of some isolates recently described (De Maayer et al. 2014).

5.7 Horizontal Gene Transfer in Bacteria

*H*orizontal *g*ene *t*ransfer (HGT) is defined as the transmission of genetic material between related and relatively unrelated microorganisms. Most of these transfers remain silent if the material can't be replicated after transfer. However, successful transfer events in which the elements can be replicated may allow for expression of novel functions in the recipient organism. HGT may eventually result in improved fitness of the recipient cell compared with the rest of the population. HGTs play an important role modulating evolutionary processes in prokaryotes (Boto 2010). Classically, three mechanisms for HGT have been described: conjugation, transformation, and transduction (Fig. 5.2). Some elements involved in HGT have been very well characterized, including plasmids, viruses, and mobile elements such as transposons. Besides these components, integrons are included in this group. Integrons are non-itinerant structures, but they are usually associated with mobile elements and contain gene cassettes and also have a high rate of mobility (Cambray et al. 2010). The detection of HGT in bacteria is usually done by bioinformatic analysis and by culture-dependent methods (Martinez-Rosales et al. 2015).

Recently, a new mechanism for transfer of DNA was described in archaeal organisms isolated from Antarctica. This mechanism involves the transfer of a plasmid mediated by a structure that resembles a virus termed vesicle and which is discussed below (Erdmann et al. 2017).

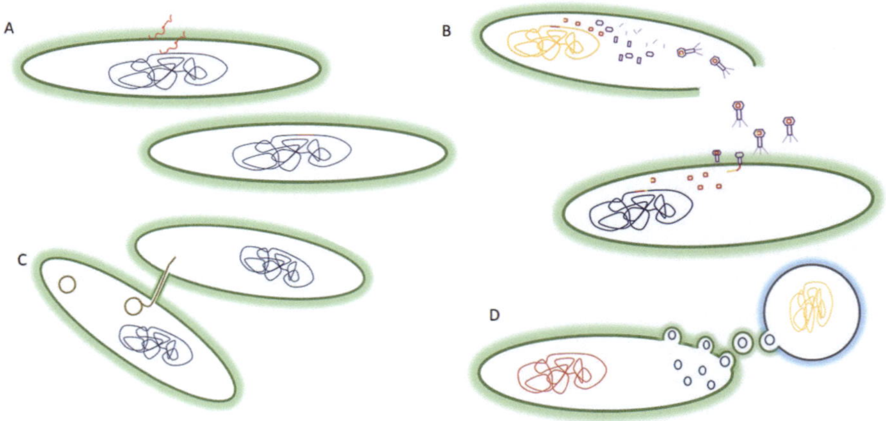

Fig. 5.2 Mechanisms for horizontal gene transfer. (**a**) Transformation consists in the acquisition of free DNA which first gets attached to the cell wall, and then it is internalized as nucleoprotein. Transformation requires a competence state, which is not very common in environmental bacteria. (**b**) Transduction involves the participation of phages. DNA of these elements can get incorporated into the chromosome as prophages. When they excise, part of host's chromosome can be packaged in their capsids, which are able to infect and transfer part of this foreign DNA to a new host cell. (**c**) Conjugative transfer of a plasmid. Conjugation is one of the most important HGT mechanism in the environment. Conjugative plasmids encode the proteins necessary for the formation of a conjugative pilus and DNA processing. Several conjugative plasmids isolated from Antarctic environment have been described. (**d**) Newly discovered mechanism of HGT in Antarctic archaeal isolates. Plasmid R1S1 encodes proteins involved in membrane metabolism. These plasmids can be transferred in plasmids vesicles which get fussed with the membrane of an acceptor cell

5.8 Plasmids

Plasmids are extrachromosomal genetic elements able to be autonomously replicated. Their contribution to evolution and diversity of bacterial genomes is well recognized. Usually, plasmids carry genes that provide adaptive advantages to host cells such as metal or antibiotic resistance, as well as degradation pathways for complex organic molecules. Besides these adaptive functions, plasmids contain regions organized in modules to assure their replication, maintenance, and transfer. Some of these modules can be absent and are extremely diverse when different plasmids are compared.

The replicon is the minimal portion of a plasmid necessary for stable self-replication. This module gives a plasmid the ability to replicate independently of the host's chromosome. Different mechanisms for replication have been identified and are associated with their corresponding replicon genes and genetic elements. The iteron-mediated replication mechanism is the most studied and involves expression of a replication initiator protein, which binds, *in cis*, to repeated sequence elements, termed iterons. This is the first step in the formation of a replication initiation complex. Plasmids also regulate their copy number by processes linked to replication. Copy number results from the balance between up- and downregulation of replication processes. RNAs also play an important role in regulation of certain plasmid's maintenance functions. For instance, in ColE1 replicons, a trans-acting RNA molecule, named RNAI, represses action of the gene that triggers plasmid replication, by pairing with its mRNA molecule. This mechanism of regulation, involving antisense RNA, is widely distributed in prokaryotes and eukaryotes.

Besides having a replicon, plasmids can have a wide array of stabilization systems. They include genes that play a role in preventing plasmid loss during the cell cycle. Stabilization systems are not always present. These functions have been linked to high molecular weight plasmids, which may become a heavy metabolic burden for host cells, and thus are generally present in low copy number. For instance, partition systems specify segregation of at least one replicon to each daughter cell during bacterial cell division. There are also other stabilization mechanisms such as addiction systems, also known as post-segregational killing processes. These mechanisms mediate plasmid maintenance in a population by encoding genes for a stable toxic protein and an unstable antidote molecule, which prevents the action of the stable toxic protein. When a cell loses the plasmid, the stable toxic gene product is inherited, but the antidote is rapidly degraded and can no longer be generated, which will result in post-segregational killing of the cell.

Plasmids can also be disseminated within a bacterial population by different mechanisms. Conjugative transfer is the prevailing HGT process in soils. Plasmids that encode for the complete conjugation machinery are termed conjugative plasmids. Other plasmids have evolved to take advantage of other co-resident conjugative plasmid's apparatus, to be transferred from one cell to another. Specifically, these so-called mobilizable plasmids have a transfer origin structure (OriT) and a gene coding for a relaxase and in some cases can also have a type IV coupling protein. These three elements are known as the MOB module, which gives plasmids the capacity to be mobilized. Plasmids are called conjugative or self-mobilizable when they also have the mating pore formation module. This consists of a set of 12–30 proteins that form a type IV secretion system that acts as a physical bridge between the donor and recipient cells. Relaxases are important proteins in plasmid biology; they function to recognize the OriT and cut one DNA strand relaxing plasmid topology for the transfer process. Relaxases are large multi-domain proteins, and their amino-terminal domains have been used for plasmid classification. Phylogenetic analyses demonstrated that conjugative and mobilizable plasmids could be classified into 6 families by comparative analysis of the 300 amino acid N-terminal domain sequences.

5.9 Plasmids in Antarctic Environments

Study of plasmids in Antarctic microorganisms is done in a similar way to study of plasmids from other environments. Culture-dependent and culture-independent strategies have been used, and in some cases, presence of plasmids becomes evident when the complete genome of a microorganism is analyzed. Detection of plasmids has been mainly associated with biotechnological or basic research objectives (Martínez-Rosales et al. 2012). In some cases, detection of plasmids is linked with the presence of genetic elements for antibiotic or heavy metal resistance or those involved in the degradation of pollutants such as complex hydrocarbons. Other studies, however, have established the presence of plasmids without the bias of pre-selection. The greatest weakness of culture-independent techniques is increased when Antarctic metagenomes are analyzed and is due to the lack of information on replicon sequences and plasmid incompatibilities (Fig. 5.3).

5.10 Culture-Dependent Techniques for Detection of Antarctic Plasmids

In spite of the advent of culture-independent techniques, culture still can provide useful and detailed information on HGT. In this section, selected HGT studies based on culture-dependent techniques are discussed.

Kobori et al. (1984) isolated 155 psychrophilic and psychrotolerant bacteria collected from different sites, including ice, water, and sediment near McMurdo Sound (Fig. 5.1). Nearly a third of these isolates (31%) contained plasmids, with the highest presence in isolates obtained from sediments and the lowest in those from seawater. Additionally, replicons were more frequently found in those isolates obtained by growth in oligotrophic medium compared with those isolated after growth on

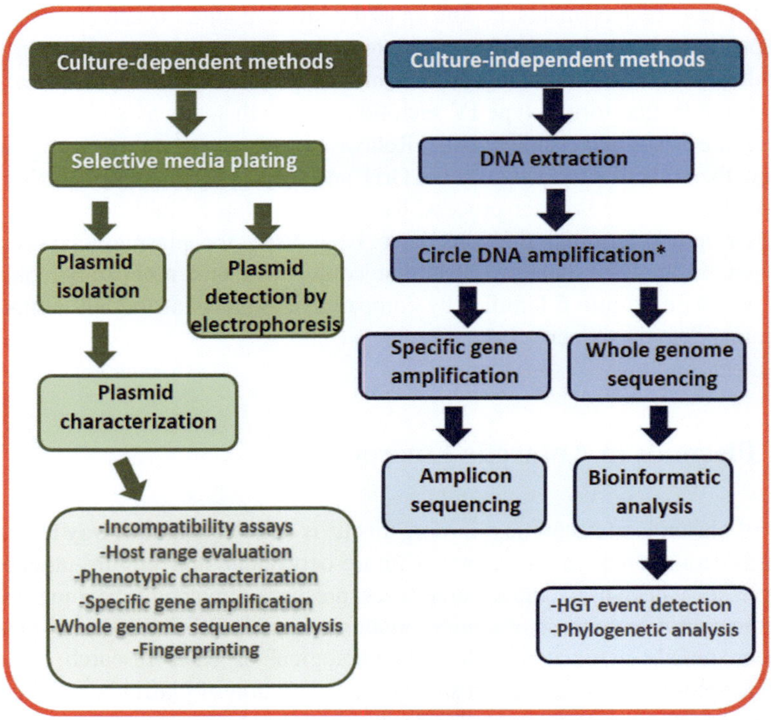

Fig. 5.3 Methods used for plasmid and HGT analyses with Antarctic samples. *Circle DNA amplification was done with a phage polymerase enzyme, phi29, coupled with an exonuclease that cleaves linear DNA. This method increases the concentration of phage and plasmid DNA. This strategy has not been used with Antarctic samples yet

rich medium. The sizes of these plasmids were lower than 10 MD, and only 7 out of 48 isolates analyzed were resistant to a set of antibiotics.

The presence of antibiotic resistance genes in plasmids is particularly alarming given the ability of some replicons to be transferred, increasing potential dispersion of those genes. Some studies on antibiotic-resistant bacteria are also oriented to look for the presence of plasmids in resistant isolates. Miller et al. (2009) obtained a collection of bacteria isolated from sites near Palmer Station (Fig. 5.1). These organisms were tested for their resistance to five common antibiotics and for the presence of plasmids. The percentage of plasmid-containing isolates was low when standard plasmid isolation methods were used, but surprisingly, this was not correlated with antibiotic resistance phenotypes. Plasmids were detected with the same frequency in drug-resistant isolates as those that were drug-sensitive. Nevertheless, these results do not diminish the importance of assessing the relatedness of antibiotic resistance genes to transferable elements such as plasmids.

Ma et al. (2006) worked with a collection of 22 bacterial isolates able to degrade polycyclic aromatic hydrocarbons (PAHs). These bacteria were isolated from Antarctic polluted soils (King George Island) (Fig. 5.1). To enrich PAHs degrading organisms, they used a mineral medium with naphthalene or phenanthrene as sole C and energy sources. Taxonomic analysis showed that 21 isolates belonged to the genus *Pseudomonas* and one to *Rahnella*. These bacteria were able to degrade naphthalene at 4 °C, and some also degraded phenanthrene. The presence of *ndo* (naphthalene dioxygenase) genes was confirmed by PCR amplification and DNA sequence analysis. Phylogenetic analysis grouped these genes into two clusters which shared 94% similarity with each other and showed about 97% similarity within a cluster. These genes were similar to those described in mesophilic bacteria. In addition, comparative analysis of 16S rRNA and *ndo* genes suggested HGT between isolates. Analysis by Southern blot hybridization, a protocol for curing of plasmids and plasmid transfer by conjugation, indicated that *ndo* genes were located on large self-transmissible plasmids, which could be transferred to mesophilic strains.

Pini et al. (2007) isolated 21 microorganisms from Ross Sea able to use diesel fuel as sole C source (Fig. 5.1). According to 16S rRNA gene sequence analysis, these isolates belonged to the genera *Rhodococcus* and *Alcaligenes*. The *alkB* gene, encoding for alkane monooxygenase, was detected in all isolates. Interestingly, *alkB* genes from these *Alcaligenes* isolates, a genus of Gram-negative bacteria, were closely related to *alkB* genes from Rhodococci. Different tree topologies resulting from *alkB* and 16S rRNA gene analyses denote that, at least for these *Alcaligenes* isolates, *alkB* genes were recently acquired by HGT. The study included a search for plasmids; however, none were found. As in other cases, not all plasmids can easily be isolated using specific standard techniques, and as such, their presence in those isolates cannot be ruled out.

Another study involved analysis of a bacterial collection isolated from soil/sediment samples collected on Deception and Galíndez Islands (Tomova et al. 2015). The 24 isolates examined exhibited diverse profiles of resistance to 13 antibiotics and 7 heavy metals. They found that 67% of sediment isolates and 92% of soil isolates had multiple antibiotic resistances, suggesting a potential anthropogenic impact at those sites. All isolates showed multiple metal resistances for two to six heavy metals, and most were resistant to lead, nickel, copper, and zinc. In addition, plasmids were detected in 21% of those isolates.

A psychrotolerant bacterium, *Pseudomonas* sp. ANT_J3, was reported to carry a plasmid named pA3J1 (Romaniuk et al. 2017) (Table 5.1). DNA sequence analysis of pA3J1 indicated the presence of 13 potential genes related with basic functions for replicons. Its characterization suggested that pA3J1 is a narrow host range replicon plasmid, maintained by partitioning functions and relBE-type toxin-antitoxin mechanisms. Sequence analysis of the relaxase (MobA) of this plasmid placed this plasmid in the MOBQ family and likely as belonging to a new subgroup in this family.

Complete genome projects can shed light on plasmid content of isolates and potential involvement of HGT. Han et al. (2016), e.g., reported the complete genome sequence of *Burkholderia* PAMC28687, a lichen-associated bacterium isolated from King George Island (Fig. 5.1). The genome of this bacterium is composed of three chromosomes and five plasmids. They identified between 9 and 277 protein coding genes in different plasmids, although they did not report if these plasmids could be transferred.

One interesting aspect of HGT by plasmid acquisition is that, hypothetically, this can result in speciation. These events are hard to track, but some evidence found in bacterial genomes can help to better understand this process. For instance, the genome of *Pseudoalteromonas haloplanktis* TAC125, a strain isolated from Maritime Antarctica, contains two chromosomes (Tutino et al. 2001) (Table 5.1). Chromosome II has a sequence pattern near its replication origin that it is consistent with unidirectional replication. This chromosome also has two genes (*kisB* and *kidB*) involved in plasmid maintenance. Almost a fifth of chromosome II genes have high similarities with plasmid-encoded genes. These findings suggest a plasmid origin for chromosome II and that it became a chromosome after acquiring one or more essential genes (Médigue et al. 2005).

In a study aimed at characterizing the diversity of psychrotolerant bacteria, 146 isolates were obtained from Terra Nova Bay (Michaud et al. 2004) (Fig. 5.1). Only 21 isolates were found to carry plasmids, with most of them containing a single plasmid. DNA fingerprinting patterns of these plasmids suggested possible HGT between isolates at different taxonomic levels.

The Antarctic environment shape microbial genotypes with unusual combinations of traits. Bacteria from the genus *Deinococcus* are typically resistant to various environmental hazards. Hirsch et al. (2004) isolated six *Deinococcus* bacteria that were classified into three new species. They were isolated from McMurdo Dry Valleys and described as tolerant to drought, UV radiation, and low temperature. Half of these isolates harbor plasmids, in at least one isolate of each species. Evidence for HGT and transferability of these plasmids was not reported.

Marine sponges are an important part of Maritime Antarctic fauna. It is known that these marine organisms are inhabited by bacterial symbionts. In a recent study, 297 endosymbionts were isolated from tissues of 5 different sponge species collected from Terra Nova Bay (Mangano et al. 2011) (Fig. 5.1). Plasmids were found in 69 of these isolates, and plasmid-containing bacteria were present in each sponge species. Significantly, 72% of these plasmid-containing bacteria were resistant to more than one of the antibiotics tested. No clear conclusion on the relation of plasmid presence and antibiotic resistance could be drawn. However, authors suggested that the high frequency of antibiotic resistance in this environment could have been due to recent plasmid acquisition and HGT.

Recently, five psychrotolerant enterobacteria resistant to ampicillin, streptomycin, and trimethoprim were isolated in different sites of Fildes Peninsula (King George Island) (Antelo et al. 2018) (Table 5.1). Conjugation experiments with two of these isolates, using *Escherichia coli* DH5α as recipient strain, confirmed that drug resistance genes were contained on a large conjugative plasmid. DNA sequence analysis of one plasmid confirmed the presence of two origins of replication (IncF), as well as systems for plasmid stability and conjugation. Also, these plasmids contained a conserved fragment with class 1 integron elements of ca. 7200 bp and, almost identical to one region of plasmid pKOX105, obtained from a clinical isolate of *Klebsiella pneumoniae*. Bioinformatics analysis suggests that this plasmid is a mosaic of plasmid fragments previously characterized in mesophilic enterobacteria (Carattoli et al. 2010).

Table 5.1 Plasmids of Antarctic bacteria partially described in previous studies

Name	Host	Size	Replication	Stabilization/ segregation	Conjugation	Characteristics	Traits	Sample location	Reference
pMtBL	*Pseudoalteromonas haloplanktis* strain TAC 125	4081 bp	n.d.	n.d.	n.d.	High copy number, broad host range, without transcriptional activity under tested conditions	n.d.	Sea water next Dumont d'Urville Antarctic station (66°40′S, 40°01′E)	Tutino et al. (2001)
pTAUp	*Psychrobacter* sp. strain TA144	1921 bp	Rep-coding gene (Psyrep) suggests rolling-circle replication mechanism	n.d.	n.d.	n.d.	n.d.	Sea water near Dumont d'Urville, Antarctic station (66°40′S, 40°01′E)	Tutino et al. (2000)
pTADw		1313 bp	n.d.	n.d.	n.d.	n.d.	n.d.		
pA3J1	*Pseudomonas* sp. strain ANT_J3	9794 bp	Replication initiation gene (pA3J1_p07)	Partitioning (pA3J1_p05-p06), toxin-antitoxin (pA3J1_p03-p04) and multimer resolution system (pA3J1_p02) genes	Conjugative transfer module genes (pA3J1_p09-p11)	n.d.	n.d.	Soil from the depth of 15–20 cm, collected at the Jardine Peak, King George Island (62°09′S, 58°29′W)	Romaniuk et al. (2017)

(continued)

Table 5.1 (continued)

Name	Host	Size	Replication	Stabilization/ segregation	Conjugation	Characteristics	Traits	Sample location	Reference
pGIAK1	Bacillaceae bacterium JMAK1 (obligate alkaliphilic, aerobic and halotolerant spore-forming bacterium)	38,362 bp	ORFs encoding a Replic_ Relax family protein	pGIAK1_40 encodes a putative 30 kDa ParA nucleotide binding domain protein. ParA is a membrane-associated ATPase involved in the segregation of plasmid and bacterial chromosome	pGIAK1_2 protein has VirD4 conserved domain at NH2-terminus. pGIAK1_3 and pGIAK1_5 involved in conjugation. pGIAK1_6 protein has a AAA-like domain, similar to VirB4 family protein, required for conjugation in *Agrobacterium tumefaciens*	n.d.	Cadmiun (pGIAK1_32–34; pGIAK1_39) and arsenicum resistance, regulator and transport genes (pGIAK1_25–29; pGIAK1_38)	Stairway step inside the Antarctic Concordia station	Guo and Mahillon (2013)
pOA307_63	*Octadecabacter antarcticus* strain 307	62.9 kb	n.d.	n.d.	n.d.	n.d.	Transporters of unknown function, peptide transporters; toxin/antitoxin plasmid stabilization system	McMurdo Sound	Vollmers et al. (2013)

pMWHK1	*Pedobacter cryoconitis* strain BG5	6206 bp	The ori region is located upstream of an AT-rich region of orf4	n.d.	n.d.	Potential Θ-type replicating plasmid, with genes for toxin-antitoxin proteins (PemK, Pem-I like), plasmid mobilization protein, a replicase and an entry exclusion protein (Exc1)	n.d.	Soil collected next to Belen Lake (62°13′45.4′′S, 58°58′53.9′′W), King George Island	Wong et al. (2013)
pKW1	*Pseudoalteromonas* sp. strain 643A	4583 bp	ORF1 has 87% sequence homology to the replicase protein of *Psychrobactercryohalolentis* K5. Similar to replicase proteins of the θ-replicating *E. coli* plasmids pColE2 and pColE3	ORF-4 has amino acid sequence homology of 46% with the putative relaxase/ mobilization nuclease domain of MobA protein from the *Hafnia alvei* and *Neisseria lactamica*	The oriT (origin of transfer) sequences are absent. P-type, Q-type, and F-type of the oriT relaxase system could not be found	n.d.	n.d.	Krill *Euphasia superba*; Admiralty Bay waters (King George Island, (62°10′S, 58°28′W)	Cieśliński et al. (2008)

(continued)

Table 5.1 (continued)

Name	Host	Size	Replication	Stabilization/ segregation	Conjugation	Characteristics	Traits	Sample location	Reference
pGLE121P1	*Pseudomonas* sp. strain GLE121	6899 bp	Region shows homology to the REP regions of plasmids of IncP-9 incompatibility group. It should be located in a new subgroup of IncP-9 group	n.d.	n.d.	Narrow host range limited to the genus *Pseudomonas*	n.d.	n.d.	Dziewit et al. (2013)
pGLE121P2		8330 bp	IncP-7 incompatibility group. RepA shows the highest level of aa sequence identity (86%) with the replication initiator of *P. syringae* pv. savastanoi PS93 plasmid pPS10	Contain TA and multimer resolution systems (MRS). TA system comprises two overlapping (23-bp) ORFs: orf4 encoding a predicted protein similar to RelE toxins and orf3 encoding a predicted antitoxin with a HTH-XRE domain typical of some transcriptional regulators. The predicted MRS encodes a putative serine resolvase	n.d.		n.d.	n.d.	
pGLE121P3		39,583 bp	Replication protein shares 62% aa sequence identity with RepA of plasmid pVS1 of *P. aeruginosa* PAO	n.d.	n.d.		Contains *rulAB* genes that confer tolerance to UV radiation	n.d.	

PsyG_26	*Psychrobacter* sp. strain G	26,087 bp	n.d.	n.d.	n.d.	n.d.	n.d.	King George Island (62°12′39′′S, 58°54′41′′W)	Che et al. (2013)
PsyG_4		4518 bp	n.d.	n.d.	n.d.	n.d.	n.d.		
PsyG_3		3956 bp	n.d.	n.d.	n.d.	n.d.	n.d.		
pSP01	*Streptomyces* sp. strain PAMC26508	104,0 kb	n.d.	n.d.	n.d.	n.d.	n.d.	Antarctic lichen Cladonia borealis	Shin and Park
pHP19	*Enterobacter* sp. strain HP19		Two ORFs for initiation replication proteins (RepFIIA and RepFIIB)	ORFs for plasmid partitioning proteins (*parA/parB* and *sopA/sopB*)	Contains conjugative genes. The plasmid was transferred to *Escherichia coli* DH5α	Presumptively a single plasmid	Contain genes for resistance to ampicillin, trimethoprim, streptomycin, mercury, arsenic, tellurite, nickel and copper	King George Island (62°13′29′′S, 58°57′9′′W) Brown microbial mat with bryophytes located next to Half Three Point	Antelo et al. (2018)

n.d. no determined in the referred article

Some of these Antarctic plasmids, however, seem to have differences in evolutionary profiles compared with most previously described plasmids. *Pseudomonas* is a ubiquitous genus in many environments, and Antarctica is not an exception to this. *Pseudomonas* sp. GLE121 is a psychrophilic bacterium isolated from Ecology Glacier on King George Island (Dziewit et al. 2013) (Table 5.1). Three plasmids were obtained from this isolate, from sequence analysis they were presumed to be plasmids restricted to the genus *Pseudomonas.* The largest plasmid (39,583 bp) encoded for a conjugative transfer system, whereas the smallest (6899 bp) seems to be a mobilizable plasmid. The largest plasmid includes genes that may code for a DNA repair system (Sundin et al. 1996), which is involved in UV radiation tolerance, like it could be expected given the UV exposure levels in the Antarctic environment. Considering the transferability of this plasmid and the ubiquitous nature of the *Pseudomonas* genus, the potential for dispersion of this plasmid should be high. The largest plasmid was classified as IncP-9, containing a gene encoding RepA as a replication initiation protein and direct repeated sequences to which this protein should bind (Dziewit et al. 2013). This replicon could not be classified into one of the subgroups described for IncP-9 incompatibility group. These subgroups contain plasmids with high sequence similarities between their replication initiator proteins and also an overall functional similarity. All IncP-9 plasmids described could be classified within these subgroups until pGLE121P1 was found. It is not known if this could be the archetype of a new subgroup or if there are not-yet-discovered IncP-9 plasmids that simply do not follow this subgroup pattern.

5.11 Culture-Independent Techniques

Studies of metagenomes for identification and characterization of microbial plasmids has several challenges. Due to the reduced amount of information available, these difficulties are increased when samples from extreme environments such as those in Antarctica are analyzed. Classically, criteria for description and classification of plasmids were defined from available sequences of specific isolates, obtained from environmental and clinical samples.

One approach for culture-independent analysis includes the search for conserved sequences belonging to different incompatibility groups, using environmental DNA as template for PCR reactions (Imperio et al. 2007). Total DNA from Antarctic samples collected in Northern Victoria Land were used as template and the regions amplified were selected from broad-host-range plasmids. Primers used were specific for IncP (*trfA2*) and IncQ (oriV) groups. The identity of amplicons was confirmed by Southern blot hybridization using *trfA2* and oriV sequences as probes. This strategy allowed for detection of IncP-specific sequences in 8 of 15 DNA samples analyzed.

Antarctic microorganisms with a distinct evolutionary history compared with related microbes in less extreme environments were found in Deep Lake, a hypersaline system that does not freeze at temperatures of about −20 °C. DeMaere and his team (2013) studied microbial communities of this lake using a metagenomic approach. They found that these communities were dominated by haloarchaea (about 72%), exhibiting a low complexity (DeMaere et al. 2013). High rates of genetic exchange among distinct halobacteria genera, including sharing of long contiguous regions of up to 35 kb, were detected. It can be expected that HGT between different genera would tend to homogenize haloarchaea genomes in the community. Considering the low cell division rates in this extreme environment, it could be hypothesized that genomic coherence of these genera could be endangered. However, the study showed the presence of distinct haloarchaea genera, which is consistent with niche adaptation. Moreover, the study provided evidence that HGT and integration events were partitioned within genomes, thereby avoiding potential deleterious effects (Wagner et al. 2017).

Dziewit and Bartosik (2014) developed a bioinformatics approach to search for plasmid sequences in genomes of psychrophilic and psychrotolerant bacteria. This analysis allowed for the identification of 66 replicons in 39 bacterial isolates from diverse sites, including 15 plasmids from Antarctic bacteria. Most of the isolates were Gram-negative bacteria.

5.12 Alternative Mechanism for HGT

Halorubrum lacusprofundii R1S1 is an archaea isolated from Rauer Island (Erdmann et al. 2017). When cultured, this strain produced vesicles of around 80–110 nm that contained a plasmid of 50,329 bp (Erdman et al. 2017). The vesicles could transform or infect other strains of *H. lacusprofundii* isolated from the same site. Resulting transformants in turn produced similar vesicles that contained the same plasmid. This replicon, termed pR1SE, encodes for proteins associated with formation of these vesicles (plasmid vesicles). This result shows a new mechanism for HGT, in that they could propagate strain R1S1 maintaining this plasmid for 3 months. This propagation led to changes in cell morphology and proteins present in plasmid vesicles, together with a shift in the restriction pattern of the plasmid DNA. They also found that the host's chromosome had certain regions identical to pR1SE, which were also missing from the plasmid after transfer. This is consistent with integration of the plasmid in the chromosome, followed by its excision or resolution. The transfer mechanism of pR1SE resembles that of viruses, i.e., infecting plasmid-free cells via a particle made up mainly from its own encoded proteins. However, the genetic features of this replicon leave no doubt of its plasmid origin.

5.13 Conclusions

Information on HGT elements in Antarctic microorganisms is relatively limited and fragmented. There are different factors that should be considered when studying Antarctic microbial communities and their associated mobile genome, transferred in plasmids. In many Antarctic sites, endemic, native, or naturalized and other introduced organisms coexist in microbial communities. The presence of scientific bases, birds, and the wind can affect the composition of some of these communities. Some sites particularly isolated and subjected to a very extreme environment could harbor communities of endemic microorganisms that can express some unique physiological functions that have not been described previously. On the other hand, microbial communities established at sites under greater biotic influence may contain mobile elements that serve as indicators of anthropogenic or animal presence.

References

Akiyama, M., Kanda, H., & Ohyama, Y. (1986). Allelopathic effect of penguin excrements and guanos on the growth of Antarctic soil algae. *Memoirs of National Institute of Polar Research. Series E Biology and Medical Science, 37*, 11–16.

Alcántara-Hernández, R. J., Centeno, C. M., Ponce-Mendoza, A., et al. (2014). Characterization and comparison of potential denitrifiers in microbial mats from King George Island, Maritime Antarctica. *Polar Biology, 37*, 403–416. https://doi.org/10.1007/s00300-013-1440-3.

Antelo, V., Guerout, A. M., Mazel, D., et al. (2018). Bacteria from Fildes Peninsula carry class 1 integrons and antibiotic resistance genes in conjugative plasmids. *Antarctic Science, 30*, 22–28. https://doi.org/10.1017/S0954102017000414.

Boto, L. (2010). Horizontal gene transfer in evolution: Facts and challenges. *Proceedings of the Royal Society B: Biological Sciences, 277*, 819–827. https://doi.org/10.1098/rspb.2009.1679.

Callejas, C., Gill, P. R., Catalán, A. I., et al. (2011). Phylotype diversity in a benthic cyanobacterial mat community on King George Island, maritime Antarctica. *World Journal of Microbiology and Biotechnology, 27*, 1507–1512. https://doi.org/10.1007/s11274-010-0578-1.

Callejas, C., Azziz, G., Souza, E. M., et al. (2018). Prokaryotic diversity in four microbial mats on the Fildes Peninsula, King George Island, maritime Antarctica. *Polar Biology*. https://doi.org/10.1007/s00300-018-2256-y.

Cambray, G., Guerout, A.-M., & Mazel, D. (2010). Integrons. *Annual Review of Genetics, 44*, 141–166. https://doi.org/10.1146/annurev-genet-102209-163504.

Carattoli, A., Aschbacher, R., March, A., et al. (2010). Complete nucleotide sequence of the IncN plasmid pKOX105 encoding VIM-1, QnrS1 and SHV-12 proteins in Enterobacteriaceae from Bolzano, Italy compared with IncN plasmids encoding KPC enzymes in the USA. *The Journal of Antimicrobial Chemotherapy, 65*, 2070–2075. https://doi.org/10.1093/jac/dkq269.

Che, S., Song, L., Song, W., Yang, M., Guiming, L., & Lin, X. (2013). Complete genome sequence of Antarctic bacterium *Psychrobacter* sp. strain G. *Genome, 1*, 2012–2013. https://doi.org/10.1093/nar/gkm321.2.

Cieśliński, H., Werbowy, K., Kur, J., & Turkiewicz, M. (2008). Molecular characterization of a cryptic plasmid from the psychrotrophic antarctic bacterium *Pseudoalteromonas* sp. 643A. *Plasmid, 60*, 154–158. https://doi.org/10.1016/j.plasmid.2008.06.002.

Convey, P. (2010). Terrestrial biodiversity in Antarctica – Recent advances and future challenges. *Polar Science, 4*, 135–147. https://doi.org/10.1016/j.polar.2010.03.003.

Cowan, D. A. (2014). *Antarctic terrestrial microbiology*. Berlin: Springer.
Darling, C. A., & Siple, P. A. (1941). Bacteria of Antarctica. *Journal of Bacteriology, 42*, 83–98.
de los Rios, A., Ascaso, C., Wierzchos, J., et al. (2004). Microstructural characterization of cyanobacterial mats from the McMurdo Ice Shelf, Antarctica. *Applied and Environmental Microbiology, 70*, 569–580. https://doi.org/10.1128/AEM.70.1.569.
De Maayer, P., Anderson, D., Cary, C., & Cowan, D. A. (2014). Some like it cold: Understanding the survival strategies of psychrophiles. *EMBO Reports, 15*, 508–517. https://doi.org/10.1002/embr.201338170.
DeMaere, M. Z., Williams, T. J., Allen, M. A., et al. (2013). High level of intergenera gene exchange shapes the evolution of haloarchaea in an isolated Antarctic lake. *Proceedings of the National Academy of Sciences, 110*, 16939–16944. https://doi.org/10.1073/pnas.1307090110.
Dziewit, L., & Bartosik, D. (2014). Plasmids of psychrophilic and psychrotolerant bacteria and their role in adaptation to cold environments. *Frontiers in Microbiology, 5*, 1–14. https://doi.org/10.3389/fmicb.2014.00596.
Dziewit, L., Grzesiak, J., Ciok, A., et al. (2013). Sequence determination and analysis of three plasmids of *Pseudomonas* sp. GLE121, a psychrophile isolated from surface ice of Ecology Glacier (Antarctica). *Plasmid, 70*, 254–262. https://doi.org/10.1016/j.plasmid.2013.05.007.
Erdmann, S., Tschitschko, B., Zhong, L., et al. (2017). A plasmid from an Antarctic haloarchaeon uses specialized membrane vesicles to disseminate and infect plasmid-free cells. *Nature Microbiology, 2*, 1446–1455. https://doi.org/10.1038/s41564-017-0009-2.
Guo, S., & Mahillon, J. (2013). pGIAK1, a heavy metal resistant plasmid from an obligate alkaliphilic and halotolerant bacterium isolated from the Antarctic Concordia Station confined environment. *PLoS One, 8*, 1–8. https://doi.org/10.1371/journal.pone.0072461.
Han, S. R., Yu, S. C., Ahn, D. H., et al. (2016). Complete genome sequence of *Burkholderia* sp. strain PAMC28687, a potential octopine-utilizing bacterium isolated from Antarctica lichen. *Journal of Biotechnology, 226*, 16–17. https://doi.org/10.1016/j.jbiotec.2016.03.043.
Helmke, E., & Weyland, H. (2004). Psychrophilic versus psychrotolerant bacteria – Occurrence and significance in polar and temperate marine habitats. *Cellular and Molecular Biology (Noisy-le-Grand, France), 50*, 553–561. https://doi.org/10.1170/T545.
Hirsch, P., Gallikowski, C. A., Siebert, J., et al. (2004). Deinococcus frigens sp. nov., *Deinococcus saxicola* sp. nov., and *Deinococcus marmoris* sp. nov., low temperature and draught-tolerating, UV-resistant bacteria from continental Antarctica. *Systematic and Applied Microbiology, 27*, 636–645. https://doi.org/10.1078/0723202042370008.
Imperio, T., Bargagli, R., & Marri, L. (2007). Detection of IncP replicon-specific regions in DNA from Antarctic microbiota. *Open Life Sciences, 2*, 378–384. https://doi.org/10.2478/s11535-007-0025-y.
Jungblut, A. D., & Neilan, B. A. (2010). *nifH* gene diversity and expression in a microbial mat community on the McMurdo Ice Shelf, Antarctica. *Antarctic Science, 22*, 117–122. https://doi.org/10.1017/S0954102009990514.
Kobori, H., Sullivan, C. W., & Shizuya, H. (1984). Bacterial plasmids in antarctic natural microbial assemblages. *Applied and Environmental Microbiology, 48*, 515–518.
Komárek, O., & Komárek, J. (1999). Diversity of freshwater and terrestrial habitats and their oxyphototroph microflora in the Arctowski Station region, South Shetland Islands. *Polish Polar Research, 20*, 259–282.
Komárek, O., & Komárek, J. (2003). Diversity and ecology of cyanobacterial microflora of Antarctic seepage habitats: Comparison of King George Island, Shetland Islands, and James Ross Island, Nw Weddell Sea, Antarctica. In *Microbial mats. Modern and ancient microorganisms in stratified systems* (pp. 517–539). Springer: Dordrecht.
Konrad, H., Shepherd, A., Gilbert, L., et al. (2018). Net retreat of Antarctic glacier grounding lines. *Nature Geoscience*. https://doi.org/10.1038/s41561-018-0082-z.
Ma, Y., Wang, L., & Shao, Z. (2006). Pseudomonas, the dominant polycyclic aromatic hydrocarbon-degrading bacteria isolated from Antarctic soils and the role of large plas-

mids in horizontal gene transfer. *Environmental Microbiology, 8*, 455–465. https://doi.org/10.1111/j.1462-2920.2005.00911.x.

Mangano, S., Caruso, C., Michaud, L., et al. (2011). Incidence of plasmid and antibiotic resistance in psychrotrophic bacteria isolated from Antarctic sponges. *AAPP Atti della Accademia Peloritana dei Pericolanti, Classe di Scienze Fisiche Matematiche e Naturali, 89*, 1–9. https://doi.org/10.1478/C1A8901003.

Martínez-Rosales, C., Fullana, N., Musto, H., & Castro-Sowinski, S. (2012). Antarctic DNA moving forward: Genomic plasticity and biotechnological potential. *FEMS Microbiology Letters, 331*, 1–9. https://doi.org/10.1111/j.1574-6968.2012.02531.x.

Martinez-Rosales, C., Marizcurrena, J. J., Iriarte, A., et al. (2015). Characterizing proteases in an Antarctic *Janthinobacterium* sp. isolate: Evidence of a protease horizontal gene transfer event. *Advances in Polar Science, 1*, 012. https://doi.org/10.13679/j.advps.2015.1.00088.

Médigue, C., Krin, E., Pascal, G., et al. (2005). Coping with cold: The genome of the versatile marine Antarctica bacterium *Pseudoalteromonas haloplanktis* TAC125. *Genome Research, 15*, 1325–1335. https://doi.org/10.1101/gr.4126905.

Michaud, L., Di Cello, F., Brilli, M., et al. (2004). Biodiversity of cultivable psychrotrophic marine bacteria isolated from Terra Nova Bay (Ross Sea, Antarctica). *FEMS Microbiology Letters, 230*, 63–71. https://doi.org/10.1016/S0378-1097(03)00857-7.

Miller, R. V., Gammon, K., & Day, M. J. (2009). Antibiotic resistance among bacteria isolated from seawater and penguin fecal samples collected near Palmer Station, Antarctica. *Canadian Journal of Microbiology, 55*, 37–45. https://doi.org/10.1139/W08-119.

Morita, R. Y. (1975). Psychrophilic Bacteria. *Bacteriological Reviews, 39*, 144–167.

Peeters, K., Verleyen, E., Hodgson, D. A., et al. (2012). Heterotrophic bacterial diversity in aquatic microbial mat communities from Antarctica. *Polar Biology, 35*, 543–554. https://doi.org/10.1007/s00300-011-1100-4.

Pini, F., Grossi, C., Nereo, S., et al. (2007). Molecular and physiological characterisation of psychrotrophic hydrocarbon-degrading bacteria isolated from Terra Nova Bay (Antarctica). *European Journal of Soil Biology, 43*, 368–379. https://doi.org/10.1016/j.ejsobi.2007.03.012.

Ramsay, A. J. (1983). Bacterial biomass in ornithogenic soils of Antarctica. *Polar Biology, 1*, 221–225. https://doi.org/10.1007/BF00443192.

Romaniuk, K., Krucon, T., Decewicz, P., et al. (2017). Molecular characterization of the pA3J1 plasmid from the psychrotolerant Antarctic bacterium *Pseudomonas* sp. ANT_J3. *Plasmid, 92*, 49–56. https://doi.org/10.1016/j.plasmid.2017.08.001.

Sancho, L. G., Pintado, A., Navarro, F., et al. (2017). Recent warming and cooling in the Antarctic Peninsula region has rapid and large effects on lichen vegetation. *Scientific Reports, 7*, 1–8. https://doi.org/10.1038/s41598-017-05989-4.

Shepherd, A., Ivins, E. R., Geruo, A., et al. (2012). A reconciled estimate of ice-sheet mass balance. *Science, 338*, 1183–1189. https://doi.org/10.1126/science.1228102.

Smith, R. I. L. (1994). Vascular plants as bioindicators of regional warming in Antarctica. *Oecologia, 99*, 322–328.

Straka, R. P., & Stokes, J. L. (1960). Psychrophilic bacteria in Antarctica. *Journal of Applied Microbiology, 80*, 622–625.

Sundin, G. W., Kidambi, S. P., Ullrich, M., & Bender, C. L. (1996). Resistance to ultraviolet light in *Pseudomonas syringae*: Sequence and functional analysis of the plasmid-encoded *rulAB* genes. *Gene, 177*, 77–81. https://doi.org/10.1016/0378-1119(96)00273-9.

Terauds, A., & Lee, J. R. (2016). Antarctic biogeography revisited: Updating the Antarctic Conservation Biogeographic regions. *Diversity and Distributions, 22*, 836–840. https://doi.org/10.1111/ddi.12453.

Tomova, I., Stoilova-Disheva, M., Lazarkevich, I., & Vasileva-Tonkova, E. (2015). Antimicrobial activity and resistance to heavy metals and antibiotics of heterotrophic bacteria isolated from sediment and soil samples collected from two Antarctic islands. *Front Life Science, 8*, 348–357. https://doi.org/10.1080/21553769.2015.1044130.

Tutino, M. L., Duilio, A., Moretti, M. A., et al. (2000). A rolling-circle plasmid from *Psychrobacter* sp. TA144: Evidence for a novel Rep subfamily. *Biochemical and Biophysical Research Communications, 274*, 488–495. https://doi.org/10.1006/bbrc.2000.3148.

Tutino, M. L., Duilio, A., Parrilli, E., et al. (2001). A novel replication element from an Antarctic plasmid as a tool for the expression of proteins at low temperature. *Extremophiles, 5*, 257–264. https://doi.org/10.1007/s007920100203.

Velázquez, D., López-Bueno, A., Aguirre De Cárcer, D., et al. (2016). Ecosystem function decays by fungal outbreaks in Antarctic microbial mats. *Scientific Reports, 6*, 1–7. https://doi.org/10.1038/srep22954.

Vollmers, J., Voget, S., Dietrich, S., et al. (2013). Poles apart: Arctic and Antarctic *Octadecabacter* strains share high genome plasticity and a new type of Xanthorhodopsin. *PLoS One*. https://doi.org/10.1371/journal.pone.0063422.

Wagner, A., Whitaker, R. J., Krause, D. J., et al. (2017). Mechanisms of gene flow in archaea. *Nature Reviews Microbiology, 15*, 492–501. https://doi.org/10.1038/nrmicro.2017.41.

Wong, C. M. V. L., Tam, H. K., Ng, W. M., et al. (2013). Characterisation of a cryptic plasmid from an Antarctic bacterium *Pedobacter cryoconitis* strain BG5. *Plasmid, 69*, 186–193. https://doi.org/10.1016/j.plasmid.2012.12.002.

Chapter 6
Diversity and Ecological Roles of Prokaryotes in the Changing Antarctic Marine Environment

Angelina Lo Giudice and Maurizio Azzaro

Abstract The Antarctic marine ecosystem is particularly vulnerable to climate change, with further exacerbation mainly deriving from the potential impacts of human activities at research stations. Anthropogenic changes in Antarctica pose a serious questioning about the ability of microbial communities to respond to environmental stresses in this extreme and fragile environment. Establishing the baselines of Antarctic prokaryotic population composition and ecophysiological activities becomes essential to monitor the functioning of ecosystems and the effects of climate change. In this chapter, we present an overview of the prokaryotic communities in the Antarctic marine environment and the potential/current influence of climate change, mainly related to rising temperatures, on their composition and activities. A focus will be done on the role of prokaryotes in the changing polar carbon cycle in seawater, sea-ice and sediments.

Keywords Marine ecosystem · Carbon cycling · Chemoheterotrophs · Organic matter sink · Climate change

6.1 Introduction

Marine microbial communities are strictly involved in nutrient cycling of both the water column and the deep biosphere (Arrigo 2005; Edwards et al. 2012) and are greatly sensitive to environmental changes and human perturbations.

Climate change is negatively impacting Antarctica, which is experiencing a fast rising of temperatures, already threatening benthic and pelagic food webs. The reduction of the ice mass around the continent has a consequent influence on water temperature, salinity and nutrient cycling (Learman et al. 2016). Primary productivity is stimulated by ice melting due to the release into seawater of organic matter,

A. Lo Giudice (✉) · M. Azzaro
Institute for the Biological Resources and Marine Biotechnologies, National Research Council, U.O.S. Messina, Messina, Italy
e-mail: angelina.logiudice@cnr.it

S. Castro-Sowinski (ed.), *The Ecological Role of Micro-organisms in the Antarctic Environment*, Springer Polar Sciences,
https://doi.org/10.1007/978-3-030-02786-5_6

which then moves downward from the water column to sediments. Moreover, both the quantity and quality of carbon sources in pelagic and benthic environments are altered by terrestrial inputs from the continental ice sheet melting. Thus, the flow of organic matter and nutrients to benthic sediments may be altered by climate change in Antarctica. Further, the vast retreat of coastal ice sheets and melting of the glacial ice caps, as the most dramatic consequences of climate warming, are opening shallow-water coastal environments of highly fragmented bathymetry and substrate type (e.g. rocks and bolder fields but also sandy and muddy areas) to new colonization (Abele et al. 2017). Since climate change in Antarctica will alter the flow of organic matter and nutrients of benthic sediments, understanding how these changes will impact diversity is essential to predict change in ecosystem functions. Due to such climate change-driven alterations, establishing the base-lines of microbial population composition and ecophysiological activities in Antarctica becomes essential to understand the impact that these changes could have on ecosystem functioning.

In contrast to the terrestrial realm, and despite the extremely low water temperatures, the Antarctic marine environment is extremely productive and diverse due to the abundance of nutrients that supports life. Within the Antarctic ecosystem, the marine one is the most stable, and this is reflected in the biodiversity of the organisms found there (Pearce 2012). According to Pearce (2012), heterogeneity exists at microscale levels (such as across physical discontinuities of the sea-ice or seawater column or in animal or detritus surfaces), and changes in microbial biodiversity may induce a cascade of changes through the marine food chain.

The following sections will deal with the structure and roles of the prokaryotic community inhabiting main Antarctic marine compartments (seawater, sea-ice and sediment), mainly focusing on the continental shelf (Fig. 6.1). We aimed at identifying the possible effects of climate change on ecosystem functioning.

6.2 Prokaryotic Communities and Climate Changes

The analysis of the prokaryotic communities in Antarctic marine environments has been frequently carried out mainly by using culture-independent techniques, with only a few reports that deal with bacterial isolates (mainly focused on biodiscovery and taxonomic fields). Even if we are living in the metagenomics era, the isolation and characterization of marine microorganisms remain a fundamental approach. These may allow achieving a better comprehension of their physiology and ecology, thus providing information about communities that cannot be obtained directly from culture-independent methods alone (Lo Giudice and Fani 2015). In the following sections, the prokaryotic community structure and composition in different Antarctic habitats, such as seawater, sediment and sea-ice, will be discussed (Fig. 6.2). A summary of the information provided in the following section are found in Table 6.1.

Fig. 6.1 Coastal sites in Antarctica (Terra Nova Bay, Ross Sea)

6.2.1 Antarctic Seawater

In Antarctica, the diversity of marine prokaryotic communities in seawater correlates to both physical (Wilkins et al. 2013) and chemical oceanographic variations (Luria et al. 2014; Signori et al. 2014). In particular, in the absence of primary production during the winter period, the prokaryotic heterotrophic production is supported by the dissolved organic carbon (DOC) pool produced during the previous summer, and it becomes the pivotal food source for the heterotrophic grazer communities (Azam et al. 1991; Abele et al. 2017).

Gaining knowledge of marine prokaryotes can be achieved by establishing their taxonomic affiliation, functional roles and relationships in a given habitat. Studies describing the structure and composition of the prokaryotic communities in Antarctic waters are quite rare and have been mainly focused on the open ocean, with emphasis in communities thriving at the coastal Antarctic and subantarctic areas that remain poorly characterized (Piquet et al. 2010).

Fig. 6.2 Typical Antarctic habitats: up and down the ice

In coastal Antarctica intensive spring and early summer blooms are found, except in areas that receive a major input from glacial melt waters and eroded sediments (Schloss et al. 2014). Studies on the prokaryotic communities in Antarctic coastal areas span from the Antarctic Peninsula to the Ross Sea and mainly focussed on water column, surface seawater and ice-affected seawater, as follows.

Recently, Abele et al. (2017) reviewed the knowledge of shallow-water community composition and species-specific functional responses to Antarctic coastal change in Potter Cove on King George Island (Isla 25 de Mayo, South Shetland Archipelago). This is a shallow fjord, strongly impacted by a glacial melting (sea surface temperature has increased ca. 0.36 °C per decade) that accelerated the microbial and heterotrophic turnover rates. *Proteobacteria* (mainly *Alphaproteobacteria* and *Gammaproteobacteria*) and *Bacteroidetes* were the two dominant phyla found in Potter Cove coastal waters (Landone Vescovo et al. 2014). Among *Alphaproteobacteria*, *Rhodobacteraceae* were abundant. Interestingly, this bacterial group has the ability to switch between photosynthetic and chemoheterotroph energy production under anoxic and oxic conditions, respectively (Abele et al. 2017). A peculiar adaptation of *Rhodobacteria* to an extremely light regime was suggested by Ghiglione and Murray (2012), as their occurrence was strongly linked to seasonal variations, with greater dominance in summer compared to the low-light

Table 6.1 Relative percentages (%) of main phylogenetic groups in some Antarctic coastal sites

		Phylogenetic group[a]											
Antarctic area	Matrix	ALF	BET	GAM	DEL	EPS	PLA	CFB	VER	ACT	FIR	Method[b]	Reference(s)
Cape Hallett	Seawater	25.7	0–1	19.9				29.6				DGGE	Celussi et al. (2009a)
Potter Cove	Seawater	50	3	31				5				Clone	Landone Vescovo et al. (2014)
Terra Nova Bay	Seawater	1–11	18–31	48–81						0–9	<1	Clones	Yakimov et al. (2004)
Terra Nova Bay	Seawater	0–2	0–33	43–79		0–5		12–14				Clones	Gentile et al. (2006)
Terra Nova Bay	Seawater	9.4	6.8	11.3				14.8		11.8	9.1	FISH	Lo Giudice et al. (2012)
Terra Nova Bay	Seawater	9.4		67.8				5.3		16.7	<1	Isolates	Lo Giudice et al. (2012)
Potter Cove	Sediment	~10 to 20	<2	~20	~8 to 9		~6 to 12	>30	~8 to 12	~5		Clones	Hernandez et al. (2015)
Terra Nova Bay	Sediment	<1	<1	23–24	19–23			21.5–32.6		5–6	<1	Clones	Baldi et al. (2010)
Terra Nova Bay	Sediment	1		65.3				24.5		9.2		Isolates	Lo Giudice et al. (2013)
Amundsen Sea	Sea-ice	~1 to 23		~12 to 78				~11 to 68				NGS	Torstensson et al. (2015)
Weddell Sea	Sea-ice	31–48		12–19				23–32				NGS	Eronen-Rasimus et al. (2017)

[a]**ALF** *Alphaproteobacteria*, **BET** *Betaproteobacteria*, **GAM** *Gammaproteobacteria*, **DEL** *Deltaproteobacteria*, **EPS** *Epsilonproteobacteria*, **PLA** *Planctomycetes*, **CFB** CF group of *Bacteroidetes*, **VER** *Verrucomicrobia*, **ACT** *Actinobacteria*, **FIR** *Firmicutes*
[b]*NGS* next-generation sequencing

regime occurring during the winter season. Thus, *Rhodobacteria* may play an important role as primary producers in low-light coastal areas (e.g. *Roseobacter* and *Erythrobacter* members are involved in anoxygenic photosynthesis at low-light levels and at wavelengths in the near-infrared spectrum).

Yakimov et al. (2004) and Gentile et al. (2006) analysed surface seawater samples in coastal sites at Terra Nova Bay (Ross Sea), reporting the strong influence of both sea-ice and human activity on the composition of the bacterial communities. In details, Yakimov et al. (2004) analysed seawater samples collected under pack-ice (2 m) at Road Bay and Adelie Cove. The microbial community structure observed at both sites was similar and characterized by a great abundance of *Gammaproteobacteria*. The prevalence and relative abundance of *Alpha-* and *Betaproteobacteria* depended on the sampling site, but bacteria from the CFB group (5% of total bacterial sequences) were always a small fraction of the communities. As determined by clone library sequencing, several *Gammaproteobacteria* corresponding to chemoheterotrophic species and genera inhabiting cold marine ecosystems, including members of *Colwellia*, *Halomonas*, *Marinobacter*, *Methylophaga*, *Pseudoalteromonas*, *Shewanella* and *Pseudomonas*, were found. A single phylotype group from the Adelie Cove (Terra Nova Bay) library was placed in *Betaproteobacteria* and was closely related to the methylotrophic species *Methylophilus methylotrophus*. Altogether, these results reflect a very common bacterial community composition that can be found in the Southern Ocean. This microbial composition might be influenced by the thaw of sea-ice that releases particles and dissolved organic matter derived from the diatom blooming in the ice bottom. Among *Betaproteobacteria* from the Road Bay, many clones with high similarity to different hydrocarbon-degrading soil *Burkholderia* spp. were detected. Moreover, Gram-positive bacteria, atypical for marine bacterioplankton communities, were present in the Road Bay library, covering 10% of total clones. This finding found an explanation in the proximity of the Road Bay to the Italian research station and, therefore, in the direct influence of human activities.

In order to have a clear vision of the microbial community composition, Gentile et al. (2006) analysed the composition of bacterioplankton assemblages in surface seawater at Road Bay and Evans Cove (Terra Nova Bay) after ice melting, sequencing clone libraries from 16S rDNA and 16S rRNA. The majority of clones were related to psychrophilic or psychrotrophic microorganisms, previously detected in permanently cold Arctic and Antarctic marine environments. Samples were dominated by *Gammaproteobacteria* (mainly affiliated to the psychrophilic genera *Colwellia* and *Psychrobacter*) and *Bacteroidetes* (mainly *Polaribacter* and *Flectobacillus* genera) and showed a low abundance of *Alphaproteobacteria* (typical from Antarctic sea-ice microbial compartments). The high degradation rates of aerobic heterotrophs, embedded in the ice bottom, might explain the occurrence of an anoxic microflora, including strictly anaerobic *Epsilonproteobacteria*. In addition, sequences related to *Betaproteobacteria* (genera *Aquaspirillum* and *Curvibacter*, typical in the activated sludge and freshwater environments) were abundant in the Road Bay but only when searching in 16S rDNA clone libraries (thus suggesting that they were not part of the metabolically active community).

Experiments aimed at analysing the bacterial community structure by denaturing gradient gel electrophoresis (DGGE) fingerprinting in an under-ice area, characterized by permanent low temperatures and a thin but extensive sea-ice coverage, showed a temporal shift in *Flavobacteria* (within the CF group of *Bacteroidetes*) (Celussi et al. 2009b). In particular, differences between daytime and night were observed, suggesting the possible influence of cyclic sea-ice melting and freezing in shaping *Flavobacteria* assemblages, which are specialized in colonizing organic matter particles released by sea-ice thawing. The sequencing of bacterial DGGE bands showed a classical taxonomic composition of the assemblages dominated by *Flavobacteria*, *Alpha-* and *Gammaproteobacteria*, confirming previous observation by Gentile et al. (2006).

Structurally different prokaryotic assemblages are strictly dependent on the characteristics of the water masses where they develop. Celussi et al. (2009a) analysed bacterial assemblages by DGGE in two areas of the Ross Sea, differing mainly in their productivity regime: Terra Nova Bay polynya (highly productive during summer) and Cape Adare (low phytoplankton biomass and activity). A pronounced stratification of bacterial assemblages was identified, with a dominance of epipelagic communities over the mesopelagic and the bathypelagic communities. Generally, the dominant phylotypes were typical from marine and Antarctic habitats, with the dominance of the CF group of *Bacteroidetes*, *Alphaproteobacteria* and *Gammaproteobacteria*. Celussi et al. (2009a) found *Actinobacteria* but only in the meso- and the bathypelagic communities. Further insights on the structure of the bacterioplankton inhabiting different water masses in the Ross Sea were reported by the same authors (Celussi et al. 2010) who also analysed samples from Cape Adare, Terra Nova Bay polynya and southernmost locations along the Ross Ice Shelf coast by DGGE. No substantial differences in the phylogenetic composition of assemblages were found between different water masses, with a dominance of *Alpha-* and *Gammaproteobacteria*. Surprisingly, *Bacteroidetes* did not strongly contribute to the bacterial community. Celussi et al. (2010) also detected members of *Betaproteobacteria*, which are typical of freshwater, confirming previous observations from the Ross Sea area (Gentile et al. 2006; Celussi et al. 2009b). They suggested that *Betaproteobacteria* could constitute a very small fraction of the bacterial community in the whole basin.

Among the Archaea domain, the occurrence of Thaumarchaeota has been frequently reported for the Southern Ocean and Arctic surface waters (Amano-Sato et al. 2013). A similar result was observed in Antarctic sediments (Signori et al. 2014; Learman et al. 2016), in sea-ice (Cowie et al. 2011) and even in the symbiont population colonizing Antarctic sponges (Rodríguez-Marconi et al. 2015), suggesting that Thaumarchaeota may have an important role in the Antarctic ecosystems (Abele et al. 2017). Hernández et al. (2015) also reported the predominance of members of the phylum Thaumarchaeota within the archaeal communities of Potter Cove (King George Island, Barton Peninsula, South Shetland Islands), with a high occurrence of the well-known *Nitrosopumilus maritimus* and other ammonium-oxidizing Archaea, indicating a role of Thaumarchaeota in the biogeochemical cycling of carbon and nitrogen in both the water column and surface sediments. The same authors also retrieved members from the order Thermoplasmatales, which includes iron oxidizing Archaea likely involved in the oxidation of ferrous iron derived from subglacial erosion or oozing ferruginous sediments in Potter Cove (Abele et al. 2017).

Most microbiological studies on Antarctic have been carried out in summer and spring periods, when photoautotrophic production supplies organic carbon sources for prokaryotic growth. But there is a strong seasonal variation in primary production, mainly driven by the seasonal cycle of irradiance, vertical mixing and ice covering of polar oceans. This raises the question of how the marine food web, and especially the marine microbial food web, functions during wintertime. Chemoautotrophy supported by nitrification has been suggested as an important contribution to prokaryotic production during the polar winter (Manganelli et al. 2009; Grzymski et al. 2012; Williams et al. 2012). On such basis, Tolar et al. (2016) determined seasonal ammonia oxidation rates, gene and transcript abundance in continental shelf waters west of the Antarctic Peninsula, where Thaumarchaeota strongly dominate populations of ammonia-oxidizing organisms (Murray et al. 1998; Church et al. 2003; Grzymski et al. 2012). Results demonstrated the potential importance of nitrification (9% of total nitrogen cycling) in Antarctic continental shelf to the global nitrogen cycle.

Bacteria have been frequently isolated from seawater samples collected in different Antarctic areas. First microbiological investigations, performing a culture-dependent approach, were done in the Terra Nova Bay (Maugeri et al. 1996; Bruni et al. 1999). More recently, the bacterial community along the water column in six stations at Terra Nova Bay was examined by the genotypic and phenotypic characterization of 606 bacterial isolates and FISH analysis by Lo Giudice et al. (2012). The FISH analysis again revealed a bacterioplankton composition that was typical of Antarctic marine environments, with a dominance of CF group of *Bacteroidetes*, *Actinobacteria* and *Gammaproteobacteria*, followed by *Alpha-*, *Betaproteobacteria* and *Firmicutes* (slightly less abundant) (Fig. 6.3). Again, ice melting could have an influence on the bacterioplankton composition. This finding was supported by the isolation of *Gelidibacter*, *Polaribacter* and *Psychroflexus* members (generally well represented in Antarctic sea-ice) which showed the ability to hydrolyse macromolecules, probably through the production of extracellular enzymes. A consistently great abundance of the *Gammaproteobacteria* (67.8%) was detected through the identification of cultivable isolates. Among the Gammaproteobacterial isolates, 65.4% accounted for members of the genera *Psychromonas* and *Pseudoalteromonas*. These are ubiquitous microbes, thus suggesting that they may play a key role within the analysed bacterioplankton community. In particular, *Pseudoalteromonas* isolates showed nitrate reductase activity and were able to hydrolyse substrates that indicated their ability for producing protease, esterase and β-galactosidase, suggesting that they are involved in the carbon and nitrogen cycling. Interestingly, results from this culturable-dependent approach showed a strong contribution of *Actinobacteria* to the bacterioplankton community at Terra Nova Bay. Usually, they are a minor fraction of natural bacterioplankton communities in coastal and oceanic samples. However, their occurrence in the Ross Sea was reported earlier (Lo Giudice

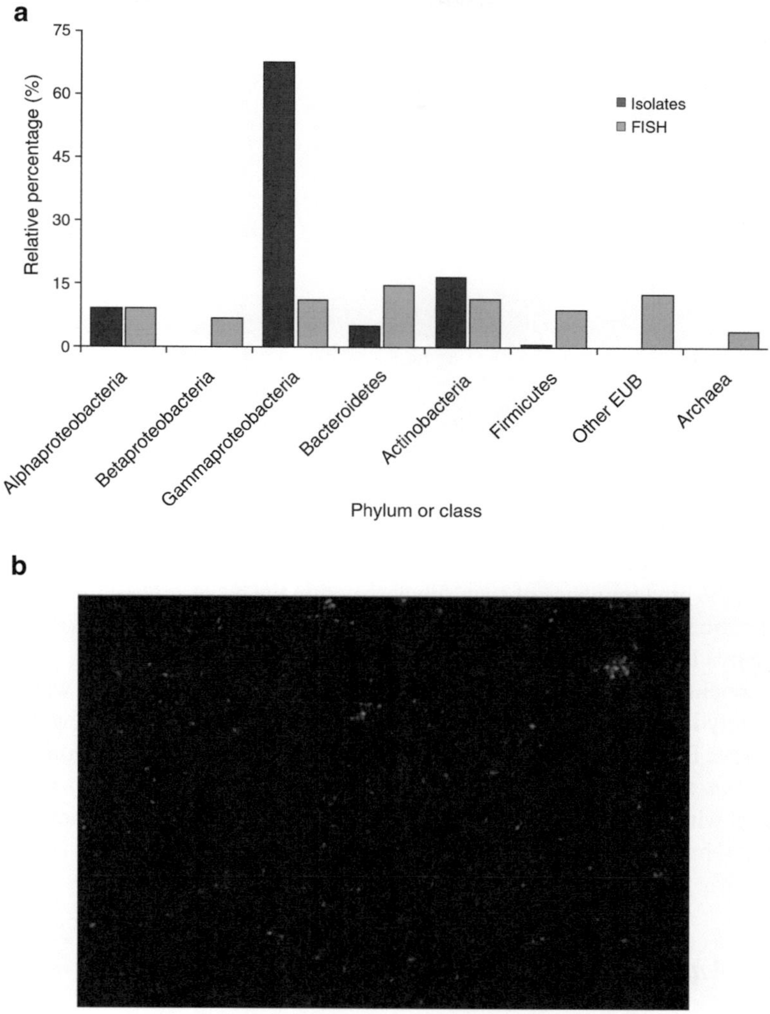

Fig. 6.3 (**a**) Relative amounts of phyla or classes detected by fluorescent in situ hybridization (*FISH*) and cultural methods (*Isolates*) in the Terra Nova Bay (Ross Sea) area (Lo Giudice et al. 2012). Data are averages across stations and depths. For FISH, abundances are related to total counts (DAPI-staining). Other EUB indicates cells detected with probe EUB338 but not with any group-specific probe (other Eubacteria); (**b**) example of an eye field under the epifluorescence microscope after the FISH analysis

et al. 2007; Celussi et al. 2009a). *Actinobacteria* isolates were able to grow in the absence of NaCl, thus suggesting a probable terrestrial origin. Terra Nova Bay is highly influenced by continental inputs, mainly deriving from frequent katabatic wind events (Fig. 6.4) or from the extension of both Campbell glacier tongue and pack-ice, which could contribute to microbial transportation. However, as observed for most marine isolates, *Actinobacteria* strains generally showed a wide range of

Fig. 6.4 A katabatic wind event in Antarctica

salt tolerance, thus indicating a strong adaptation to the polynya environment (where seawater salinity highly varies due to freezing and melting cycles of sea-ice). The low abundance of *Alphaproteobacteria* was in contrast with other studies that reported the predominance of such subclass within the bacterioplankton community in Antarctic seawater. Nevertheless, this feature may be typical of shallow waters at Terra Nova Bay, as previously reported by Gentile et al. (2006).

In conclusion, the prokaryotic communities in Antarctic coastal waters have been rarely investigated, and the information is mainly restricted to a few areas (i.e. Terra Nova Bay, Potter Cove) next to Research Bases. Overall, the predominance of *Proteobacteria* (mainly *Alpha*- and *Gammaproteobacteria*) and *Bacteroidetes* was observed within the Bacteria domain, whereas Thaumarchaeota were more frequent within Archaea.

6.2.2 Antarctic Sea-Ice

Sea-ice covers 13% of the Earth surface. It is an active environment characterized by fluctuations in salinity, temperature, light gradient and nutrient availability (Thomas and Dieckmann 2002). Unlike freshwater ice, frozen seawater forms a semisolid matrix, forming a network of channels and pores that are filled with a high salinity (up to 200‰) brine that provides a liquid-phase interface at subzero temperatures. Extreme temperature and salinity conditions, together with continuous freeze-thaw cycles, strongly affect sea-ice physical and ecological properties, damaging living cells and reducing cell viability (Thomas and Dieckmann 2002). Ice-trapped bacterial communities must adopt peculiar adaptation strategies to cope with such harsh and seasonally fluctuant environment. Despite such extreme conditions, bacterial assemblages are conspicuous and mostly derive from larger organisms that live at the surface. Their growth is fuelled by a large pool of dissolved organic matter, mainly consisting of carbohydrates released after the death and lysis of sea-ice organisms, as well as by the exudation of organic polymers by algae (mainly diatoms) and bacteria. This reach marine food web has been mainly analysed during the spring and summer seasons, when the sea-ice algal bloom provides organic compounds for bacterial growth. The available information regarding the Antarctic sea-ice bacterial community is very scarce, but it has been recently reviewed by Bowman (2015).

In both first-year ice and multiyear ice, the Antarctic bacterial communities were dominated by copiotrophic bacteria (microorganisms found in environments rich in nutrients), mainly *Gammaproteobacteria* (e.g. *Glaciecola* and *Colwellia*), *Alphaproteobacteria* (e.g. *Octadecabacter*) and *Flavobacteria* (e.g. *Polaribacter* and *Flavobacterium*), whereas *Cyanobacteria* and Archaea were generally not detected (Bowman et al. 1997; Brown and Bowman 2001; Brinkmeyer et al. 2003; Murray and Grzymski 2007; Maas et al. 2012). In addition to copiotrophic microorganisms, Maas et al. (2012) reported 16S rDNA sequences clustering with species that encode phototrophic genes, thus raising the possibility that bacterial phototrophy may occur at the Antarctic sea-ice, as reported above for seawater. Among others, 16S rDNA clones closely related to the *Roseobacter* clade (well known to produce bacteriochlorophyll-a and to contain genes required for aerobic anoxygenic photosynthesis) and photoheterotrophic *Polaribacter* spp., which have been reported to carry the proteorhodopsin gene (a light-driven proton pump that can harvest light to generate ATP), were detected, suggesting that photosynthetically competent bacteria are also abundant in the Antarctic sea-ice. But also sulphate-reducing bacteria (e.g. *Desulforhopalus*, *Desulfofrigus* and *Sulfurospirillum*) and the presence of H_2S were observed in a thick apparently anoxic ice sheet, suggesting that the development of an anaerobic bacterial community and the reduction of sulphur compounds may occur under suitable conditions in sea-ice (Eronen-Rasimus et al. 2017).

Recently, it has showed that the bacterial community in the Antarctic sea-ice can stay active throughout the winter period, and as a consequence of a future warming, the increase in bacterial production may lead to changes in bacteria-mediated processes, affecting the biogeochemical cycles and food webs of ice-covered seas (Eronen-Rasimus et al. 2017). According to Grossmann and Dieckmann (1994) and Helmke and Weyland (1995), within newly formed Antarctic sea-ice, bacterial activity is temporarily suppressed and later restored after consolidation of the sea-ice. At this stage, psychrophilic bacteria predominated over psychrotolerant bacteria, indicating changes in the community structure during winter. Sea-ice bacterial genera (such as *Octadecabacter*, *Polaribacter* and *Glaciecola*) are often abundant in spring and summer during the sea-ice algal bloom but also predominated in the winter communities. This variability in the bacterial community composition may be correlated with the different concentrations of chlorophyll-a, suggesting that sea-ice bacteria and algae may also be coupled during the Antarctic winter (Eronen-Rasimus et al. 2017). Overall, the results from Eronen-Rasimus et al. (2017) described an active sea-ice bacterial community also throughout the winter period, suggesting that future warming of sea-ice, with the consequent increase in bacterial production, may impact food web dynamics and elemental cycling in the Antarctic sea-ice zones. At this regard, Torstensson et al. (2015) reported a sea-ice bacterial community (in the Amundsen and Ross seas) that was strongly dependent on sea-ice temperature and brine salinity. According to the authors, warmer and less saline brines, deriving from climate and seasonal changes, could reduce the diversity, richness and composition of psychrophilic and halophilic sea-ice bacteria.

6.2.3 *Antarctic Sediment*

Polar shelves are regions of intense biological activity and biogeochemical cycling. In sediments, microbial communities play a major role in carbon, sulphur, nitrogen and phosphorus cycles (Edwards et al. 2012). Sediments are characterized by seasonally high primary production, large drawdowns of pCO_2 and intense sedimentation rates (Fonda Umani et al. 2005). An integrated approach, examining both the available carbon sources and indicators of microbial abundance, diversity and metabolism is required to understand the effects that the organic matter (OM) quality and quantity have on the resident microbial communities (Carr et al. 2013). For example, in the Terra Nova Bay (Ross Sea), the particulate organic matter was dominated by proteins associated with large phytoplankton cells or phytodetritus. Such protein pool was characterized by a lower digestibility (due to the low temperature and low microbial activities) than that from oligotrophic waters, thus resulting in an inefficient recycling along the water column (Fabiano and Pusceddu 1998). This leads to the accumulation of organic detritus in the coastal belt and, in turn, potentially supports high microbial biomass and activity in the sediments (Fabiano and Danovaro 1998).

Marine sediments of the Ross Sea harbour microbial communities that play a significant role in the decomposition, mineralization and recycling of organic carbon matter (Carr et al. 2013). The composition of this organic matter is modulated by a pulsed organic supply of particulate matter, a phenomenon restricted to a short period of the spring-summer time, when primary production in blooms and sea-ice release their solid load, rich in labile compounds (Fabiano and Pusceddu 1998). Further, the global warming leads to an increase in the amount of melt water entering the coast at the ice shelf, impacting both coastal and open ocean water chemical and biological composition (Depoorter et al. 2013). This phenomenon could lead to the increase in the occurrence of phytoplankton blooms in polar areas (Ducklow et al. 2006) and, therefore, could increase the transport of organic matter to the seafloor, thus changing the community from lithotrophic organisms to ones often associated with degradation of organic matters (Learman et al. 2016).

A number of studies have examined microbial diversity in Antarctic sediments, by using various sequencing methods and recording different geochemical parameters (e.g. Bowman and McCuaig 2003; Bowman et al. 2003; Powell et al. 2003; Baldi et al. 2010; Carr et al. 2013, 2015; Jamieson et al. 2013; Ruff et al. 2014). For example, Carr et al. (2013) and Ruff et al. (2014), analysing sediments from the Ross Sea and the Drake Passage at the Antarctic Polar Front, reported an increase in the microbial abundance in a direct relationship with the level of the organic matter. However, Jamieson et al. (2013), examining sediments from the Southern Ocean, reported that the organic matter did not influence the species richness.

Baldi et al. (2010) analysed the relationship between the level of labile organic matter and the bacterial communities in Antarctic shallow sediments along the Terra Nova Bay coast, between Gerlache Inlet and Adelie Cove. The results showed that the organic carbon represented the major fraction of total carbon, at similar or lower concentrations than those previously found in Antarctic bottom sediments (Isla et al. 2002, 2004, 2006; Masquè et al. 2002). The biopolymeric carbon molecules ranged from 4.1% to 19.9% and showed a wide trophic range (65–834 $\mu g\ g^{-1}$ of dry weight). On average, proteins represented the main biochemical molecules contributing to the labile organic carbon pool, followed by lipids and carbohydrates.

As previously observed in sediments from other Antarctic areas (Bowman et al. 2003; Bowman and McCuaig 2003), phylogenetic analyses carried out on two 16S rDNA clone libraries (samples from Tethys Bay and Gerlache Inlet) revealed that *Proteobacteria* and *Bacteroidetes* were over-represented. Almost half of the proteobacterial clones were allocated in the Delta subclass (about 20% of the clone libraries). They were probably responsible for the sulphur, iron and phosphorous precipitations which coated the siliceous and alumino-silicate sediments analysed throughout the study (Baldi et al. 2010). *Deltaproteobacteria* sequences were mainly affiliated with the genera *Desulforhopalus* and *Desulfosarcina.* The occurrence of such strictly anaerobic bacteria suggests that sediments might also undergo anoxic conditions that could favour the accumulation of proteins and the fermentation of monosaccharides from polysaccharide decomposition. The potential anoxia occurring at surface sediments was also strengthened by the high abundance of sequences affiliated to the *Ectothiorhodospiraceae* within *Chromatiales*, typically retrieved in anoxic samples.

The conspicuous presence of sequences affiliated within the *Flavobacteriaceae* suggested that in vivo degradation of organic polymers in shallow-water Antarctic sediments may be mainly carried out by these bacteria, which are known for their ability to efficiently degrade many types of polysaccharides (Bowman and Nichols 2005). The complex population structure, consisting of aerobic and anaerobic heterotrophic bacteria with the occurrence of chemolithotrophs, strongly suggested that surface sediments at Terra Nova Bay may alternate from oxic to anoxic conditions and vice versa.

A 16S rRNA gene pyrosequencing analysis from a soft sediment core (15 depths) collected in Ross Ice Shelf, the world's largest ice shelf (ca. 560,000 km^2), allowed the identification of a diverse community (50 bacterial phyla), including heterotrophs that utilize organic carbon sources such as amino acids, oligosaccharides and lactose (Carr et al. 2013). The community at the surface of the sediment was more diverse than at deeper sediment locations and was dominated by members of the phyla *Actinobacteria*, *Bacteroidetes*, *Chloroflexi* and *Firmicutes* and the subphyla *Beta-*, *Delta-* and *Gammaproteobacteria*. However, a minor representation (1%) of bacteria belonging to *Acidobacteria*, Candidate Division OP3, Candidate Division OP9, *Cyanobacteria*, *Deferribacteres*, *Deinococcus-Thermus*, *Planctomycetes* and

Alphaproteobacteria was also found. Changes in the degree of diversity within the core mainly correlated with the available carbon source.

More recently, Learman et al. (2016) analysed surface sediment from 24 sites across a 5500 km region of Western Antarctica (spanning from the Ross Sea to the Weddell Sea) to examine relationships between microbial communities and sediment geochemistry. rRNA genes were sequenced via Illumina MiSeq and correlated with geochemical and nutrient data. The total dataset demonstrated a relationship between the modifications in carbon sequestration processes and the microbial community composition, highlighting a prospective mechanism that links climate change to carbon availability. The study also suggested a possible coupling between pelagic and benthic communities. Overall, the relative abundant phyla identified in these sediments were in line with those reported in other studies on Antarctic sediments, with the predominance of *Gamma-* and *Deltaproteobacteria* subclasses (Bowman and McCuaig 2003; Powell et al. 2003; Baldi et al. 2010). However, the 16S primer set used during those studies may have overestimated and underestimated the occurrence of *Gammaproteobacteria* and *Alphaproteobacteria*, respectively (Learman et al. 2016). Three of the top five documented phylotypes (i.e. Piscirickettsiaceae, *Nitrosopumilus* and OM60/NOR5clade) were present among samples, suggesting that they could play key roles within the Antarctic sedimentary prokaryotic community. The former negatively correlated with $\delta^{13}C$, NH_4^+ and Si. Among them *Cycloclasticus pugetii* degrades complex carbon. Conversely, the lithotrophic archaeal genus *Nitrosopumilus* was extremely abundant and its relative abundance statistically correlated with $\delta^{13}C$ and Si content. It was also reported as dominant in Antarctic sediment community in the Weddell Sea by Gillan and Danis (2007). As reported also by Ruff et al. (2014), OM60/NOR5 clade members are involved in the degradation of phytoplankton-derived organic matter in sediments of the Drake Passage at the Antarctic Polar Front. Additionally, samples from Western Antarctica were characterized by their relatively recalcitrant carbon content, a characteristic mainly influencing the occurrence of Thaumarchaeota within the Archaea domain, which are well known for their ability to thrive as lithotrophs. Similar results were obtained by Matos et al. (2016), who studied surface sediment samples in the inner Potter Cove and hence close to the melting glacier. Finally, sediments from the Antarctic Peninsula (Learman et al. 2016) were characterized by the presence of *Bacteroidetes*, which is generally associated with the degradation of the high molecular weight organic matter derived from sinking particles and faecal pellets from blooms and their associated grazers.

A high metabolic diversity and functional versatility of the bacterial communities (dominated by chemoheterotrophic aerobic bacteria) was reported for Potter Cove sediments, suggesting that they have the ability to respond to sudden environmental changes (Wölfl et al. 2014). As observed for the water column, *Proteobacteria* (mainly represented by *Alpha-* and *Gammaproteobacteria*) and *Bacteroidetes* (mainly *Flavobacteriaceae* and *Haliscomenobacter* spp.) dominated, followed by *Planctomycetes* (mainly members of the family *Planctomycetaceae*) and *Verrucomicrobia*. Further, differences between the sandy shallow coastal and the central muddy areas of Potter Cove were encountered within the sedimentary microbial community.

In addition to culture-independent approaches, the cultivable diversity of Antarctic bacteria in sediment samples from different areas was also studied (Yu et al. 2011; Lo Giudice et al. 2013). A comparative 16S rRNA gene sequence analysis carried out on bacterial isolates from sediment at Nella Fjord (Eastern Antarctica) and Terra Nova Bay (Ross Sea), respectively, showed sedimentary isolates mainly belonging to *Gammaproteobacteria*, followed by *Bacteroidetes*, *Actinobacteria* and *Alphaproteobacteria*, as generally observed for seawater samples.

As a general overview of Sect. 6.2, we can summarize that to date the composition of the Antarctic prokaryotic communities has been investigated in a few coastal areas, such as Potter Cove (King George Island, South Shetland Island) and Terra Nova Bay (Ross Sea). Predominant phyla (e.g. *Alpha*- and *Gammaproteobacteria* and Thaumarchaeota) in sediment, sea-ice and seawater reflect the environmental status of the sampling area, being the sea-ice a major player with a fundamental role in structuring the prokaryotic community. As investigations have been carried out mainly during the summer season, the future investigations should provide details on the wintertime prokaryotic life. Further, it is evident that our actual knowledge is still fragmentary and habitat-specific, so that a wider vision could be obtained by considering a major number of sampling sites and performing deeper phylogenetic and physiological studies of the microbial communities inhabiting such so different Antarctic environmental compartments (i.e. seawater, sediment and sea-ice).

6.3 Prokaryotes in the Changing Polar Carbon Cycle: An Overview

Microorganisms play a pivotal role in Southern Ocean habitats (sea-ice, coastal and pelagic sea, benthic zone). They control many processes, including primary production (pelagic and benthic), turnover of biogenic elements and degradation of organic matter (Fuhrman and Azam 1980; Azam et al. 1991; Karl 1993; Azzaro et al. 2006; Manganelli et al. 2009; Alder et al. 2017). Moreover, prokaryotic carbon cycling in the Southern Ocean is a major issue to understand the regulation of the Earth climate and to investigate the potential effects of climate change on cold marine habitats.

6.3.1 C-Cycle in Sea-Ice

Sea-ice covers about 50% of the Southern Ocean in winter, decreasing up to 10% in the short summer period. It represents one of the most dynamic ecosystems on Earth (Arrigo et al. 1997) and plays an important role on biogeochemical cycles, ecosystems and global climate. The sea-ice microbial community inhabits salty pore spaces and also the underside of the ice at the seawater-sea-ice interface (Bowman 2015). Despite low temperature and high salinity, the sea-ice bacterial assemblage remains relatively unchanged through the winter compared to the seawater assemblage (Collins et al. 2010). Sea-ice bacterial assemblages play a significant role in the Southern Ocean carbon cycle by enhancing sea-ice primary production, which is mainly exported to the water column (Fernández-Méndez et al. 2014) and ocean floor (Morata et al. 2010). The sea-ice microbial communities are evidenced by high rates of prokaryotic heterotrophic production, but their metabolic diversity extends far beyond heterotrophy, and their functionality encompasses much more than carbon turnover (Bowman 2015). In fact, the investigations performed during the last three decades have identified an active role of sea-ice microorganisms in nitrogen and phosphate cycling, mutualistic partnerships with ice algae and even prokaryotic carbon fixation. In the last 50 years, the loss of ice shelves and retreat of coastal glaciers around the Antarctic Peninsula has exposed a large marine area (at least 2.4×10^4 km^2) and produced new carbon sinks in Antarctica acting as a negative feedback to climate change (Peck et al. 2010). This phenomenon may be partially attributed to the high concentration of cryoprotective exopolymers commonly found within sea-ice (Krembs et al. 2002; Underwood et al. 2010).

6.3.2 C-Cycle in Seawater

In the seawater of the Southern Ocean, prokaryotic abundance and activity change significantly over the annual cycle as sea-ice melts and phytoplankton blooms develop (Ducklow et al. 2001; Pearce et al. 2007). Microbial food chains develop even in regions where large euphausiids are abundant and these chains include small metazoans, and in the northern open waters are predominated by multiple trophic levels (copepods, chaetognaths, amphipods, myctophids, fish and birds), in contrast with the classical short chain of diatoms – krill – vertebrates (Atkinson et al. 2012; Alder et al. 2017). Genomic studies, trying to decipher the microbial metabolic capabilities using Southern Ocean water samples, are currently limited (Béja et al. 2002; Grzymski et al. 2006). However, Grzymski et al. (2012) found that the bacterioplanktonic assemblages in nearshore surface water samples of the Antarctic Peninsula were dominated by chemolithoautotrophic organisms in the winter season, but they were not detected in the summer season, when incident solar irradiance and the primary productivity are high. If chemolithoautotrophy is widespread in the Southern Ocean at winter, this process may be a previously unidentified carbon sink. Currently little is known about biogeochemical dynamics during the long dark winter of the southern oceans, particularly in the deeper part of the ocean. Manganelli et al. (2009) demonstrated that Bacteria and Archaea become the major producers of biogenic particles during austral winter in the Southern Drake passage, at the expense of dissolved organic carbon drawdown. Moreover, chemoautotrophic

CO_2 fixation and heterotrophic production rates were important, also in deep water, and bacterial assemblages were controlled by viruses and protists. Thus, Southern Ocean microbial loop may substantially maintain a wintertime food web and system respiration at the expense of summer produced dissolved organic carbon, as well as must be involved in the regeneration of nutrients and Fe. Heterotrophic microbes control the extent of vertical organic carbon fluxes by respiration, excretion and decomposition (Cho and Azam 1988; Smith et al. 1992; Nagata et al. 2000). Heterotrophic remineralization in the Antarctic deep waters represent the output of the oceanic biological pump that exports organic carbon from the euphotic zone to the ocean depths (Ducklow et al. 2001). This process contributes to the sequestration of large amounts of carbon for up to 1000 year in deep waters and influences the global carbon cycle and climate (Sarmiento and Gruber 2006). In fact, the downward flux to the mesopelagic zone of particulate organic carbon (POC) is estimated to be 40–50% of the primary production (1949 ± 70.1 Tg C $year^{-1}$ – mean value between 1998 and 2006; Arrigo et al. 2008). In this context, an investigation about the impact of subtropical waters on the microbial community of the subantarctic zone (SAZ) of the southeast of Tasmania (Australia) and at stations in the Polar Fontal zone (PFZ) showed that the arrival of subtropical waters impacted on the microbial community, resulting in communities similar to those from subtropical waters. As a consequence, a future shift of SAZ microbial communities from particulate carbon export from the photic zone towards an increased production by smaller cells may significantly increase the microbial loop and viral lysis (Evans et al. 2011). These changes would promote carbon recycling within the euphotic zone, thereby potentially decreasing the capacity of the future subantarctic zone to sequester CO_2.

6.3.3 C-Cycle in Sediment

Microbial communities from sediments also play a key role in carbon cycling (Mayor et al. 2012), mineralizing the organic matter sinking from the water column to sediments (Ducklow et al. 2006) and representing the main energy source for microbial life in the seafloor. Since climate change will alter the flow of organic matter and nutrients to benthic sediments in the Southern Ocean, understanding how these changes will impact on microbial metabolism is essential to predict change in ecosystem functions (Learman et al. 2016). Studies of the marine sediments from the Ross Sea (Carr et al. 2013) and the Drake Passage at the Antarctic Polar Front (Ruff et al. 2014) have shown that the increase in organic matter can positively influence the microbial community abundance and composition at local scale. The increase of the melt water from the continental ice sheet would introduce more terrestrial carbon into marine and benthic environments, altering the quantity and type of carbon. Thus, a greater understanding of Antarctic ecosystems is essential to predict how changing climate will influence organic fluxes between benthic and pelagic communities (Learman et al. 2016).

6.4 Concluding Remarks

Prokaryotes responds to environmental stresses, including climate warming, and could be used as bioindicators of changes from human perturbations. As the Antarctic environment remains sufficiently no-impacted, the study of the microbial communities may allow the identification of responses of these communities to the climate changes. In general, sampling was performed during summer, when the prokaryotic communities experience organic matter inputs from glacial melt waters and algal blooms. As a consequence of increased ice melting due to global change-driven rising temperatures, such conditions will become more prominent and frequent in the Antarctic coastal environments in the near future. As a general conclusion, it seems evident that this could result in the predominance of chemoheterotrophs, microbes involved in the degradation of phytoplankton-derived organic matter (Fig. 6.5). Such increase in prokaryotic production may finally impact food web dynamics and elemental cycling in Antarctic coastal areas. In our opinion, this review highlights a main aspect which merits to be better elucidated in the near future, i.e. the impact of sea-ice as a sink of organic compounds, which are released in the marine environment during melting, thus impacting on the food web. Comparative studies including different sampling sites will be certainly needed to gain a more valuable picture of the response of the Antarctic prokaryotic communities to environmental changes. Finally, it could be useful to do a monitoring work over time (e.g. from the initial ice break-up to the total ice melting) and to analyse the influence of sea-ice on the activity and composition of the prokaryotic community, instead of focusing the attention on punctual and isolated changes environmental conditions.

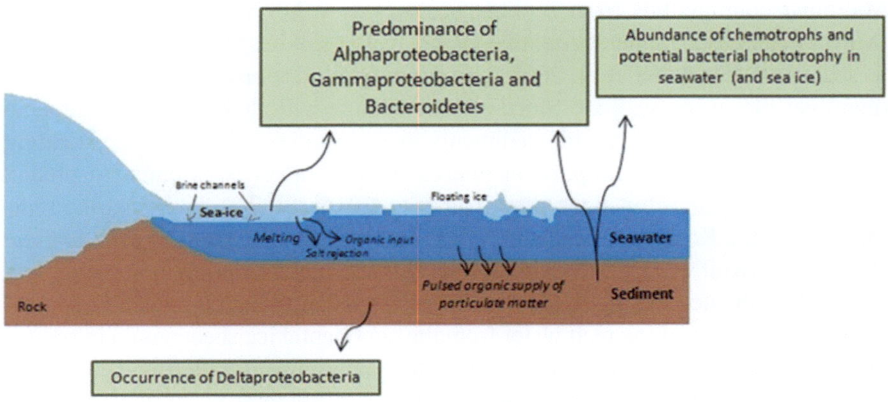

Fig. 6.5 Schematic view of the main processes and bacterial phyla occurring in Antarctic seawater, sea-ice and sediment

References

Abele, D., Vazquez, S., Buma, A. G. J., Hernandez, E., Quiroga, C., Held, C., Frickenhaus, S., Harms, L., Lopez, J. L., Helmke, E., & Mac Cormack, W. P. (2017). Pelagic and benthic communities of the Antarctic ecosystem of Potter Cove: Genomics and ecological implications. *Marine Genomics, 33*, 1–11.

Alder, V., Azzaro, M., Hucke-Gaete, R., Mosetti, R., Orgeira, J. L., Quartino, L., Rey, A. R., Schejter, L., Secchione, M., Marschoff, E. R. (2017). Southern Ocean. In *First Global Marine Assessment, Chapter: 36H, Oceans and Law of the Sea, United Nations* (p. 41).

Amano-Sato, C., Akiyama, S., Uchida, M., Shimada, K., & Utsumi, M. (2013). Archaeal distribution and abundance in water masses of the Arctic Ocean, Pacific sector. *Aquatic Microbial Ecology, 69*, 101–112.

Arrigo, K. R. (2005). Marine microorganisms and global nutrient cycles. *Nature, 437*, 349–355.

Arrigo, K. R., Worthen, D. L., Lizotte, M. P., Dixon, P., & Dieckmann, G. (1997). Primary production in Antarctic sea ice. *Science, 276*, 394–397.

Arrigo, K. R., van Dijken, G. L., & Bushinsky, S. (2008). Primary production in the Southern Ocean, 1997–2006. *Journal of Geophysical Research, 113*, C08004.

Atkinson, A., Ward, P., Hunt, B. P. V., Pakhomov, E. A., & Hosie, G. W. (2012). An overview of southern ocean zooplankton data: Abundance, biomass, feeding and functional relationships. *CCAMLR Science, 19*, 171–218.

Azam, F., Smith, D. C., & Hollibaugh, J. T. (1991). The role of the microbial loop in Antarctic pelagic ecosystems. *Polar Research, 10*, 239–244.

Azzaro, M., La Ferla, R., & Azzaro, F. (2006). Microbial respiration in the aphotic zone of the Ross Sea (Antarctica). *Marine Chemistry, 99*, 199–209.

Baldi, F., Marchetto, D., Pini, F., Fani, R., Michaud, L., Lo Giudice, A., Berto, D., & Giani, M. (2010). Biochemical and microbial features of shallow marine sediments along the Terra Nova Bay (Ross Sea, Antarctica). *Continental Shelf Research, 30*, 1614–1625.

Béja, O., Koonin, E. V., Aravind, L., Taylor, L. T., Seitz, H., Stein, J. L., Bensen, D. C., Feldman, R. A., Swanson, R. V., & Delong, E. F. (2002). Comparative genomic analysis of archaeal genotypic variants in a single population and in two different oceanic provinces. *Applied and Environmental Microbiology, 68*, 335–345.

Bowman, J. S. (2015). The relationship between sea ice bacterial community structure and biogeochemistry: A synthesis of current knowledge and known unknowns. *Elementary Science Anthology, 3*, 72.

Bowman, J. P., & McCuaig, R. D. (2003). Biodiversity, community structural shifts, and biogeography on prokaryotes within Antarctic continental shelf sediment. *Applied and Environmental Microbiology, 69*, 2463–2483.

Bowman, J. P., & Nichols, D. S. (2005). Novel members of the family Flavobacteriaceae from Antarctic maritime habitats including *Subsaximicrobium wynnwilliamsii* gen. nov., sp. nov., *Subsaximicrobium saxinquilinus* sp. nov., *Subsaxibacter broadyi* gen. nov., sp. nov., *Lacinutrix copepodicola* gen. nov., sp. nov., and novel species of the genera *Bizionia*, *Gelidibacter* and *Gillisia*. *International Journal of Systematic and Evolutionary Microbiology, 55*, 1471–1486.

Bowman, J. P., McCammon, S. A., Brown, M. V., Nichols, D. S., & McMeekin, T. A. (1997). Diversity and association of psychrophilic bacteria in Antarctic sea ice. *Applied and Environmental Microbiology, 63*, 3068–3078.

Bowman, J. P., McCammon, S. A., Gibson, J. A. E., Robertson, L., & Nichols, P. D. (2003). Prokaryotic metabolic activity and community structure in Antarctic continental shelf sediments. *Applied and Environmental Microbiology, 69*, 2448–2462.

Brinkmeyer, R., Knittel, K., Jurgens, J., Weyland, H., Amann, R., & Helmke, E. (2003). Diversity and structure of bacterial communities in Arctic versus Antarctic pack ice. *Applied and Environmental Microbiology, 69*, 6610–6619.

Brown, M. V., & Bowman, J. P. (2001). A molecular phylogenetic survey of sea-ice microbial communities (SIMCO). *FEMS Microbiology Ecology, 35*, 267–275.

Bruni, V., Gugliandolo, C., Maugeri, T., & Allegra, A. (1999). Psychrotrophic bacteria from a coastal station in the Ross Sea (Terra Nova Bay, Antarctica). *Microbiologica, 22*, 357–363.

Carr, S. A., Vogel, S. W., Dunbar, R. B., Brandes, J., Spea, J. R., Levy, R., Naish, T. R., Powell, R. D., Wakeham, S. G., & Mandernack, K. W. (2013). Bacterial abundance and composition in marine sediments beneath the Ross Ice Shelf, Antarctica. *Geobiology, 11*, 377–395.

Carr, S. A., Orcutt, B. N., Mandernack, K. W., & Spear, J. R. (2015). Abundant *Atribacteria* in deep marine sediment from the Adelie Basin, Antarctica. *Frontiers in Microbiology, 6*, 872.

Celussi, M., Cataletto, B., Fonda Umani, S., & Del Negro, P. (2009a). Depth profiles of bacterioplankton assemblages and their activities in the Ross Sea. *Deep Sea Research, 56*, 2193–2205.

Celussi, M., Paoli, A., Crevatin, E., Bergamasco, A., Margiotta, F., Saggiomo, V., Fonda Umani, S., & Del Negro, P. (2009b). Short-term under-ice variability of prokaryotic plankton communities in coastal Antarctic waters (Cape Hallett, Ross Sea). *Estuarine Coastal and Shelf Science, 81*, 491–500.

Celussi, M., Bergamasco, A., Cataletto, B., Fonda Umani, S., & Del Negro, P. (2010). Water masses' bacterial community structure and microbial activities in the Ross Sea, Antarctica. *Ant Science, 22*, 361–370.

Cho, B. C., & Azam, F. (1988). Major role of bacteria in biogeochemical fluxes in the ocean's interior. *Nature, 332*, 441–443.

Church, M. J., DeLong, E. F., Ducklow, H. W., Karner, M. B., Preston, C. M., & Karl, D. M. (2003). Abundance and distribution of planktonic Archaea and Bacteria in the waters west of the Antarctic Peninsula. *Limnology and Oceanography, 48*, 1893–1902.

Collins, R. E., Rocap, G., & Deming, J. W. (2010). Persistence of bacterial and archaeal communities in sea ice through an Arctic winter. *Environmental Microbiology, 12*, 1828–1841.

Cowie, R. O., Maas, E. W., & Ryan, K. G. (2011). Archaeal diversity revealed in Antarctic sea ice. *Antarctic Science, 23*, 531–536.

Depoorter, M. A., Bamber, J. L., Griggs, J. A., Lenaerts, J. T., Ligtenberg, S. R., van den Broeke, M. R., & Moholdt, G. (2013). Calving fluxes and basal melt rates of Antarctic ice shelves. *Nature, 502*, 89–92.

Ducklow, H., Carlson, C., Church, M., Kirchman, D., Smith, D., & Steward, G. (2001). The seasonal development of the bacterioplankton bloom in the Ross Sea, Antarctica, 1994–1997. *Deep Sea Research II, 48*, 4199–4221.

Ducklow, H. W., Fraser, W., Karl, D. M., Quetin, L. B., Ross, R. M., Smith, R. C., Stammerjohn, S. E., Vernet, M., & Daniels, R. M. (2006). Water-column processes in the West Antarctic Peninsula and the Ross Sea: Interannual variations and food web structure. *Deep Sea Research II, 53*, 834–852.

Edwards, K. J., Becker, K., & Colwell, F. (2012). The deep, dark energy biosphere: Intraterrestrial life on earth. *Annual Review of Earth and Planetary Sciences, 40*, 551–568.

Eronen-Rasimus, E., Luhtanen, A. M., Rintala, J. M., Delille, B., Dieckmann, G., Karkman, A., & Tison, J. L. (2017). An active bacterial community linked to high *chl*-a concentrations in Antarctic winter-pack ice and evidence for the development of an anaerobic sea-ice bacterial community. *The ISME Journal, 11*, 2345–2355.

Evans, C., Thomson, P. G., Davidson, A. T., Bowie, A. R., Van Den Enden, R., Witte, H., & Brussaard, C. P. D. (2011). Potential climate change impacts on microbial distribution and carbon cycling in the Australian Southern Ocean. *Deep Sea Research II, 58*, 2150–2161.

Fabiano, M., & Danovaro, R. (1998). Enzymatic activity, bacterial distribution, and organic matter composition in sediments of the Ross Sea (Antarctica). *Applied and Environmental Microbiology, 64*, 3838–3845.

Fabiano, M., & Pusceddu, A. (1998). Total and hydrolyzable particulate organic matter (carbohydrates, proteins and lipids) at a coastal station in Terra Nova Bay (Ross Sea, Antarctica). *Polar Biology, 19*, 125–132.

Fernández-Méndez, M., Wenzhöfer, F., Peeken, I., Sørensen, H. L., Glud, R. N., et al. (2014). Composition, buoyancy regulation and fate of ice algal aggregates in the Central Arctic Ocean. *PLoS One, 9*, e107452.

Fonda Umani, S., Montia, M., Bergamasco, A., Cabrini, M., De Vittor, C., Burba, N., & Del Negro, P. (2005). Plankton community structure and dynamics versus physical structure from Terra Nova Bay to Ross Ice Shelf (Antarctica). *Journal of Marine Systems, 55*, 31–46.

Fuhrman, J. A., & Azam, F. (1980). Bacterioplankton secondary production estimates for coastal waters of British Columbia, Antarctica, and California. *Applied and Environmental Microbiology, 39*, 1085–1095.

Gentile, G., Giuliano, L., D'Auria, G., Smedile, F., Azzaro, M., De Domenico, M., & Yakimov, M. M. (2006). Study of bacterial communities in Antarctic coastal waters by a combination of 16S rRNA and 16S rDNA sequencing. *Environmental Microbiology, 8*, 2150–2161.

Ghiglione, J., & Murray, A. (2012). Pronounced summer to winter differences and higher wintertime richness in coastal Antarctic marine bacterioplankton. *Environmental Microbiology, 14*, 617–629.

Gillan, D. C., & Danis, B. (2007). The archaebacterial communities in Antarctic bathypelagic sediments. *Deep Sea Research, I54*, 1682–1690.

Grossmann, S., & Dieckmann, G. S. (1994). Bacterial standing stock, activity, and carbon production during formation and growth of sea ice in Weddell Sea, Antarctica. *Applied and Environmental Microbiology, 60*, 2746–2753.

Grzymski, J. J., Carter, B. J., DeLong, E. F., Feldman, R. A., Ghadiri, A., & Murray, A. E. (2006). Comparative genomics of DNA fragments from six Antarctic marine planktonic bacteria. *Applied and Environmental Microbiology, 72*, 1532–1541.

Grzymski, J. J., Riesenfeld, C. S., Williams, T. J., Dussaq, A. M., Ducklow, H., Erickson, M., et al. (2012). A metagenomic assessment of winter and summer bacterioplankton from Antarctica Peninsula coastal surface waters. *The ISME Journal, 6*, 1901–1915.

Helmke, E., & Weyland, H. (1995). Bacteria in sea ice and underlying water of the eastern Weddel sea in midwinter. *Marine Ecology Progress Series, 117*, 269–287.

Hernández, E. A., Piquet, A. M.-T., Lopez, J. L., Buma, A. G., & Mac Cormack, W. P. (2015). Marine archaeal community structure from Potter Cove Antarctica: High temporal and spatial dominance of the phylum Thaumarchaeota. *Polar Biology, 38*, 117–130.

Isla, E., Masqué, P., Palanques, A., Sanchez-Cabeza, J. A., Bruach, J. M., Guillen, J., & Puig, P. (2002). Sediment accumulation rates and carbon burial in the bottom sediment in a high-productivity area: Gerlache Strait (Antarctica). *Deep Sea Research, 49*, 3275–3287.

Isla, E., Masqué, P., Palanques, A., Guillén, J., Puig, P., & Sanchez-Cabeza, J. A. (2004). Sedimentation of biogenic constituents during the last century in western Bransfield and Gerlache Straits, Antarctica: A relation to currents, primary production, and sea floor relief. *Marine Geology, 209*, 265–277.

Isla, E., Rossi, S., Palanques, A., Gili, J. M., Gerdes, D., & Amtz, W. (2006). Biochemical composition of marine sediment from the eastern Weddell Sea (Antarctica): High nutritive value in a high benthic-biomass environment. *Journal of Marine Systems, 60*, 255–267.

Jamieson, R. E., Heywood, J. L., Rogers, A. D., Billett, D. S. M., & Pearce, D. A. (2013). Bacterial biodiversity in deep-sea sediments from two regions of contrasting surface water productivity near the Crozet Islands, Southern Ocean. *Deep Sea Research I, 75*, 67–77.

Karl, D. M. (1993). Microbial processes in the southern oceans. In E. I. Friedmann (Ed.), *Antarctic microbiology* (pp. 1–63). New York: Wiley.

Krembs, C., Eicken, H., Junge, K., & Deming, J. (2002). High concentrations of exopolymeric substances in Arctic winter sea ice: Implications for the polar ocean carbon cycle and cryoprotection of diatoms. *Deep Sea Research Part I: Oceanographic Research Papers, 49*, 2163–2181.

Landone Vescovo, I. A., Golemba, M. D., Di Lello, F. A., Culasso, A. C., Levin, G., Ruberto, L., MacCormack, W. P., & López, J. L. (2014). Rich bacterial assemblages from Maritime Antarctica (Potter Cove, South Shetlands) reveal several kinds of endemic and undescribed phylotypes. *Revista Argentina de Microbiología, 46*, 218–230.

Learman, D. R., Henson, M. W., Thrash, J. C., Temperton, B., Brannock, P. M., Santos, S. R., Mahon, A. R., & Halanych, K. M. (2016). Biogeochemical and microbial variation across

5500 km of Antarctic surface sediment implicates organic matter as a driver of benthic community structure. *Frontiers in Microbiology, 7*, 284.
Lo Giudice, A., & Fani, R. (2015). Cold-adapted bacteria from the coastal Ross Sea (Antarctica): Linking microbial ecology to biotechnology. *Hydrobiologia, 761*, 417–441.
Lo Giudice, A., Brilli, M., Bruni, V., De Domenico, M., Fani, R., & Michaud, L. (2007). Bacterium-bacterium inhibitory interactions among psychrotrophic bacteria isolated from Antarctic seawaters (Terra Nova Bay, Ross Sea). *FEMS Microbiology Ecology, 60*, 383–396.
Lo Giudice, A., Caruso, C., Mangano, S., Bruni, V., De Domenico, M., & Michaud, L. (2012). Marine bacterioplankton diversity and community composition in an Antarctic coastal environment. *Microbial Ecology, 63*, 210–223.
Lo Giudice, A., Casella, P., Bruni, V., & Michaud, L. (2013). Response of bacterial isolates from Antarctic shallow sediments towards heavy metals, antibiotics and polychlorinated biphenyls. *Ecotoxicology, 22*, 240–250.
Luria, C. M., Ducklow, H. W., & Amaral-Zettler, L. A. (2014). Marine bacterial, archaeal and eukaryotic diversity and community structure on the continental shelf of the western Antarctic Peninsula. *Aquatic Microbial Ecology, 73*, 107–121.
Maas, E. W., Simpson, A. M., Martin, A., Thompson, S., Koh, E. Y., Davy, S. K., Ryan, K. G., & O'Toole, R. F. (2012). Phylogenetic analyses of bacteria in sea ice at Cape Hallett, Antarctica. *New Zealand Journal of Marine and Freshwater Research, 46*, 3–12.
Manganelli, M., Malfatti, F., Samo, T. J., Mitchell, B. G., Wang, H., & Azam, F. (2009). Major role of microbes in carbon fluxes during austral winter in the southern Drake Passage. *PLoS One, 4*, e6941.
Masquè, P., Isla, E., Sanchez-Cabeza, J. A., Palanques, A., Bruach, J. M., Puig, P., & Guillèn, J. (2002). Sediment accumulation rates and carbon fluxes to bottom sediments at the Western Bransfield Strait (Antarctica). *Deep Sea Research, 49*, 921–933.
Matos, M. N., Lozada, M., Anselmino, L. E., Musumeci, M. A., Henrissat, B., Jansson, J. K., Mac Cormack, W. P., Carroll, J., Sjöling, S., Lundgren, L., & Dionisi, H. M. (2016). Metagenomics unveils the attributes of the alginolytic guilds of sediments from four distant cold coastal environments. *Environmental Microbiology, 18*, 4471–4484.
Maugeri, T. L., Gugliandolo, C., & Bruni, V. (1996). Heterotrophic bacteria in the Ross Sea (Terra Nova Bay, Antarctica). *New Microbiologica, 19*, 67–76.
Mayor, D. J., Thornton, B., Hay, S., Zuur, A. F., Nicol, G. W., McWilliam, J. M., & Witte, U. F. M. (2012). Resource quality affects carbon cycling in deep-sea sediments. *The ISME Journal, 6*, 1740–1748.
Morata, N., Poulin, M., & Renaud, P. E. (2010). A multiple biomarker approach to tracking the fate of an ice algal bloom to the sea floor. *Polar Biology, 34*, 101–112.
Murray, A. E., & Grzymski, J. J. (2007). Diversity and genomics of Antarctic marine microorganisms. *Philosophical Transactions of the Royal Society B, 362*, 2259–2271.
Murray, A. E., Preston, C. M., Massana, R., Taylor, L. T., Blakis, A., Wu, K., et al. (1998). Seasonal and spatial variability of bacterial and archaeal assemblages in the coastal waters near Anvers Island, Antarctica. *Applied and Environmental Microbiology, 64*, 2585–2595.
Nagata, T., Fukuda, H., Fukuda, R., & Koike, I. (2000). Bacterioplankton distribution and production in deep Pacific waters: Large-scale geographic variations and possible coupling with sinking particle fluxes. *Limnology and Oceanography, 45*, 426–435.
Pearce, D. A. (2012). Extremophiles in Antarctica: Life at low temperatures. In H. Stan-Lotter & S. Fendrihan (Eds.), *Adaption of microbial life to environmental extremes* (pp. 87–118). Vienna: Springer.
Pearce, I., Davidson, A. T., Bell, E. M., & Wright, S. (2007). Seasonal changes in the concentration and metabolic activity of bacteria and viruses at an Antarctic coastal site. *Aquatic Microbial Ecology, 47*, 11–23.
Peck, L. S., Barnes, D. K. A., Cook, A. J., Fleming, A. H., & Clarke, A. (2010). Negative feedback in the cold: Ice retreat produces new carbon sinks in Antarctica. *Global Change Biology, 16*, 2614–2623.

Piquet, A. M., Bolhuis, H., Davidson, A. T., & Buma, A. G. (2010). Seasonal succession and UV sensitivity of marine bacterioplankton at an Antarctic coastal site. *FEMS Microbiology Ecology, 73*, 68–82.

Powell, S. M., Bowman, J. P., Snape, I., & Stark, J. S. (2003). Microbial community variation in pristine and polluted nearshore Antarctic sediments. *FEMS Microbiology Ecology, 45*, 135–145.

Rodríguez-Marconi, S., De la Iglesia, R., Díez, B., Fonseca, C. A., Hajdu, E., & Trefault, N. (2015). Characterization of bacterial, archaeal and eukaryote symbionts from Antarctic sponges reveals a high diversity at a three-domain level and a particular signature for this ecosystem. *PLoS One, 10*, e0138837.

Ruff, S. E., Probandt, D., Zinkann, A. C., Iversen, M. H., Klaas, C., Wurzberg, L., et al. (2014). Indications for algae-degrading benthic microbial communities in deep-sea sediments along the Antarctic Polar Front. *Deep Sea Research I, 108*, 6–16.

Sarmiento, J. L., & Gruber, N. (2006). *Ocean biogeochemical dynamics*. Princeton: Princeton University Press.

Schloss, I. R., Wasilowska, A., Dumont, D., Almandoz, G. O., Hernando, M. P., Michaud-Tremblay, C.-A., Saravia, L., Rzepecki, M., Monien, P., Monien, D., Kopczyńska, E. E., Bers, A. V., & Ferreyra, G. A. (2014). On the phytoplankton bloom in coastal waters of southern King George Island (Antarctica) in January 2010: An exceptional feature? *Limnology and Oceanography, 59*, 195–210.

Signori, C. N., Thomas, F., Enrich-Prast, A., Pollery, R. C., & Sievert, S. M. (2014). Diversity and community structure across environmental gradients in Bransfield Strait, Western Antarctic Peninsula. *Frontiers in Microbiology, 5*, 647.

Smith, D. C., Simon, M., Alldredge, A. L., & Azam, F. (1992). Intense hydrolytic enzyme activity on marine aggregates and implications for rapid particle dissolution. *Nature, 359*, 139–142.

Thomas, D. N., & Dieckmann, G. S. (2002). Antarctic sea ice-a habitat for extremophiles. *Science, 295*, 641.

Tolar, B. B., Ross, M. J., Wallsgrove, N. J., Liu, Q., Aluwihare, L. I., Popp, B. N., & Hollibaugh, J. T. (2016). Contribution of ammonia oxidation to chemoautotrophy in Antarctic coastal waters. *The ISME Journal, 10*, 2605–2619.

Torstensson, A., Dinasquet, J., Chierici, M., Fransson, A., Riemann, L., & Wulff, A. (2015). Physicochemical control of bacterial and protist community composition and diversity in Antarctic sea ice. *Environmental Microbiology, 17*, 3869–3881.

Underwood, G. J. C., Fietz, S., Papadimitriou, S., et al. (2010). Distribution and composition of dissolved extracellular polymeric substances (EPS) in Antarctic sea ice. *Marine Ecology Progress Series, 404*, 1–19.

Wilkins, D., Lauro, F. M., Williams, T. J., Demaere, M. Z., Brown, M. V., Hoffman, J. M., Andrews-Pfannkoch, C., McQuaid, J. B., Riddle, M. J., Rintoul, S. R., & Cavicchioli, R. (2013). Biogeographic partitioning of Southern Ocean microorganisms revealed by metagenomics. *Environmental Microbiology, 15*, 1318–1333.

Williams, T. J., Long, E., Evans, F., DeMaere, M. Z., Lauro, F. M., Raftery, M. J., et al. (2012). A metaproteomic assessment of winter and summer bacterioplankton from Antarctic Peninsula coastal surface waters. *The ISME Journal, 6*, 1883–1900.

Wölfl, A. C., Lim, C. H., Hass, H. C., Lindhorst, S., Tosonotto, G., Lettmann, K. A., Kuhn, G., Wolff, J.-O., & Abele, D. (2014). Distribution and characteristics of marine habitats in a subpolar bay based on hydroacoustics and bed shear stress estimates—Potter Cove, King George Island, Antarctica. *Geo-Marine Letters, 34*, 435–446.

Yakimov, M. M., Gentile, G., Bruni, V., Cappello, S., D'Auria, G., Golyshin, P. N., & Giuliano, L. (2004). Crude oil-induced structural shift of coastal bacterial communities of Rod Bay (Terra Nova Bay, Ross Sea) and characterization of cultured cold-adapted hydrocarbonoclastic bacteria. *FEMS Microbiology Ecology, 49*, 419–432.

Yu, Y., Li, H.-R., Zeng, Y.-X., & Chen, B. (2011). Bacterial diversity and bioprospecting for cold-active hydrolytic enzymes from culturable bacteria associated with sediment from Nella Fjord, Eastern Antarctica. *Marine Drugs, 9*, 184–195.

Chapter 7
Antarctic Soil Microbial Communities in a Changing Environment: Their Contributions to the Sustainability of Antarctic Ecosystems and the Bioremediation of Anthropogenic Pollution

Cecilia G. Flocco, Walter P. Mac Cormack, and Kornelia Smalla

Abstract Antarctica, one of the most harsh and isolated environments on Earth, is concurrently facing the shaping forces of global environmental changes and the impact of human activities in the continent, the latter represented by the operation of scientific stations and touristic expeditions which have significantly expanded over the last decades. Although still largely preserved, different types and levels of contamination of soils, sediments, and coastal waters were detected, being petroleum-derived hydrocarbons the predominant pollutants. Since the environmental protocol to the Antarctic Treaty prohibits the application of site-disruptive cleanup methods and the introduction of foreign species, studying the natural capacity of indigenous microbial communities to degrade pollutants (bioremediation) emerges as a key step in the design of suitable in situ pollution remediation strategies. This chapter describes the interactions of environmental and anthropogenic

C. G. Flocco (✉)
Julius Kühn-Institut (JKI), Federal Research Centre for Cultivated Plants, Braunschweig, Germany

Instituto Antártico Argentino (IAA), Buenos Aires, Argentina

Instituto de Nanobiotecnología (NANOBIOTEC), Universidad de Buenos Aires- Consejo Nacional de Investigaciones Científicas y Técnicas (CONICET), Buenos Aires, Argentina

W. P. Mac Cormack
Instituto Antártico Argentino (IAA), Buenos Aires, Argentina

Instituto de Nanobiotecnología (NANOBIOTEC), Universidad de Buenos Aires- Consejo Nacional de Investigaciones Científicas y Técnicas (CONICET), Buenos Aires, Argentina

K. Smalla
Julius Kühn-Institut (JKI), Federal Research Centre for Cultivated Plants, Braunschweig, Germany

S. Castro-Sowinski (ed.), *The Ecological Role of Micro-organisms in the Antarctic Environment*, Springer Polar Sciences,
https://doi.org/10.1007/978-3-030-02786-5_7

sources of disturbance to Antarctic ecosystems and the recent advances in our understanding of the ecological contributions of indigenous soil prokaryote communities, delving into their capacity to remediate polluted terrestrial environments. The aims are to highlight the crucial contributions of soil microbes to the functioning and sustainability of Antarctic ecosystems and enrich the body of knowledge supporting the engineering of bioremediation strategies for the restoration of terrestrial environments affected by anthropogenic sources of pollution.

Keywords Soil microbial communities · Anthropogenic disturbances · Chemical contamination · Cold-temperature bioremediation · Gene flow

7.1 Introduction

Unlike the northern polar regions, in which permanent settlements, mining, trade, and fluid intra- and intercontinental transit are established (Poland et al. 2003), Antarctica is still a largely isolated and inhabited region. The human presence is restricted to scientific and logistic crews temporarily living in research stations and visitors arriving with touristic expeditions. The Antarctic Treaty and its instruments regulate the international relations and activities in the region, prohibiting the militarization and the exploitation of natural resources south of 60 °S (Convey et al. 2012).

Antarctica is covered by a kilometer-deep ice layer, with only 0.34% (44,000 km^2) of its continental area not permanently layered by snow or ice (Cowan et al. 2014). These comparatively small sections of exposed soil concentrate most of the terrestrial biota and are mainly located on coastal strips around the Antarctic Peninsula, its offshore islands, and some scattered continental regions, comprising as well the nunataks (isolated mountain peaks or ridges) emerging from the circumventing ice masses (Bottos et al. 2014; Terauds and Lee 2016). Figure 7.1 shows an example of a coastal landscape located in the Maritime Antarctic region. Although Antarctic terrestrial environments are by no means homogeneous (Cowan et al. 2014; Terauds and Lee 2016), the predominant environmental conditions (extremely cold temperatures, strong winds, sharp seasonality in light provision, frequent freeze-thaw cycles, and low water and nutrient availability) preclude the existence of most plant and animal species, defining the reduced complexity that terrestrial Antarctic ecosystems exhibit (Wall 2007; Chan et al. 2013; Convey 2013). The vegetation is composed of mosses, lichens, liverworts, and just two vascular plant species (Fig. 7.1), *Deschampsia antarctica* (Antarctic grass) and the cushion-forming *Colobanthus quitensis* (Antarctic pearlwort), which are found along the ice-free areas of the Antarctic peninsula and subantarctic islands (Day et al. 1999; Lewis Smith 2003; Flocco et al. 2009; Teixeira et al. 2010; Yergeau 2013). Microorganisms, adapted to the extreme and fluctuating environmental conditions, are the dominant biota and play a key role in the ecology and sustainability of Antarctic terrestrial ecosystems (Vishniac 1993; Wynn-Williams 1996; Cowan et al. 2014).

Fig. 7.1 Landscape and some vegetation specimens of the Maritime Antarctica, King George Island (Isla 25 de Mayo), South Shetland Islands, Antarctica. (**a**) Panoramic view of Carlini Research Station located in Potter Peninsula, facing the ice debris-sprinkled waters of the Potter Cove; the Three Brothers Hill, the remnant neck of an extinct volcano, is seen on the background. (**b**) The cushion-forming *Colobanthus quitensis* and (**c**) the Antarctic grass *Deschampsia antarctica*, the only two vascular plants growing in Antarctica (bars represent 1 cm). (Photos: **a**, courtesy of the icebreaker vessel ARA Irizar`s crew; **b** and **c**, C.G. Flocco)

Although still largely preserved, the natural environment is changing rapidly in some Antarctic regions due to concomitant factors: global environmental changes and the increasing impact of human activities on-site. The diverse impacts of those shaping forces on Antarctic habitats and ecosystems were documented over the last decades (Bargagli 2005, 2008; Frenot et al. 2005; Convey and Smith 2006), but their long-term consequences are largely unknown and represent a source of growing concern. This chapter reviews the main sources and interactions of anthropogenic and environmental disturbance to Antarctic ecosystems and delves into the long-hidden biodiversity of soil microbial communities and their crucial contributions to the sustainability of Antarctic ecosystems, focusing on the microbe-driven remediation (bioremediation) of chemically polluted terrestrial environments.

7.2 Antarctic Soil Microbial Communities

7.2.1 Contributions to Ecosystems, Biotechnological Applications, and Bioprospecting

Antarctic microbes developed a wide range of highly specialized strategies that allow them to survive and thrive under the harsh and changing conditions that shape life in the continent. Those adaptation mechanisms include habitat selection, life cycle strategies, changes in cellular composition and enzyme activity, and the production of extracellular polymeric substances (an in-depth review on the subject is provided by Vincent 1988 and recent advances by De Maayer et al. 2014).

Ongoing research efforts (reviewed in Sect. 7.2.3), progressing from phylogenetic description to functional and ecological analysis, advance our understanding of the crucial contributions of soil microorganisms to the sustainability of Antarctic terrestrial habitats and unravel their vital links to the health of sediment and marine ecosystems (Vishniac 1993; Wynn-Williams 1996; Wall 2007; Vincent 2002; Howard-Williams et al. 2010; Quartino et al. 2015; Cowan et al. 2011, 2014; De Maayer et al. 2014; Abele et al. 2017; Thompson et al. 2017). Antarctic microorganisms are energy producers, drive biogeochemical cycles, and sustain niche ecosystems by creating microhabitat conditions (Cowan et al. 2014 and references therein). For example, *Cyanobacteria* and algae are often the primary energy producers (Vincent 2002; Pandey et al. 2004), a crucial contribution given the scarcity or absence of vegetation characteristic of the harshest Antarctic terrestrial ecosystems (Wall 2007). Some members of this phylum (*Nostoc* spp.) express photosynthetic activity at subzero temperature and are involved in nitrogen cycling (Pandey et al. 2004). Several cold-adapted microorganisms are producers of enzymes and polysaccharides crucial to their survival under extreme environmental stress, which also support the growth of other organisms by creating an enabling habitat (De Maayer et al. 2014). A detailed description of the contributions of Antarctic microorganisms to biogeochemical cycles and niche-habitat strategies is presented in other chapters included in this book.

Due to their remarkable adaptive features, Antarctic terrestrial and marine microbial communities represent an invaluable and largely untapped source of novel microorganisms and biomolecules of biotechnological interest, which are the target of numerous bioprospecting studies (Lohan and Johnston 2005; Bercovich et al. 2008; Berlemont et al. 2011; de Pascale et al. 2012; Godinho et al. 2013; Aran et al. 2014; Bosi et al. 2017; Martorell et al. 2017; Musumeci et al. 2017). For example, the production of antibiotics by members of Actinomycetes, a feature often described in temperate environments, is also a frequent finding associated to Antarctic representatives of this taxon (Bruntner et al. 2005; Tian et al. 2017). Marine bacteria belonging to the genus *Pseudoalteromonas* are known for their production of biomolecules with marked antimicrobial and antitumor activity (Bosi et al. 2017), with promising applications in human health. The microbial enzymatic machinery and highly specialized molecular mechanisms of tolerance to cold and chilling are intensively studied for applications in the food industry (Vázquez et al. 2004; Pulicherla et al. 2011; Martínez-Rosales et al. 2012). In addition, several cold-adapted indigenous microorganisms pose the metabolic capacity to degrade environmental contaminants rendering them a central component of bioremediation strategies, one of the few remediation options compatible with the stringent environmental and regulatory conditions pertaining to Antarctica, as it is further discussed in this chapter.

7.2.2 *Links to the Sustainability of Antarctic Ecosystems: A Dual Role*

Considering the concepts reviewed in the preceding section, it becomes apparent that the crucial contributions of microbial communities to the sustainability of Antarctic ecosystems, at a landscape level, are exerted in a dual way: (1) supporting life under polar conditions through the vital contributions to the functioning of ecosystems and (2) restoring areas affected by anthropogenic sources of contamination, by providing the key catabolic capacity to bioremediation protocols (Fig. 7.2).

In summary, the relatively unexplored microbial diversity thriving in Antarctic soils is essential to the functioning and sustainability of local and linked ecosystems and represents as well a source of unique microorganisms and biomolecules of industrial and pharmacological interest. Therefore, it becomes essential to deploy substantial efforts to characterize these communities to support the design of conservation areas aimed at preserving such a vital and irreplaceable resource. Significant advances in our understanding of the biodiversity and ecology of these unique communities were made over the last decades, enabled by the fast-evolving molecular biology technology applied to microbial research, as reviewed in the next section.

7.2.3 *Unraveling the Hidden Soil Microbial Communities*

Different methods have been used to study the microbial communities of Antarctic soils, and their resolution is constantly improving – although no method is free of a certain degree of bias (Tindall 2004; Logares et al. 2012). The very early efforts to describe microbial communities of Antarctic soils (for a historical perspective see Shivaji and Reddy 2010) retrieved none or only a few microorganisms prompting researchers to depict some environments as predominantly sterile (Horowitz et al. 1969, 1972), a description that fitted to the barren visuals of several Antarctic landscapes. The pioneering microbiological studies employed cultivation-based methods, which are based on techniques aiming to grow microorganisms under controlled laboratory conditions for their isolation and further morphological and

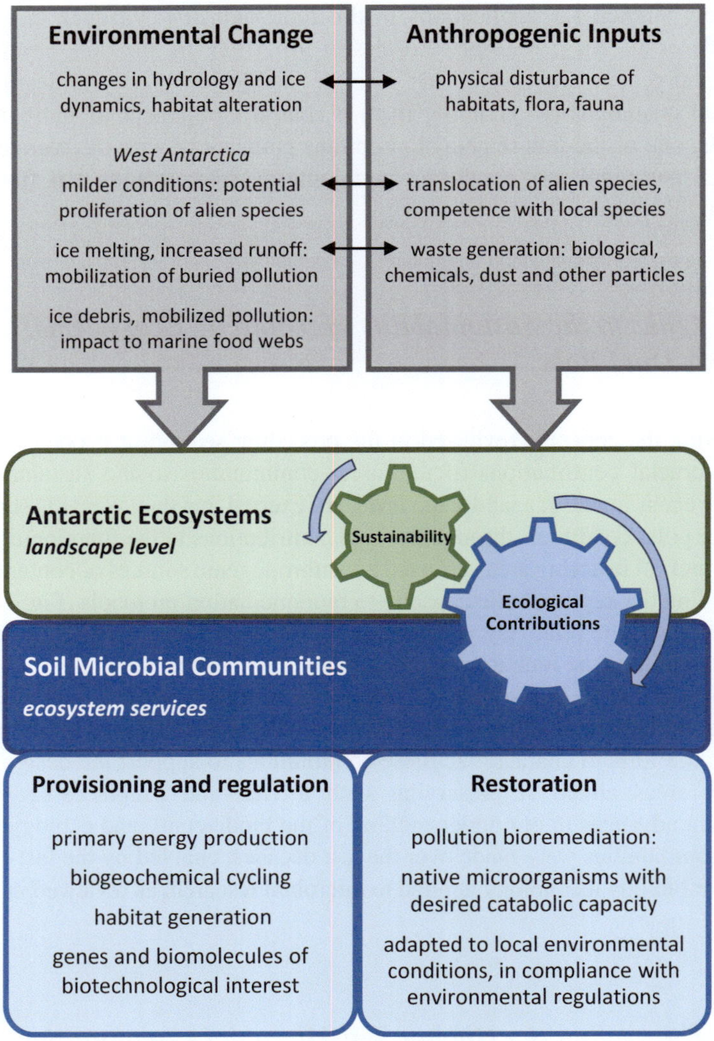

Fig. 7.2 Schematic representation illustrating the ecological contributions of soil microbial communities to the functioning and sustainability of Antarctic ecosystems and the main environmental and anthropogenic shaping forces impacting Antarctic ecosystems; the double-headed black arrows indicate events that can synergistically facilitate impacts to ecosystems. West Antarctica, a region that is facing anthropogenic influence and environmental change forces at a faster pace

physiological characterization (Fry 2004). Although the cultivation techniques have substantially improved over time (Ferrari et al. 2008; Pudasaini et al. 2017) and several Antarctic prokaryotic microorganisms were successfully isolated and characterized (mostly belonging to the phyla *Proteobacteria*, *Actinobacteria*, *Bacteroidetes*, *and Firmicutes*, as summarized by Shivaji and Reddy 2010), the approach still presents several challenges given (1) the slow growth rate of many

Antarctic microorganisms and (2) the difficulties in reproducing in the laboratory the extreme conditions of the selective habitats in which those microorganisms grow (Ferrari et al. 2008).

With the breakthrough and expansion of cultivation-independent molecular methods (Smalla 2004; Schreiter et al. 2015) over the last two decades (such as polymerase chain reaction (PCR) amplification, fingerprinting, hybridization, Sanger sequencing) and the subsequent development of high-throughput sequencing (HTS) technologies (Logares et al. 2012), it was possible to overcome cultivation-associated obstacles and further unveil the rich taxonomic and functional microbial diversity "hidden" in Antarctic habitats (Tindall 2004; Chong et al. 2015 and references therein). These advances are largely based on constantly improving sequencing technologies (454 pyrosequencing, Illumina platform and other HTS methods) applied either to the analysis of gene fragments (e.g., PCR products of specific phylogenetic markers or functional genes) or used in whole approaches capturing the complete genetic information present in a microbial isolate or environmental sample (metagenomics, shotgun sequencing). HTS methods, combined with powerful bioinformatics tools, produced a substantial gain in technology accessibility, processing speed and resolution power (Logares et al. 2012), greatly advancing our understanding of microbial communities, their ecological roles and evolution mechanisms.

Currently, different cultivation-based and cultivation-independent methods and combinations of those are used in polyphasic approaches to study the microbial taxonomic and functional diversity of Antarctic soils (Vázquez et al. 2004, 2017; Flocco et al. 2009; Yergeau et al. 2007, 2012; Shivaji and Reddy 2010; Howard-Williams et al. 2010; Chong et al. 2012; Stomeo et al. 2012; Cowan et al. 2014; Pudasaini et al. 2017). Although the methodology across different studies varied and the different Antarctic bioregions (Terauds and Lee 2016) were not homogeneously covered, some consistent global observations can be retrieved. The array of dominant phyla or classes resembles that of soil communities from other parts of the world (Janssen 2006), with the exception of mineral desert soils (Dry Valleys region) which showed a relatively low biodiversity and a high proportion of uncultured phylotypes (Smith et al. 2006). Bottos et al. (2014) reviewed a large number of microbiological studies carried out in different terrestrial Antarctic environments and concluded that the predominant prokaryotic (bacterial) types in the different Antarctic bioregions include representatives of the phyla *Proteobacteria*, *Actinobacteria*, *Bacteroidetes*, *Firmicutes*, *Acidobacteria*, and *Cyanobacteria* with *Gemmatimonadetes* and *Deinococcus-Thermus* frequently represented in the drier soils of the Dry Valleys region. *Proteobacteria* and *Actinobacteria* are usually the most prominent groups, likely reflecting the metabolic versatility of members belonging to the phyla (Cowan et al. 2014). Other frequent findings were the detection of an important proportion of unclassified phylotypes, suggesting the presence of novel species (Smith et al. 2006; Peeters et al. 2011a; de Pascale et al. 2012) and the existence of high levels of endemism (Cowan et al. 2010; Peeters et al. 2011b), with marked differences in the microbial community composition across relatively close locations within a region (Bottos et al. 2014).

Our preliminary analysis of the soil prokaryote taxonomic diversity (16S rRNA amplification and 454 multiplex sequencing, Flocco et al. unpublished) in different areas of the Maritime Antarctic (Carlini Station and ASPA 132, King George Island), presented in Fig. 7.3, reflects the above described trends, in terms of dominant phyla (*Proteobacteria*, *Actinobacteria*, *Bacteroidetes*, *Firmicutes*, with a significant proportion of unclassified items) and a location effect on the community distribution (differential phyla distribution among sites with similar environmental conditions and soil physicochemical properties, sites PI and PV) but separated by terrain features (Flocco et al. 2009).

7.3 Antarctica, a Changing Environment

7.3.1 Anthropogenic and Environmental Drivers of Change: Interacting Forces

Although still largely preserved, the Antarctic region is concurrently experiencing shaping forces that can affect its highly specialized and vulnerable ecosystems (Frenot et al. 2005) in the short term and potentially produce disturbances – with unforeseen consequences – in the long term.

On the one hand, global driving forces represented by atmospheric and environmental changes are significantly altering local physicochemical conditions of several bioregions (Frenot et al. 2005; Convey and Smith 2006; Yergeau et al. 2012; Yergeau 2013; Quartino et al. 2013; Seefeldt et al. 2017). The different manifestations of environmental change are visible over a relatively recent time span (Convey and Smith 2006; Martín-Español et al. 2016). For example, the altered temperature patterns documented over the last few decades in Antarctica triggered changes in the dynamics of glaciers and other ice masses, with some regions, such as the West Antarctic Peninsula (WAP) registering changes (ice mass loss) at a faster pace, in comparison to other parts of the continent (Cook et al. 2005; Turner et al. 2010; Martín-Español et al. 2016). Contrasting with this, but being part of the same global atmospheric phenomenon, increased snowfall in East Antarctica has contributed to the thickening of the ice layer, balancing in part the total ice-mass losses in the continent (Martín-Español et al. 2016). The described environmental changes produce profound habitat alterations and consequently, affect the dynamics of the ecosystems they host in different ways (Frenot et al. 2005; Convey and Smith 2006; Wall 2007; Yergeau et al. 2012).

On the other hand, the expanding anthropogenic influence on-site, represented by the activities at the scientific stations and a steadily growing tourism industry (IAATO 2017), unavoidably translates into an increased risk of disturbance to Antarctic ecosystems, with consequences that are not yet fully elucidated (Frenot et al. 2005; Tin et al. 2009; Hughes et al. 2015), representing a source of growing concern (Bargagli 2008; Hughes and Bridge 2010b; Hughes et al. 2015).

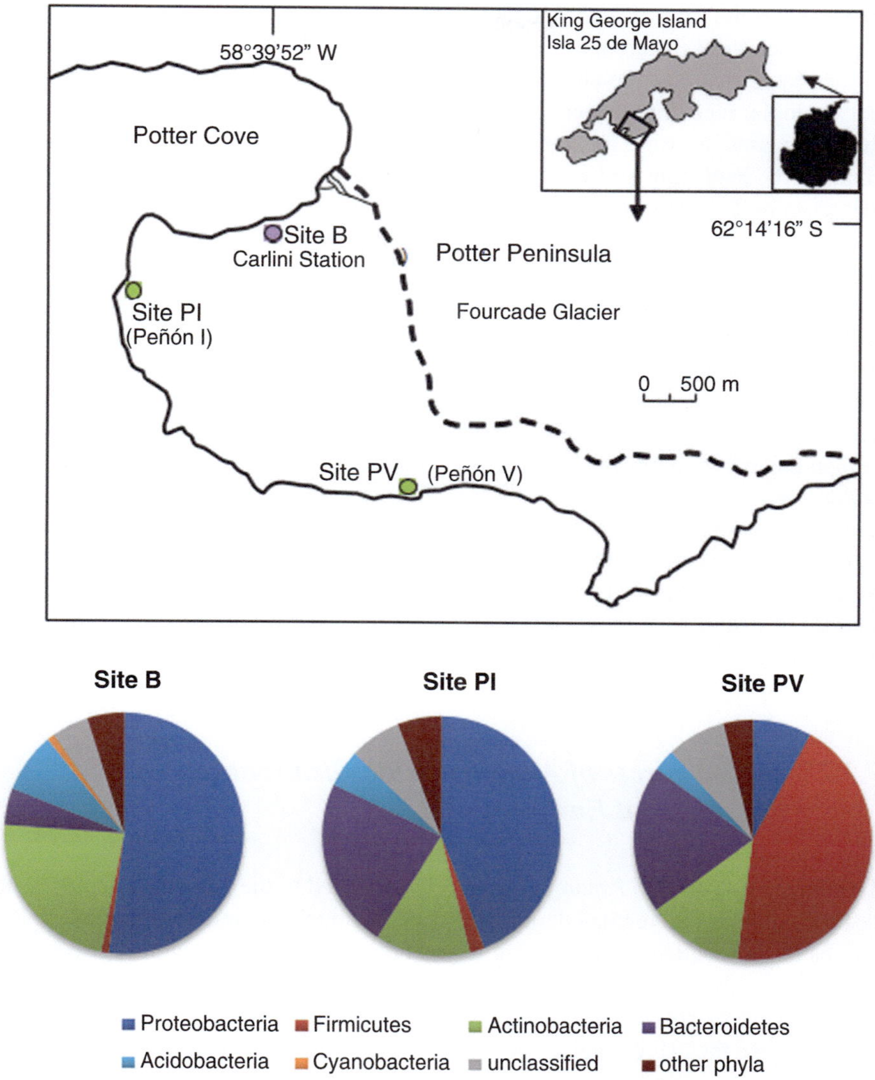

Fig. 7.3 Bacterial community composition of Antarctic soils from Potter Peninsula and Carlini station (King George Island/Isla 25 de Mayo, South Shetland Islands, Maritime Antarctic) at the phylum level (analyzed by multiplex pyrosequencing). Experimental sites are indicated as follows: (B) fuel-contaminated site within the area of the research station; (PI) and (PV) are non-contaminated sites within the adjacent protected area ASPA 132. Other phyla include: *Chlamydiae, Chloroflexi, Deinococcus-Thermus, Gemmatimonadetes, Nitrospirae, Planctomycetes; Spirochaetes, and Verrucomicrobia*

Furthermore, on-site anthropogenic activities and global environmental forces can produce local disturbances *synergistically*, since changing environmental conditions can exacerbate the challenges to ecosystems generated by human activities. For example, receding glaciers can expose contamination, otherwise buried below the ice masses. Melting ice, combined with increased precipitation, enhances the chances of displacing soil contamination toward the water bodies with the augmented runoff (Curtosi et al. 2007; Cabrerizo et al. 2013). The presence of pollutants in the water can negatively affect the pelagic primary producers (Azam et al. 1991; Curtosi et al. 2010; Quartino et al. 2013) and consequently the equilibrium of the marine food web. Physical habitat disturbance can also occur through the generation of ice debris produced by receding glaciers (Fig. 7.1) which increases the turbidity of the water bodies, affecting the amount of light that primary producers can receive (Quartino et al. 2013; Pasotti et al. 2015). Increased temperature and water availability generate more permissive conditions for the survival of non-indigenous microorganisms introduced by human activities, migratory birds, or air masses, with potential effects on the local ecosystems (Cowan and Ah Tow 2004 and references therein). Figure 7.2 summarizes the interactions of the global environmental and on-site anthropogenic shaping forces and the main ecosystem disturbances they may cause. In the following section, we focus on the anthropogenic sources of disturbance to Antarctic terrestrial ecosystems and the role of microorganisms in the remediation of contaminated environments.

7.3.2 *Main Sources of Anthropogenic Disturbance to Terrestrial Environments*

The human presence in Antarctica is tightly restricted to the personnel deployed at research stations and regulated touristic visits. Although the first adventurous explorations of the continent and hunting expeditions started in the eighteenth century, both the establishment of research stations and the initiation of touristic activities are relatively recent, dating back to the early 1950s (Tin et al. 2009), expanding steadily since then (Frenot et al. 2005; IAATO 2017). Past expeditions (Blanchette et al. 2004) and current activities (Kriwoken and Rootes 2000; Ah Tow and Cowan 2005; Jabour 2009; Tin et al. 2009; IAATO 2017) leave their footprint in Antarctica, affecting the environment in several ways: (1) physically disturbing habitats, fauna and flora, with vehicular and pedestrian transit; (2) generating human waste and chemical contamination; (3) producing dust and particulate material during the construction of Antarctic bases and through the open air rubbish incineration (executed in early expeditions prior to regulations); and (4) importing foreign microorganisms and propagules, as summarized in Fig. 7.2 (for a comprehensive review, see Cowan and Ah Tow 2004; Tin et al. 2009). Due to its impact and persistence, chemical contamination represents one of the most significant anthropogenic sources of disturbance to terrestrial Antarctic environments (Bargagli 2008) and is the focus of the next sections of this chapter.

7.3.2.1 Chemical Contamination

Petroleum-derived hydrocarbons (refined petroleum products) are some of the most common pollutants in the Antarctic, since fossil fuels power the operation of Antarctic bases, including heating and electricity generation, and are essential for sea and land transportation. The Special Antarctic Blend (SAB) diesel is commonly used at Antarctic research stations, and it is composed mainly of aliphatic hydrocarbons (lineal hydrocarbon chains) and a fraction containing polycyclic aromatic hydrocarbons (PAHs), heavy waxes, resins, and tars (Payne et al. 2014). SAB and other fuels are transported to Antarctica in vessels and transferred to storage tanks, typically located around the research stations. During these logistic operations and further manipulations for fuel use, accidental spillages do occur (Aislabie et al. 2004; Vázquez et al. 2017). Depending on the type of fuel, the contamination may contain different levels of aliphatic compounds, which are relatively easier to clean up, and PAHs, which may arise as well from the combustion of fossil fuels and natural sources, such as volcanic activity and wood fires (Shuttleworth and Cerniglia 1995). PAHs contain two or more aromatic rings and are a source of environmental concern since they are particularly toxic and recalcitrant compounds, with many of them labeled as carcinogenic (Achten and Andersson 2015). Due to their physicochemical properties, these compounds may bind to sediments and tissues of living organisms and accumulate in the food chain (bioaccumulation), representing a risk to the sustainability of the fragile Antarctic ecosystems (Goerke et al. 2004; Bargagli 2008; Curtosi et al. 2010; Quartino et al. 2013). Their fate in Antarctic environments depends mostly on the contaminant's physicochemical properties, the soil type, environmental conditions, and the presence of ice (Aislabie and Foght 2010). Their persistence in polar regions is exacerbated by the low rates of natural dissipation that are observed at high latitudes, in comparison to warmer regions of the planet (Delille and Pelletier 2002). Hence, chronically contaminated sites have been documented at several Antarctic locations (Mazzera et al. 1999; Aislabie et al. 2000a, b; Delille and Pelletier 2002; Martins et al. 2004; Vázquez et al. 2017), and degrading microbial populations were detected several years after a contamination event (Ruberto et al. 2005; Flocco et al. 2009; Aislabie and Foght 2010; Vázquez et al. 2017).

7.4 Remediation of Polluted Antarctic Terrestrial Environments

7.4.1 Microbial Communities and Their Catabolic Potential Are Key

Under the strict environmental conditions and regulatory framework inherent to the Antarctic territory, exploiting the natural capacity of microorganisms to remediate contamination (bioremediation) emerges as one of the few suitable and environment-friendly cleanup approaches (Aislabie et al. 2006; Atlas 2010). Several microorganisms have been successfully used for the remediation of contaminated areas in temperate and subantarctic conditions (Aislabie et al. 2006 and references therein). However, those cannot be transferred to the Antarctic since (1) the microbial degraders may not survive under the extreme environmental stress and (2) may retain viability, but do not produce any significant degradation under very low temperatures, since that factor may inhibit or drastically reduce the enzymatic activity, and (3) the introduction of foreign microorganisms as inoculants is prohibited by the environmental protocol to the Antarctic Treaty (Convey et al. 2012). Therefore, the bioremediation path narrows down to studying the catabolic capacity of the indigenous microbial populations adapted to survive and function under Antarctic environmental conditions.

Due to the significant environmental impact of petroleum-derived hydrocarbons, substantial research efforts have focused on understanding the biodegradation of that group of contaminants under Antarctic conditions. The microbial populations with capacity to degrade fuel-derived contaminants detected so far in the Antarctic environment mostly belong to the bacterial genera *Pseudomonas*, *Sphingomonas*, *Rhodococcus*, *Acinetobacter*, and *Polaromonas* (Aislabie et al. 2006; Luz et al. 2010; Vázquez et al. 2017) together with less frequent genera such as *Pedobacter* and *Brevundimonas* (Vázquez et al. 2013).

Also some Antarctic soil fungi, predominantly non-ligninolytic species, have been described as able to tolerate or degrade oil hydrocarbons (for a review, see Hughes and Bridge 2010a). It is important to point out that microbial consortia, rather than single strains, are better equipped to biodegrade complex mixtures (as is the case of fuel-derived contamination) due to the broader catabolic capacity and synergy resulting from the association (Mrozik and Piotrowska-Seget 2010; Festa et al. 2016).

The main gene classes involved in petroleum–hydrocarbon catabolism are the alkane monooxygenase (*alk*) genes related to the biodegradation of alkanes and the aromatic ring hydroxylating dioxygenases (ARHD) genes, a multicomponent and diverse enzyme family related to the degradation of different aromatic compounds (Nam et al. 2001). Both classes of hydrocarbon-degrading genes have been described in several bacterial taxa mostly belonging or closely related to the genera *Pseudomonas*, *Acinetobacter*, *Rhodococcus*, and *Actinomycetes*, detected in both polar and mesophilic environments (Bej et al. 2000; Gomes et al. 2007; van Beilen and Funhoff 2007; Flocco et al. 2009; Luz et al. 2010; Jurelevicius et al. 2012; de Sousa et al. 2017; Musumeci et al. 2017). Table 7.1 presents some examples of the Antarctic prokaryotic microbial taxa studied for their use in bioremediation.

7.4.2 Pollution Shaping Microbial Communities

In addition to the harsh environmental determinants, the presence of contamination can impact the abundance and structure of microbial communities in different ways. For example, in a cultivation-independent survey of the composition of

Table 7.1 Some examples of the Antarctic prokaryotic microbial taxa studied for their use in bioremediation

Microorganism	Source	Site of isolation	Growth substrate	Reference
Rhodococcus sp. strain ADH	Fuel-contaminated soil	Potter Peninsula, 25 de Mayo (King George Island), South Shetlands	Antarctic diesel fuel, aliphatics C12–C16	Ruberto et al. (2005)
Rhodococcus spp. strains 5/14 and 7/1	Oil-contaminated soil	Scott Base, Pram Point, Ross Island	n-alkanes (C6-C20)	Bej et al. (2000)
Consortium M10: *Pseudomonas, Stenotrophomonas, Sphingobacterium*	Fuel-contaminated soil	Marambio Station, Marambio (Seymour) Island	Phenanthrene	Ruberto et al. (2009b)
Pseudomonas spp. strain group LCY	Contaminated soil	Great Wall Station, Fildes Peninsula, 25 de Mayo (King George) Island, South Shetlands	Naphthalene, phenanthrene	Ma et al. (2006)
Pseudomonas sp. strain J3	Contaminated soil	Potter Peninsula, 25 de Mayo (King George) Island, South Shetlands	Diesel fuel	Shukor et al. (2009)
Pseudomonas extremaustralis DSM 17835	Pristine temporary pond	Primavera Station, Cierva Point, on the Danco Coast (WAP) West Peninsula	Diesel fuel	Tribelli et al. (2012)
Stenotrophomonas DM 1-41 and *Stenotrophomonas* MP 2-4	Contaminated soil	Marambio Station, Marambio (Seymour) Island	Phenanthrene, anthracene, fluorene, and dibenzothiophene	Ruberto et al. (2008)
Pantoea sp. strain A-13	Ornithogenic soil	Dewart Island, Frazier Islands	n-paraffins and kerosene	Vasileva-Tonkova and Gesheva (2007)
Sphingobium xenophagum D43FB	Contaminated soil	King George Island, South Shetland Islands	Phenanthrene	Gran-Scheuch et al. (2017)
Sphingomonas spp. (strains Ant17, Ant20) and *Pseudomonas* spp. (strains Ant 5, Ant 9, 7/22)	Contaminated soil	Scott Base, Pram Point, Ross Island	JP8 jet fuel and aromatic hydrocarbons	Aislabie et al. (2000a, b)
Sphingomonas sp. strain Ant 17	Contaminated soil	Scott Base, Pram Point, Ross Island	Aromatic hydrocarbons	Baraniecki et al. (2002)
Nocardioides sp. strain A-8	Contaminated soil	Dewart Island, Frazier Islands	*n*-paraffins	Vasileva-Tonkova and Gesheva (2005)

soil microbial communities of Carlini base and surrounding areas in Potter peninsula (by 454 pyrosequencing of 16S rRNA genes PCR-amplified from total community DNA), we found out that the main phyla detected in contaminated and non-contaminated soils were *Proteobacteria*, *Firmicutes*, *Actinobacteria*, and *Bacteroidetes*, as shown in Fig. 7.3 (Flocco et al. unpublished), which are the usual phyla found in soils from both polar and temperate regions (Janssen 2006). However, when analyzing the cluster formation of those communities, we could observe that those corresponding to a site with a chronical low-level contamination with petroleum hydrocarbons (mostly PAHs) clustered in a group (Fig. 7.4, samples with prefix B) that separated from the other two non-contaminated sites (Fig. 7.4, samples with prefix PV and PI), although all soils shared similar physicochemical properties, vegetation type, and exposure to environmental conditions (Flocco et al. 2009). Dias et al. (2015) also reported a clear clustering of the communities from contaminated soil samples from Carlini station after biostimulation with different nutrient sources. Those changes in the composition of microbial communities in response to contamination can be observed after a long-term exposure or detected within a relative short period following a

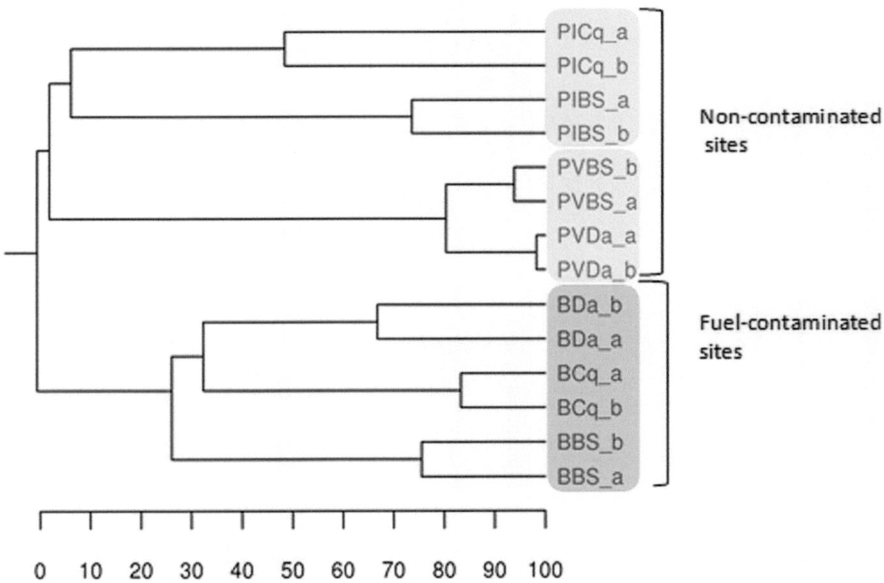

Fig. 7.4 Bacterial community structure of Antarctic soils from Potter Peninsula and Carlini station (King George Island/Isla 25 de Mayo, South Shetland Islands, Maritime Antarctic), analyzed by multiplex pyrosequencing. Experimental sites are indicated by the prefixes: (B) fuel-contaminated site within the area of the research station; (PI) and (PV) are non-contaminated sites within the adjacent protected area ASPA 132. Soil types are indicated by the additional prefixes: (BS) bulk soil, (Cq) and (Da) rhizospheric soils of the vascular plants *Colobanthus quitensis* and *Deschampsia antarctica*, respectively; a and b indicate replicas

pollution event. For example, in a bioremediation experiment at Carlini station, Ruberto et al. (2009a) described a shift during the treatment (50 days), with a significant increase in the proportion of *Actinobacteria*. Vázquez et al. (2009) working at the same location with in situ biopiles bioaugmented with two different hydrocarbon-degrading consortia or amended with nutrients also observed a fast and a significant increase in the number of hydrocarbon degraders, in parallel to a significant decrease in the soil hydrocarbon content, as compared to untreated soils.

Shifts in the structure of microbial communities in response to fuel contamination indicate an enrichment in degrading organisms, a pattern also reported for other polar regions (Aislabie and Foght 2010 and references therein) and temperate environments (Bordenave et al. 2007; Gomes et al. 2010; de Sousa et al. 2017). However, there are cases in which the growth of the degrading population is arrested, for example, by the presence of co-contaminants that are toxic for the hydrocarbon-degrading organisms. Examples of that phenomenon have been early documented in Antarctica by Aislabie et al. (1998), being the culprit organic lead compounds present in a gasoline spill. Heavy metals are frequent co-contaminants that can inhibit microbial activity (Aislabie and Foght 2010; Olaniran et al. 2013) and also freeze protection agents, such as glycols, which can reduce the efficiency of hydrocarbon biodegradation by becoming the preferred substrate (Wyrwas et al. 2011). The described inhibitory effect of co-contaminants is an important factor, among many, that must be taken into account when designing bioremediation strategies.

7.4.3 Optimizing Cold Bioremediation

Bioremediation processes imply an intricate array of biotic and abiotic factors interacting at a complexity level that has challenged many early implementation attempts (Cases and de Lorenzo 2005), prompting researchers to take a broader approach that combines bioengineering with systems biology (de Lorenzo 2008; Dvořák et al. 2017). Bioremediation in polar environments requires considering as well aspects inherent to the dynamics of biotic and abiotic pollution dissipation processes under extreme cold temperature and sharp fluctuations of environmental conditions. Harnessing cold-adapted microorganisms with the desired genetic catabolic capacity active under the given environmental conditions, assessing the type of pollutants and their location (ice, permafrost, soil layer, sediments), and defining the acceptable time frame for a defined decontamination strategy, given that the bioremediation processes in cold environments are slower compared to those in temperate regions (Aislabie et al. 2004; Yang et al. 2009; Atlas 2010), are crucial elements to factor into the design of a bioremediation strategy. The desired remediation end-points should be defined in terms of acceptable contaminant concentration and toxicity level (Atlas 2010), considering the different sensitivity of Antarctic species to pollutants as compared to that of closely related species from nonpolar habitats (Payne et al. 2014) and the potential generation of toxic metabolic intermediates during the enzymatic breakdown of PAHs (Flocco et al. 2002).

The natural attenuation of contaminants in the environment (driven by physicochemical process and indigenous degrading microbes) can be enhanced through two broadly defined approaches: (1) biostimulation, which implies adjusting physicochemical conditions and/or the addition of nutrients and electron acceptors in order to enhance the natural biodegradation processes, and (2) bioaugmentation, which involves reinforcing the biodegradation process by inoculating microorganisms with the required catabolic capacity (Tyagi et al. 2010). After many years of own laboratory and in situ studies with Antarctic soils contaminated with hydrocarbons, we outlined a number of conclusions in relation to the optimization of bioremediation strategies, which are in consonance with observations collected by several other authors focused in the same field of research. First, the biostimulation of Antarctic soils with nitrogen and phosphorus sources is key to achieving good levels of decontamination. This feature was repeatedly observed by our research group (Vázquez et al. 2009; Dias et al. 2012; Martínez Alvarez et al. 2015, 2017) and many other researchers (Delille and Coulon 2008; Delille et al. 2009; Cury et al. 2015). Second, when bioremediation processes are applied to soils with a long history of hydrocarbon exposure (chronic contamination), the bioaugmentation strategies do not produce a significant enhancement of the pollutant catabolization, as compared to biostimulated soils. Considering the complexity inherent to inocula preparation as well as the increase in treatment costs when a bioaugmentation step is involved, it becomes apparent that the use of microbial inoculants is not a suitable option for enhancing the bioremediation of chronically contaminated polar soils (Thomassin-Lacroix et al. 2002; Ruberto et al. 2008, 2009a, 2010). Third, noncontaminated soils receiving the impact of a pollution event (acute contamination) do not have a developed microbial community adapted to tolerate and degrade the contaminants. In this situation, the convenience of applying bioaugmentation remains an open question. The strategy could be used for enhancing the microbial catabolic activity and reducing the time required for reaching adequate degradation levels (Ruberto et al. 2003, 2006; Stallwood et al. 2005) and should be evaluated on a case-by-case basis.

7.4.4 *A Focus on Functional Genes and Rhizosphere Communities*

Plants can play a role in the degradation of pollutants (phytoremediation) either directly, through plant-mediated dissipation processes (Marmiroli et al. 2006; Flocco and Giulietti 2007), or indirectly through the release of exudates, which can serve as a carbon and energy source for microbial communities and as inducers of their catabolic activity (Siciliano et al. 2003; Gao et al. 2011; Berg et al. 2017; Vergani et al. 2017). In this context, we investigated the occurrence and diversity of

PAH catabolic genes in the soil strictly associated to the roots (rhizospheric soil) of the native Antarctic vascular plants *Deschampsia antarctica* and *Colobanthus quitensis* (shown in Fig. 7.1) growing at both contaminated and non-contaminated sites located in Potter Peninsula, since those pollutants are the predominant ones around Antarctic bases (Curtosi et al. 2007; Dauner et al. 2015). We surveyed both bulk soils and specific rhizospheres at Carlini base and the surrounding Antarctic protected area (ASPA132), tracing the occurrence of a class of ARHD genes, the naphthalene dioxygenase (*ndo*) genes, a well-studied molecular indicator of the microbial PAH degradation potential (Gomes et al. 2007; Flocco et al. 2009 and references therein). When we screened the samples by fingerprinting methods (denaturant gradient gel electrophoresis, DGGE, of *ndo* gene PCR fragments), we were able to detect the presence of *ndo* genes in all sample types, with the PAH-polluted sites exhibiting additional DGGE bands and apparently higher levels of *ndo* amplicons. Both the bulk and rhizosphere soils affected by a low but chronical presence of PAHs showed an increased level of a predominant *ndo* gene type (*nahAc* gene) of approximately two orders of magnitude (quantified by real-time PCR of *nahAc* gene), when compared with the respective soil samples from non-polluted sites. The presence of plants did not significantly alter the abundance and type of *nahAc* genes at polluted nor at non-polluted sites (Flocco et al. 2009), although we detected a rhizosphere effect (being it plant specific in some cases) in the bacteria domain community profile (Flocco et al. unpublished). In spite of the increased levels of PAH catabolic genes at contaminated sites, no significant differences in the total bacterial abundance (estimated by real-time PCR of 16S rRNA gene) across all sites and soil types were found (approximately 10^8 gene copies per gram of soil on dry weight basis, Flocco et al. 2009). The observed *ndo* and 16S rRNA gene level ratios strongly indicate a shift in the community structure, which is depicted by the clustering pattern of the total bacterial community (resulting from multiplex sequencing analysis, Flocco et al. unpublished), in which samples from the polluted site form a segregated group (Fig. 7.4); a similar clustering pattern was observed at the class and order levels within the main phyla detected (*Actinobacteria, Bacteroidetes, Firmicutes* and *Proteobacteria*; Flocco et al. unpublished). Taken together, these results indicate that the chronic history of a low-level PAH pollution was a major factor shaping the abundance and distribution of PAH-degrading genes in the soil microbial communities analyzed. A prevailing effect of a long-term hydrocarbon contamination on the structure of microbial communities was observed as well in temperate environments and microcosm conditions (Gomes et al. 2007; Sutton et al. 2013; Barbato et al. 2016). Other studies have assessed the influence of vegetation on Antarctic soil microbial communities across latitudinal gradients and sites affected by animal and anthropogenic influence (Yergeau 2013 and references therein) and revealed different levels of a vegetation effect on microbial communities. However, those studies did not encompass the specific evaluation of the effect of hydrocarbon pollution, as to compare with the results obtained for Potter Peninsula (Flocco et al. 2009) regarding the prevailing effect of long-term pollution in shaping soil microbial communities.

7.4.5 Gene Flow in the Cold: Mobile Genetic Elements and Bioremediation

The occurrence of mobile genetic elements (MGEs) in Antarctica (first studied by Kobori et al. 1984) is key to the adaptation of microbial species to the polar environmental conditions and the presence of pollutants, since it facilitates the dissemination of enabling genes among indigenous communities (de Lorenzo 2010; Martínez-Rosales et al. 2012; Smalla et al. 2015). The genus *Pseudomonas* has been extensively studied, given the metabolic versatility and capacity to catabolize aromatic compounds harbored by microorganisms belonging to this vast taxonomic group (de Lorenzo et al. 2013; Jun et al. 2015; Nikel et al. 2016 and references therein). It was demonstrated that the genes involved in the degradation of aromatic hydrocarbons are located in chromosomes (Li et al. 2012; Phale et al. 2013) or associated to mobile genetic elements (Yen and Gunsalus 1982; Sota et al. 2006; Ladino-Orjuela et al. 2016 and references therein). Within this context, we investigated the total community DNA of hydrocarbon-polluted and non-polluted soils from Carlini station and Potter Peninsula (Flocco et al. 2009) and detected a microdiversity of *ndo* genes (*nahAc*-like) closely related to the archetypal PAH degradation pathway (Yen and Gunsalus 1982). The most abundant genotypes were related to homologs located in the plasmid pNAH7 (Sota et al. 2006) and pDTG1 (Dennis and Zylstra 2004), affiliated to the IncP9 plasmid group (IncP standing for incompatibility group, according to the plasmid classification system for the *Pseudomonas* group; Thomas and Haines 2004). Using a combination of molecular methods (hybridization with specific probes, quantitative PCR and cloning and sequencing), we detected the occurrence of *Pseudomonas*-related plasmids IncP-1 and IncP-9 at PAH-contaminated sites. These plasmids were either not detected or produced faint hybridization signals when non-contaminated control soil samples were tested (Flocco et al. unpublished). Interestingly, the plasmid distribution pattern mirrored that of the *ndo* genes (*nahAc*) detected in the same soil samples (Flocco et al. 2009), strongly indicating a role of horizontal gene transfer (HGT) in the transmission of PAH catabolic genes. Similar *ndo* genotypes, located in large plasmids contained in *Pseudomonas* isolates (that could transfer to mesophilic representatives), were described by Ma et al. (2006) at both, a nearby Antarctic base (Great Wall Station, King George Island) and a distant one in East Antarctica (Zhongshan Station). A recent global study of the evolutionary relationships and distribution of ARDH genes across several temperate and polar sites showed a connectivity between Antarctica and South America which could be facilitated by the anthropogenic or airborne transportation of microorganisms across these locations (de Sousa et al. 2017 and references therein) and the subsequent propagation through HGT.

Gaining insight into the pool of mobile genetic elements, the *mobilome* (Jørgensen et al. 2015), and the underlying mechanisms of gene transfer in the environment are key to the optimization of biodegradation methods through the enhancement of the catabolic activity of microbial communities (Heuer and Smalla 2012). The facilitation of natural processes carried out by native microbial communities is of particular interest for application in polluted Antarctic areas given the strict environmental and regulatory conditions pertaining to the territory. At the same time, the contrasting scenario provided by relatively untouched Antarctica habitats represents a unique opportunity for studying the processes regulating gene transfer and their ecological implications without interference (or at least a minimal one) of anthropogenic inputs. Recent advances in molecular biology techniques and sequencing methods (Jørgensen et al. 2015; Smalla et al. 2015) boost our understanding of the *mobilome* and feed the development of environmental applications.

7.5 Concluding Remarks and Outlook

Microbial communities are key components of Antarctic terrestrial environments and central to their functioning and sustainability since microorganisms are primary energy producers, drive biogeochemical cycles, create enabling microhabitats, and help remediate environmental pollution. Their contributions and value transcend the boundaries of the Antarctic region, since they constitute as well a largely untapped source of unique genes and biomolecules with important biotechnological applications, ranging from industrial processes to human health. Habitat conditions in some Antarctic regions are changing rapidly due to concomitant factors: global environmental changes and the increasing impact of human activities on-site. We are starting to understand how these changes could affect the composition and functioning of the microbial communities in the short term, potentially altering the cycles and processes they sustain, but unforeseeable consequences are hovering in the long-term horizon. Further understanding of the biodiversity and ecological interactions of Antarctic microbial communities prior to the occurrence of disturbances introduced by environmental and anthropogenic drivers is a necessary step toward the preservation and sustainable management of this unique natural resource. As recently highlighted by Hughes et al. (2015), surveying the microscopic biodiversity of Antarctic terrestrial habitats and creating an inventory supporting the design of environmental management protocols represents a challenging task, compounded by the inherent complexity of the data (in comparison to macroscopic forms of life, which are easier to visualize), the vast territorial extension and harsh environmental conditions, and the enormous logistical resources needed to conduct in situ surveys.

Notwithstanding, collaborative international projects such as Imconet (http://www.imconet.eu) and the Earth Microbiome (http://www.earthmicrobiome.org) are making significant advances in this avenue of research. The first one was an international research network that followed an interdisciplinary approach to understand the consequences of climate change in coastal Western Antarctica. The microbial communities of the Antarctic sites assessed during this recently concluded project represented a central research target. The second one represents an international collaborative effort to characterize the microbial communities of the planet, with Antarctica as one of the focal points. The first communal catalogue performed by this global consortium was recently published and revealed the Earth's multiscale microbial diversity (Thompson et al. 2017). In perspective, boosting large-scale, collaborative research efforts, as the examples mentioned above, will be essential to gain a deeper and global understanding of the microbial communities of Antarctica or any other region of our planet. International cooperation, strategic long-term planning, and continued support for survey programs are key to succeed in this challenging and, at the same time, fascinating endeavor.

Acknowledgments C.G. Flocco is very thankful to the Alexander von Humboldt Foundation, supporting the creation of this chapter through a visiting scientist grant, and to the host institution, the Julius Kühn-Institut in Braunschweig, Germany. Our fieldwork at the Antarctic base Carlini was supported by Instituto Antártico Argentino, the University of Buenos Aires, and Consejo Nacional de Investigaciones Científicas y Técnicas (CONICET), Argentina. Many thanks go to Carlini station and the icebreaker ARA Irizar crews for transportation and logistic support.

References

Abele, D., Vazquez, S., Buma, A. G. J., et al. (2017). Pelagic and benthic communities of the Antarctic ecosystem of Potter Cove: Genomics and ecological implications. *Marine Genomics, 33*, 1–11. https://doi.org/10.1016/j.margen.2017.05.001.

Achten, C., & Andersson, J. T. (2015). Overview of polycyclic aromatic compounds (PAC). *Polycyclic Aromatic Compounds, 35*, 177–186.

Ah Tow, L., & Cowan, D. A. (2005). Dissemination and survival of non-indigenous bacterial genomes in pristine Antarctic environments. *Extremophiles, 9*, 385–389. https://doi.org/10.1007/s00792-005-0452-5.

Aislabie, J., & Foght, J. M. (2010). Response of polar soil bacterial communities to fuel spills. In A. K. Bej, J. Aislabie, & R. M. Atlas (Eds.), *Polar microbiology: The ecology, biodiversity and bioremediation potential of microorganisms in extremely cold environments* (pp. 215–230). Boca Raton: Taylor & Francis.

Aislabie, J., McLeod, M., & Fraser, R. (1998). Potential for biodegradation of hydrocarbons in soil from the Ross Dependency, Antarctica. *Applied Microbiology and Biotechnology, 49*, 210–214.

Aislabie, J., Foght, J., & Saul, D. (2000a). Aromatic hydrocarbon-degrading bacteria from soil near Scott Base, Antarctica. *Polar Biology, 23*, 183–188.

Aislabie, J., Fraser, R., Duncan, S., et al. (2000b). Effects of oil spills on microbial heterotrophs in Antarctic soils. *Polar Biology, 24*, 308–313.

Aislabie, J. M., Balks, M. R., Foght, J. M., et al. (2004). Hydrocarbons spills on Antarctic soils: Effects and management. *Environmental Science & Technology, 38*, 1265–1247.

Aislabie, J., Saul, D. J., & Foght, J. M. (2006). Bioremediation of hydrocarbon-contaminated polar soils. *Extremophiles, 10*, 171–179.

Aran, M., Smal, C., Pellizza, L., et al. (2014). Solution and crystal structure of BA42, a protein from the Antarctic bacterium *Bizionia argentinensis* comprised of a stand-alone TPM domain. *Proteins: Structure, Function, and Bioinformatics, 82*, 3062–3078. https://doi.org/10.1002/prot.24667.

Atlas, R. M. (2010). Microbial bioremediation in polar environments: Current status and future directions. In A. K. Bej, J. Aislabie, & R. M. Atlas (Eds.), *Polar microbiology: The ecology, biodiversity and bioremediation potential of microorganisms in extremely cold environments* (pp. 373–391). Boca Raton: Taylor & Francis.

Azam, F., Smith, D. C., & Hollibaugh, J. T. (1991). The role of the microbial loop in Antarctic pelagic ecosystems. *Polar Research, 10*, 239–244. https://doi.org/10.1111/j.1751-8369.1991.tb00649.x.

Baraniecki, C. A., Aislabie, J., & Foght, J. M. (2002). Characterization of *Sphingomonas* sp. Ant 17, an aromatic hydrocarbon-degrading bacterium isolated from Antarctic soil. *Microbial Ecology, 43*, 44–54.

Barbato, M., Mapelli, F., Magagnini, M., et al. (2016). Hydrocarbon pollutants shape bacterial community assembly of harbor sediments. *Marine Pollution Bulletin, 104*, 211–220. https://doi.org/10.1016/j.marpolbul.2016.01.029.

Bargagli, R. (2005). Antarctic ecosystems: environmental contaminants, climate change and human impact. In *Ecological studies* (Vol. 175). Berlin: Springer, 395pp.

Bargagli. (2008). Environmental contamination in Antarctic ecosystems. *Science of the Total Environment, 400*, 212–226.

Bej, A., Saul, D., & Aislabie, J. (2000). Cold-tolerant alkane-degrading *Rhodococcus* species from Antarctica. *Polar Biology, 23*, 100–105. https://doi.org/10.1007/s003000050014.

Bercovich, A., Vazquez, S. C., Yankilevich, P., et al. (2008). *Bizionia argentinensis* sp. nov., isolated from surface marine waters, Antarctica. *International Journal of Systematic Bacteriology, 58*, 2363–2367.

Berg, G., Köberl, M., Rybakova, D., et al. (2017). Plant microbial diversity is suggested as the key to future biocontrol and health trends. *FEMS Microbial Ecology, 93*, 5. https://doi.org/10.1093/femsec/fix050.

Berlemont, R., Pipers, D., Delsaute, M., et al. (2011). Exploring the Antarctic soil metagenome as a source of novel cold-adapted enzymes and genetic mobile elements. *Revista Argentina de Microbiología, 43*, 94–103. https://doi.org/10.1590/S0325-75412011000200005.

Blanchette, R., Held, B. W., Jurgens, J. A., et al. (2004). Environmental pollution from the Scott and Shackleton expeditions during the Heroic Age of Antarctic exploration. *Polar Record, 40*, 143–151.

Bordenave, S., Goñi-Urriza, M. S., Caumette, P., et al. (2007). Effects of heavy fuel oil on the bacterial community structure of a pristine microbial mat. *Applied and Environmental Microbiology, 73*, 6089–6097. https://doi.org/10.1128/AEM.01352-07.

Bosi, E., Fondi, M., Orlandini, V., et al. (2017). The pangenome of Antarctic *Pseudoalteromonas* bacteria: Evolutionary and functional insights. *BMC Genomics, 18*, 93. https://doi.org/10.1186/s12864-016-3382-y.

Bottos, E., Scarrow, J., Archer, S., et al. (2014). Bacterial community structures of Antarctic soils. In D. A. Cowan (Ed.), *Antarctic terrestrial microbiology: Physical and biological properties of Antarctic soils* (pp. 9–33). Heidelberg: Springer. https://doi.org/10.1007/978-3-642-45213-0_2.

Bruntner, C., Binder, T., Pathom-Aree, W., et al. (2005). Frigocyclinone, a novel angucyclinone antibiotic produced by a *Streptomyces griseus* strain from Antarctica. *The Journal of Antibiotics, 58*, 346–349.

Cabrerizo, A., Dachs, J., Barceló, D., et al. (2013). Climatic and biogeochemical controls on the remobilization and reservoirs of persistent organic pollutants in Antarctica. *Environmental Science & Technology, 47*, 4299–4306. https://doi.org/10.1021/es400471c.

Cases, I., & de Lorenzo, V. (2005). Genetically modified organisms for the environment: Stories of success and failure and what we have learned from them. *International Microbiology, 8*, 213–222.

Chan, Y., Van Nostrand, J. D., Zhou, J., et al. (2013). Functional ecology of an Antarctic Dry Valley. *Proceedings of the National Academy of Sciences, 110*, 8990–8995. https://doi.org/10.1073/pnas.1300643110.

Chong, C.-W., Pearce, D., Convey, P., et al. (2012). Patterns in the distribution of soil bacterial 16S rRNA gene sequences from different regions of Antarctica. *Geoderma, 181–182*, 45–55. https://doi.org/10.1016/j.geoderma.2012.02.017.

Chong, C.-W., Pearce, D. A., & Convey, P. (2015). Emerging spatial patterns in Antarctic prokaryotes. *Frontiers in Microbiology, 6*, 1058. https://doi.org/10.3389/fmicb.2015.01058.

Convey, P. (2013). Antarctic ecosystems. In S. A. Levin (Ed.), *Encyclopedia of biodiversity* (Vol. 1, 2nd ed., pp. 179–188). San Diego: Elsevier.

Convey, P., & Smith, R. I. L. (2006). Responses of terrestrial Antarctic ecosystems to climate change. *Plant Ecology, 182*, 1–10.

Convey, P., Hughes, K. A., & Tin, T. (2012). Continental governance and environmental management mechanisms under the Antarctic Treaty System: Sufficient for the biodiversity challenges of this century? *Biodiversity, 13*, 1–15. https://doi.org/10.1080/14888386.2012.703551.

Cook, A. J., Fox, A. J., Vaughan, D. G., et al. (2005). Retreating glacier fronts on the Antarctic Peninsula over the past half-century. *Science, 308*, 541–544.

Cowan, D. A., & Ah Tow, L. (2004). Endangered Antarctic environments. *Annual Review of Microbiology, 58*, 649–690.

Cowan, D. A., Khan, N., Heath, C., et al. (2010). Microbiology of Antarctic terrestrial soils and rocks. In A. K. Bej, J. Aislabie, & R. M. Atlas (Eds.), *Polar microbiology: The ecology, biodiversity and bioremediation potential of microorganisms in extremely cold environments* (pp. 1–29). Boca Raton: Taylor & Francis.

Cowan, D. A., Sohm, J. A., Makhalanyane, T. P., et al. (2011). Hypolithic communities: Important nitrogen sources in Antarctic desert soils. *Environmental Microbiology Reports, 3*, 581–586. https://doi.org/10.1111/j.1758-2229.2011.00266.x.

Cowan, D. A., Makhalanyane, T. P., Dennis, P. G., et al. (2014). Microbial ecology and biogeochemistry of continental Antarctic soils. *Frontiers in Microbiology, 5*, 154. https://doi.org/10.3389/fmicb.2014.00154.

Curtosi, A., Pelletier, E., Vodopivez, C. L., et al. (2007). Polycyclic aromatic hydrocarbons in soil and surface marine sediment near Jubany Station (Antarctica). Role of permafrost as a low-permeability barrier. *Science of the Total Environment, 383*, 193–204.

Curtosi, A., Pelletier, E., Vodopivez, C., et al. (2010). Presence and distribution of persistent toxic substances in sediments and marine organisms of Potter Cove, Antarctica. *Archives of Environmental Contamination and Toxicology, 59*, 582–592. https://doi.org/10.1007/s00244-010-9509-2.

Cury, J., Jurelevicius, D. A., Villela, H. D. M., et al. (2015). Microbial diversity and hydrocarbon depletion in low and high diesel-polluted soil samples from Keller Peninsula, South Shetland Islands. *Antarctic Science, 27*, 263–273.

Dauner, A. L. L., Hernandez, E. A., Mac Cormack, W. P., et al. (2015). Molecular characterization of anthropogenic sources of sedimentary organic matter from Potter Cove, King George Island, Antarctica. *Science of the Total Environment, 502*, 408–416.

Day, T. A., Ruhland, C. T., Grobe, C. W., et al. (1999). Growth and reproduction of Antarctic vascular plants in response to warming and UV-B radiation reduction in the field. *Oecologia, 119*, 24–35.

de Lorenzo, V. (2008). Systems biology approaches to bioremediation. *Current Opinion in Biotechnology, 19*, 579–589. https://doi.org/10.1016/j.copbio.2008.10.004.

de Lorenzo, V. (2010). Genes that move the window of viability of life: Lessons from bacteria thriving at the cold extreme. *BioEssays, 33*, 38–42.

de Lorenzo, V., Pieper, D., & Ramos, J. L. (2013). From the test tube to the environment – And back. *Environmental Microbiology, 15*, 6–11. https://doi.org/10.1111/j.1462-2920.2012.02896.x.

De Maayer, P., Anderson, D., Cary, C., et al. (2014). Some like it cold: Understanding the survival strategies of psychrophiles. *EMBO Reports, 15*, 508–517. https://doi.org/10.1002/embr.201338170.

de Pascale, D., de Santi, C., Fu, J., et al. (2012). The microbial diversity of polar environments is a fertile ground for bioprospecting. *Marine Genomics, 8*, 15–22.

de Sousa, S. T. P., Cabral, L., Vieira Lacera, G., Jr., et al. (2017). Diversity of aromatic hydroxylating dioxygenase genes in mangrove microbiome and their biogeographic patterns across global sites. *Microbiology Open, 6*(4), e00490. https://doi.org/10.1002/mbo3.490.

Delille, D., & Coulon, F. (2008). Comparative mesocosm study of biostimulation efficiency in two different oil-amended sub-Antarctic soils. *Microbial Ecology, 56*, 243–252. https://doi.org/10.1007/s00248-007-9341-z.

Delille, D., & Pelletier, E. (2002). Natural attenuation of diesel-oil contamination in a subantarctic soil (Crozet Island). *Polar Biology, 25*, 682–687.

Delille, D., Pelletier, D., & Rodriguez-Blanco, A. (2009). Effects of nutrient and temperature on degradation of petroleum hydrocarbons in sub-Antarctic coastal seawater. *Polar Biology, 32*, 1521–1528. https://doi.org/10.1007/s00300-009-0652-z.

Dennis, J. J., & Zylstra, G. J. (2004). Complete sequence and genetic organization of pDTG1, the 83 kilobase naphthalene degradation plasmid from *Pseudomonas putida* strain. NCIB 9816-4. *Journal of Molecular Biology, 13*, 753–768.

Dias, R. L., Ruberto, L., Hernández, E., et al. (2012). Bioremediation of an aged diesel oil-contaminated Antarctic soil: Evaluation of the on-site biostimulation strategy using different nutrient sources. *International Biodeterioration and Biodegradation, 75*, 96–103. https://doi.org/10.1016/j.ibiod.2012.07.020.

Dias, R. L., Ruberto, L., Calabró, A., et al. (2015). Hydrocarbon removal and bacterial community structure in on-site biostimulated biopile systems designed for bioremediation of diesel contaminated Antarctic soil. *Polar Biology, 38*, 677–687.

Dvořák, P., Nikel, P. I., Damborský, J., et al. (2017). Bioremediation 3.0: Engineering pollutant-removing bacteria in the times of systemic biology. *Biotechnology Advances, 35*, 845–866. https://doi.org/10.1016/j.biotechadv.2017.08.001.

Ferrari, B. C., Winsley, T., Gillings, M., et al. (2008). Cultivating previously uncultured soil bacteria using a soil substrate membrane system. *Nature Protocols, 3*, 1261–1269. https://doi.org/10.1038/nprot.2008.102.

Festa, S., Coppotelli, B. M., & Morelli, I. S. (2016). Comparative bioaugmentation with a consortium and a single strain in a phenanthrene-contaminated soil: Impact on the bacterial community and biodegradation. *Applied Soil Ecology, 98*, 8–19.

Flocco, C. G., & Giulietti, A. M. (2007). In vitro hairy root cultures as a tool for phytoremediation research. In N. Willey (Ed.), *Phytoremediation, Methods in Biotechnology* (Vol. 23, pp. 161–173). Totowa: Humana Press. https://doi.org/10.1007/978-1-59745-098-0_14.

Flocco, C. G., Lobalbo, A., Carranza, M. P., et al. (2002). Some physiological, microbial, and toxicological aspects of the removal of phenanthrene by hydroponic cultures of alfalfa (*Medicago sativa* L.). *International Journal of Phytoremediation, 4*, 169–186. https://doi.org/10.1080/15226510208500081.

Flocco, C. G., Gomes, N. C., Mac Cormack, W., et al. (2009). Occurrence and diversity of naphthalene dioxygenase genes in soil microbial communities from the Maritime Antarctic. *Environmental Microbiology, 11*, 700–714. https://doi.org/10.1111/j.1462-2920.2008.01858.x.

Frenot, Y., Chown, S. L., Whinam, J., et al. (2005). Biological invasions in the Antarctic: Extent, impacts and implications. *Biological Reviews, 80*, 45–72.

Fry, J. (2004). Culture-dependent microbiology. In A. Bull (Ed.), *Microbial diversity and bioprospecting* (pp. 80–87). Washington, DC: ASM Press. https://doi.org/10.1128/9781555817770.ch8.

Gao, Y., Yang, Y., Ling, W., et al. (2011). Gradient distribution of root exudates and polycyclic aromatic hydrocarbons in rhizosphere soil. *Soil Science Society of America Journal, 75*, 1694–1703.

Godinho, V. M., Furbino, L. E., Santiago, I. F., et al. (2013). Diversity and bioprospecting of fungal communities associated with endemic and cold-adapted macroalgae in Antarctica. *The ISME Journal, 7*, 1434–1451. https://doi.org/10.1038/ismej.2013.77.

Goerke, H., Weber, K., Bornemann, H., et al. (2004). Increasing levels and biomagnification of persistent organic pollutants (POPs) in Antarctic biota. *Marine Pollution Bulletin, 48*, 295–302.

Gomes, N. C. M., Borges, L. R., Paranhos, R., et al. (2007). Diversity of *ndo* genes in mangrove sediments exposed to different sources of polycyclic aromatic hydrocarbon pollution. *Applied and Environmental Microbiology, 73*, 7392–7399.

Gomes, N. C. M., Flocco, C. G., Costa, R., et al. (2010). Mangrove microniches determine the structural and functional diversity of enriched petroleum hydrocarbon-degrading consortia. *FEMS Microbiology Ecology, 74*, 276–290. https://doi.org/10.1111/j.1574-6941.2010.00962.x.

Gran-Scheuch, A., Fuentes, E., Bravo, D. M., et al. (2017). Isolation and characterization of phenanthrene degrading bacteria from diesel fuel-contaminated Antarctic soils. *Frontiers in Microbiology, 8*, 1634. https://doi.org/10.3389/fmicb.2017.01634.

Heuer, H., & Smalla, K. (2012). Plasmids foster diversification and adaptation of bacterial populations in soil. *FEMS Microbiology Reviews, 36*, 1083–1104. https://doi.org/10.1111/j.1574-6976.2012.00337.x.

Horowitz, N. H., Bauman, A. J., Cameron, R. E., et al. (1969). Sterile soil from Antarctica: Organic analysis. *Science, 164*, 1054–1056.

Horowitz, N. H., Cameron, R. E., & Hubbard, J. S. (1972). Microbiology of the Dry Valleys of Antarctica. *Science, 176*, 242–245.

Howard-Williams, C., Hawes, I., & Gordon, S. (2010). The environmental basis of ecosystem variability in Antarctica: Research in the Latitudinal Gradient Project. *Antarctic Science, 22*, 591–602. https://doi.org/10.1017/S0954102010000829.

Hughes, K. A., & Bridge, P. D. (2010a). Tolerance of Antarctic soil fungi to hydrocarbons and their potential role in soil bioremediation. In A. K. Bej, J. Aislabie, & R. M. Atlas (Eds.), *Polar microbiology: The ecology, biodiversity and bioremediation potential of microorganisms in extremely cold environments* (pp. 277–300). Boca Raton: Taylor & Francis.

Hughes, K. A., & Bridge, P. D. (2010b). Potential impacts of bioprospecting and associated commercial activities upon Antarctic science and scientists. *Ethics in Science and Environmental Politics, 10*, 13–18.

Hughes, K. A., Cowan, D. A., & Wilmotte, A. (2015). Protection of Antarctic microbial communities – 'out of sight, out of mind'. *Frontiers in Microbiology, 6*, 151. https://doi.org/10.3389/fmicb.2015.00151.

IAATO: International Association for Antarctic Tour Operations. (2017). IAATO tourism statistics. Available via: https://iaato.org/de/tourism-statistics. Accessed 21 Feb 2018.

Jabour, J. (2009). National Antarctic programs and their impact on the environment. In K. R. Kerry & M. J. Riddle (Eds.), *Health of Antarctic wildlife. A challenge for science and policy* (pp. 2211–2229). Berlin: Springer.

Janssen, P. H. (2006). Identifying the dominant soil bacterial taxa in libraries of 16S rRNA and 16S rRNA genes. *Applied and Environmental Microbiology, 72*, 1719–1728.

Jørgensen, T. S., Kiil, A. S., Hansen, M. A., et al. (2015). Current strategies for mobilome research. *Frontiers in Microbiology, 22*, 750. https://doi.org/10.3389/fmicb.2014.00750.

Jun, S. R., Wassenaar, T. M., Nookaew, I., et al. (2015). Diversity of *Pseudomonas* genomes, including *Populus* associated isolates, as revealed by comparative genome analysis. *Applied and Environmental Microbiology, 82*, 375–383. https://doi.org/10.1128/AEM.02612-15.

Jurelevicius, D., Alvarez, V. M., Peixoto, R., et al. (2012). Bacterial polycyclic aromatic hydrocarbon ring-hydroxylating dioxygenases (PAH-RHD) encoding genes in different soils from King George Bay, Antarctic Peninsula. *Applied Soil Ecology, 55*, 1–9.

Kobori, H., Sullivan, C. V., & Shizuya, H. (1984). Bacterial plasmids in Antarctic natural microbial assemblages. *Applied Environmental Microbiology, 48*, 515–518.

Kriwoken, L. K., & Rootes, D. (2000). Tourism on ice: Environmental impact assessment of Antarctic tourism. *Impact Assessment and Project Appraisal, 18*, 138–150.

Ladino-Orjuela G, Gomes E, da Silva R et al (2016) Metabolic pathways for degradation of aromatic hydrocarbons by bacteria. In: de Voogt W (ed) Reviews of environmental contamination and toxicology, 237. Springer, Cham, 105–121

Lewis Smith RI (2003) The enigma of *Colobanthus quitensis* and *Deschampsia antarctica* in Antarctica. Huiskes AHL, Gieskes WWC, Rozema J, Schorno RML, van der Vies SM, Wolff WJ. Antarctic biology in a global context. Backhuys Publishers, Leiden, 234–239.

Li, S., Zhao, H., Li, Y., et al. (2012). Complete genome sequence of the naphthalene-degrading *Pseudomonas putida* strain ND6. *Journal of Bacteriology, 194*, 5154–5155. https://doi.org/10.1128/JB.01190-12.

Logares, R., Haverkamp, T. H., Kumar, S., et al. (2012). Environmental microbiology through the lens of high-throughput DNA sequencing: Synopsis of current platforms and bioinformatics approaches. *Journal of Microbiological Methods, 91*, 106–113. https://doi.org/10.1016/j.mimet.2012.07.017.

Lohan, D., & Johnston, S. (2005). United Nations University – Institute of Advanced Studies (UNU-IAS) report. Bioprospecting in Antarctica. Available via: www.ias.unu.edu/binaries2/antarctic_bioprospecting.pdf. Accessed 8 Feb 2018.

Luz, P. A., Kuhn, E., & Pellizari, V. H. (2010). Occurrence, distribution and nature of hydrocarbon-degrading genes in bio degradative microorganisms from the Antarctic environment. In A. K. Bej, J. Aislabie, & R. M. Atlas (Eds.), *Polar microbiology: The ecology, biodiversity and bioremediation potential of microorganisms in extremely cold environments* (pp. 339–355). Boca Raton: Taylor & Francis.

Ma, Y., Wang, L., & Shao, Z. (2006). *Pseudomonas*, the dominant polycyclic aromatic hydrocarbon-degrading bacteria isolated from Antarctic soils and the role of large plasmids in horizontal gene transfer. *Environmental Microbiology, 8*, 455–465.

Marmiroli, N., Marmiroli, M., & Maestri, E. (2006). Phytoremediation and phytotechnologies: A review for the present and the future. In I. Twardowska, H. E. Allen, M. M. Häggblom, & S. Stefaniak (Eds.), *Soil and water pollution monitoring, protection and remediation, NATO Science Series* (Vol. 69, pp. 403–416). Dordrecht: Springer.

Martín-Español, A., Zammit-Mangion, A., Clarke, P. J., et al. (2016). Spatial and temporal Antarctic ice sheet mass trends, glacio-isostatic adjustment, and surface processes from a joint inversion of satellite altimeter, gravity, and GPS data. *Journal of Geophysical Research – Earth Surface, 121*, 182–200.

Martínez Alvarez, L. M., Lo Balbo, A., Mac Cormack, W. P., et al. (2015). Bioremediation of a hydrocarbon contaminated Antarctic soil: Optimization of a biostimulation strategy using response-surface methodology (RSM). *Cold Regions Science and Technology, 119*, 61–67. https://doi.org/10.1016/j.coldregions.2015.07.005.

Martínez Álvarez, L. M., Ruberto, L. A. M., Balbo, L., et al. (2017). Bioremediation of hydrocarbon-contaminated soils in cold regions: Development of a pre-optimized biostimulation biopile-scale field assay in Antarctica. *Science of the Total Environment, 590–591*, 194–203. https://doi.org/10.1016/j.scitotenv.2017.02.204.

Martínez-Rosales, C., Fullana, N., Musto, H., et al. (2012). Antarctic DNA moving forward: Genomic plasticity and biotechnological potential. *FEMS Microbiology Letters, 331*, 1–9. https://doi.org/10.1111/j.1574-6968.2012.02531.x.

Martins, C. C., Bícego, M. C., Taniguchi, S., et al. (2004). Aliphatic and polycyclic aromatic hydrocarbons in surface sediments in Admiralty Bay, King George Island, Antarctica. *Antarctic Science, 16*, 117–122.

Martorell, M. M., Ruberto, L. A. M., Fernández, P. M., et al. (2017). Bioprospection of cold-adapted yeasts with biotechnological potential from Antarctica. *Journal of Basic Microbiology, 57*, 504–516. https://doi.org/10.1002/jobm.201700021.

Mazzera, D., Hayes, T., Lowenthal, D., et al. (1999). Quantification of polycyclic aromatic hydrocarbons in soil at McMurdo Station, Antarctica. *Science of the Total Environment, 229*, 65–71.

Mrozik, A., & Piotrowska-Seget, Z. (2010). Bioaugmentation as a strategy for cleaning up of soils contaminated with aromatic compounds. *Microbiological Research, 165*, 363–375.

Musumeci, M. A., Lozada, M., Rial, D. V., et al. (2017). Prospecting biotechnologically-relevant monooxygenases from cold sediment metagenomes: An in silico approach. *Marine Drugs, 15*, 114. https://doi.org/10.3390/md15040114.

Nam, J. W., Nojiri, H., Yoshida, T., et al. (2001). New classification system for oxygenase components involved in ring-hydroxylating oxygenations. *Bioscience, Biotechnology, and Biochemistry, 65*, 254–263.

Nikel, P. I., Chavarría, M., Danchin, A., et al. (2016). From dirt to industrial applications: *Pseudomonas putida* as a synthetic biology chassis for hosting harsh biochemical reactions. *Current Opinion in Chemical Biology, 34*, 20–29. https://doi.org/10.1016/j.cbpa.2016.05.011.

Olaniran, A. O., Balgobind, A., & Pillay, B. (2013). Bioavailability of heavy metals in soil: Impact on microbial biodegradation of organic compounds and possible improvement strategies. *International Journal of Molecular Sciences, 14*, 10197–10228. https://doi.org/10.3390/ijms140510197.

Pandey, K. D., Shukla, S. P., Shukla, P. N., et al. (2004). *Cyanobacteria* in Antarctica: Ecology, physiology and cold adaptation. *Cellular and Molecular Biology (Noisy-le-Grand, France), 50*, 575–584.

Pasotti, F., Saravia, L. A., De Troch, M., et al. (2015). Benthic trophic interactions in an Antarctic shallow water ecosystem affected by recent glacier retreat. *PLoS One, 10*(11), e0141742. https://doi.org/10.1371/journal.pone.0141742.

Payne, S. J., King, C., Marcus Zamora, L., et al. (2014). Temporal changes in the sensitivity of coastal Antarctic zooplankton communities to diesel fuel: A comparison between single- and multi-species toxicity tests. *Environmental Toxicology and Chemistry, 33*, 882–890.

Peeters, K., Ertz, D., & Willems, A. (2011a). Culturable bacterial diversity at the Princess Elisabeth Station (Utsteinen, Sør Rondane Mountains, East Antarctica) harbours many new taxa. *Systematic and Applied Microbiology, 34*, 360–367. https://doi.org/10.1016/j.syapm.2011.02.002.

Peeters, K., Hodgson, D. A., Convey, P., et al. (2011b). Culturable diversity of heterotrophic bacteria in Forlidas Pond (Pensacola Mountains) and Lundström Lake (Shackleton Range), Antarctica. *Microbial Ecology, 62*, 399–413. https://doi.org/10.1007/s00248-011-9842-7.

Phale, P. S., Paliwal, V., & Raju, S. C. (2013). Genome sequence of naphthalene-degrading soil bacterium *Pseudomonas putida* CSV86. *Genome Announcements, 1*, 1. https://doi.org/10.1128/genomeA.00234-12.

Poland, J. S., Riddle, M. J., & Zeeb, B. A. (2003). Contaminants in the Arctic and the Antarctic: A comparison of sources, impacts, and remediation options. *Polar Record, 39*, 369–383.

Pudasaini, S., Wilson, J., & Ji, M. (2017). Microbial diversity of Browning Peninsula, Eastern Antarctica revealed using molecular and cultivation methods. *Frontiers in Microbiology, 8*, 591. https://doi.org/10.3389/fmicb.2017.00591.

Pulicherla, K., Mrinimoy, G., Suresh Kumar, P., et al. (2011). Psychrozymes – The next generation industrial enzymes. *Journal of Marine Science: Research and Development, 1*, 102. https://doi.org/10.4172/2155-9910.1000102.

Quartino, M. L., Deregibus, D., Campana, G. L., et al. (2013). Evidence of macroalgal colonization on newly ice-free areas following glacial retreat in Potter Cove (South Shetland Islands), Antarctica. *PLoS One, 8*(3), e58223. https://doi.org/10.1371/journal.pone.0058223.

Quartino, M. L., Vazquez, S. C., Latorre, G. E., et al. (2015). Possible role of bacteria in the degradation of macro algae *Desmarestia anceps* Montagne (Phaeophyceae) in Antarctic marine waters. *Revista Argentina de Microbiología, 473*, 274–276. https://doi.org/10.1016/j.ram.2015.04.003.

Ruberto, L., Vazquez, S. C., & Mac Cormack, W. P. (2003). Effectiveness of the natural bacterial flora, biostimulation and bioaugmentation on the bioremediation of a hydrocarbon contaminated Antarctic soil. *International Biodeterioration and Biodegradation, 52*, 115–125. https://doi.org/10.1016/S0964-8305(03)00048-9.

Ruberto, L. A. M., Vazquez, S. C., LoBalbo, A., et al. (2005). Psychrotolerant hydrocarbon-degrading *Rhodococcus* strains isolated from polluted Antarctic soils. *Antarctic Science, 17*, 47–56.

Ruberto, L. A. M., Vazquez, S. C., Curtosi, A., et al. (2006). Phenanthrene biodegradation in soils using an Antarctic bacterial consortium. *Bioremediation Journal, 10*, 191–201. https://doi.org/10.1080/10889860601021449.

Ruberto, L., Dias, R., Vazquez, S. C., et al. (2008). Small-scale studies on biodegradation of hydrocarbons in acutely-contaminated Antarctic soils. *Reports on Polar and Marine Research, 571*, 340–347 TXT W2.

Ruberto, L., Dias, R., Lo Balbo, A., et al. (2009a). Influence of nutrients addition and bioaugmentation on the hydrocarbon biodegradation of a chronically contaminated Antarctic soil. *Journal of Applied Microbiology, 106*, 101–110.

Ruberto, L. A. M., Vazquez, S. C., & Mac Cormack, W. P. (2009b). Bacterial hydrocarbon-degrading consortium from Antarctic soils. *Revista Argentina de Microbiología, 41*, 262.

Ruberto, L., Vazquez, S. C., Dias, R. L., et al. (2010). Small-scale studies toward a rational use of bioaugmentation in an Antarctic hydrocarbon-contaminated soil. *Antarctic Science, 22*, 463–469. https://doi.org/10.1017/S0954102010000295.

Schreiter, S., Eltlbany, N., & Smalla, K. (2015). Microbial communities in the rhizosphere analyzed by cultivation-independent DNA-based methods. In B. Lugtenberg (Ed.), *Principles of plant-microbe interactions* (pp. 289–298). Cham: Springer.

Seefeldt, M. A., Campana, G. L., Deregibus, D., et al. (2017). Different feeding strategies in Antarctic scavenging amphipods and their implications for colonisation success in times of retreating glaciers. *Frontiers in Zoology, 14*, 59. https://doi.org/10.1186/s12983-017-0248-3.

Shivaji, S., & Reddy, G. S. N. (2010). Bacterial diversity of Antarctica. Conventional polyphasic and rRNA approaches. In A. K. Bej, J. Aislabie, & R. M. Atlas (Eds.), *Polar microbiology: The ecology, biodiversity and bioremediation potential of microorganisms in extremely cold environments* (pp. 215–230). Boca Raton: Taylor & Francis.

Shukor, M. Y., Hassan, N. A. A., Jusoh, A. Z., et al. (2009). Isolation and characterization of a *Pseudomonas* diesel-degrading strain from Antarctica. *Journal of Environmental Biology, 30*, 1–6.

Shuttleworth, K. L., & Cerniglia, C. E. (1995). Environmental aspects of PAH biodegradation. *Applied Biochemistry and Biotechnology, 1–3*, 291–302.

Siciliano, S. D., Germida, J. J., Banks, K., et al. (2003). Changes in microbial community composition and function during a polyaromatic hydrocarbon phytoremediation field trial. *Applied and Environmental Microbiology, 69*, 483–489. https://doi.org/10.1128/AEM.69.1.483-489.2003.

Smalla, K. (2004). Culture-independent microbiology. In A. Bull (Ed.), *Microbial diversity and bioprospecting* (pp. 88–99). Washington, DC: ASM Press. https://doi.org/10.1128/9781555817770.ch9.

Smalla, K., Jechalke, S., & Top, E. M. (2015). Plasmid detection, characterization, and ecology. *Microbiology Spectrum, 1*, PLAS-0038-2014. https://doi.org/10.1128/microbiolspec.

Smith, J. J., Ah Tow, L., Stafford, W., et al. (2006). Bacterial diversity in three different Antarctic Cold Desert mineral soils. *Microbial Ecology, 51*, 413–421.

Sota, M., Yano, H., Ono, A., et al. (2006). Genomic and functional analysis of the IncP-9 naphthalene-catabolic plasmid NAH7 and its transposon Tn4655 suggests catabolic gene spread by a tyrosine recombinase. *Journal of Bacteriology, 188*, 4057–4067.

Stallwood, B., Shears, J., Williams, P. A., et al. (2005). Low temperature bioremediation of oil-contaminated soil using biostimulation and bioaugmentation with a *Pseudomonas* sp. from maritime Antarctica. *Journal of Applied Microbiology, 99*, 794–802. https://doi.org/10.1111/j.1365-2672.2005.02678.x.

Stomeo, F., Makhalanyane, T. P., Valverde, A., et al. (2012). Abiotic factors influence microbial diversity in permanently cold soil horizons of a maritime-associated Antarctic Dry Valley. *FEMS Microbiology Ecology, 82*, 326–340. https://doi.org/10.1111/j.1574-6941.2012.01360.x.

Sutton, N. B., Maphosa, F., Morillo, J. A., et al. (2013). Impact of long-term diesel contamination on soil microbial community structure. *Applied and Environmental Microbiology, 79*, 619–630. https://doi.org/10.1128/AEM.02747-2712.

Teixeira, L. C., Peixoto, R. S., Cury, J. C., et al. (2010). Bacterial diversity in rhizosphere soil from Antarctic vascular plants of Admiralty Bay, Maritime Antarctica. *The ISME Journal, 4*, 989–1001. https://doi.org/10.1038/ismej.2010.35.

Terauds, A., & Lee, R. (2016). Antarctic biogeography revisited: Updating the Antarctic Conservation Biogeographic Regions. *Diversity and Distributions*, 1–5. doi:https://doi.org/10.4225/15/5729930925224.

Thomas, C. M., & Haines, A. S. (2004). Plasmids of the genus Pseudomonas. In J.-L. Ramos (Ed.), *Pseudomonas* (pp. 197–231). London: Kluwer/Plenum.

Thomassin-Lacroix, E. J. M., Eriksson, M., Reimer, K. J., et al. (2002). Biostimulation and bioaugmentation for on-site treatment of weathered diesel fuel in Arctic soil. *Applied Microbiology and Biotechnology, 59*, 551–556. https://doi.org/10.1007/s00253-002-1038-0.

Thompson, L. R., Sanders, J. G., McDonald, D., et al. (2017). A communal catalogue reveals Earth's multiscale microbial diversity. *Nature, 551*, 457–463. https://doi.org/10.1038/nature24621.

Tian, Y., Li, Y.-L., & Zhao, F.-C. (2017). Secondary metabolites from polar organisms. *Marine Drugs, 15*, 28. https://doi.org/10.3390/md15030028.

Tin, T., Fleming, Z. L., Hughes, D. G., et al. (2009). Impacts of local human activities on the Antarctic environment. *Antarctic Science, 21*, 3–33.

Tindall, B. J. (2004). Prokaryotic diversity in the Antarctic: The tip of the iceberg. *Microbial Ecology, 47*, 271–283.

Tribelli, P. M., Di Martino, C., Lopez, N. I., et al. (2012). Biofilm lifestyle enhances diesel bioremediation and biosurfactant production in the Antarctic polyhydroxyalkanoate producer *Pseudomonas extremaustralis*. *Biodegradation, 23*, 645–651.

Turner, J., Bindschadler, R., Convey, P., et al. (2010). *Antarctic climate change and the environment. A contribution to the International Polar Year 2007–2008*. Scientific Committee Antarctic Research (SCAR). Available via: http://acce.scar.org/wiki/Antarctic_Climate_Change_and_the_Environment. Accessed 8 Feb 2018.

Tyagi, M., da Fonseca, M. M. R., & de Carvalho, C. C. C. R. (2010). Bioaugmentation and biostimulation strategies to improve the effectiveness of bioremediation processes. *Biodegradation, 22*, 231–241. https://doi.org/10.1007/s10532-010-9394-4.

van Beilen, J. B., & Funhoff, E. G. (2007). Alkane hydroxylases involved in microbial alkane degradation. *Applied Microbiology and Biotechnology, 74*, 13–21. https://doi.org/10.1007/s00253-006-0748-0.

Vasileva-Tonkova, E., & Gesheva, V. (2005). Glycolipids produced by Antarctic *Nocardioides* sp. during growth on n-paraffin. *Process Biochemistry, 40*, 2387–2391. https://doi.org/10.1016/j.procbio.2004.09.018.

Vasileva-Tonkova, E., & Gesheva, V. (2007). Biosurfactant production by antarctic facultative anaerobe *Pantoea* sp. during growth on hydrocarbons. *Current Microbiology, 54*, 136–141. https://doi.org/10.1007/s00284-006-0345-6.

Vázquez, S. C., Coria, S. H., & Mac Cormack, W. P. (2004). Extracellular proteases from eight psychrotolerant Antarctic strains. *Microbiological Research, 159*, 157–166.

Vázquez, S., Nogales, B., Ruberto, L., et al. (2009). Bacterial community dynamics during bioremediation of diesel-oil contaminated Antarctic soil in a mesocosm assay. *Microbial Ecology, 57*, 598–610.

Vázquez, S., Nogales, B., Ruberto, L. C., et al. (2013). Characterization of bacterial consortia from diesel-contaminated Antarctic soils: Towards the design of tailored formulas for bioaugmentation. *International Biodeterioration and Biodegradation, 77*, 22–30. https://doi.org/10.1016/j.ibiod.2012.11.002.

Vázquez, S., Monien, P., Pepino Minetti, R., et al. (2017). Bacterial communities and chemical parameters in soils and coastal sediments in response to diesel spills at Carlini Station, Antarctica. *Science of the Total Environment, 15*, 26–37. https://doi.org/10.1016/j.scitotenv.2017.06.129.

Vergani, L., Mapelli, F., Zanardini, E., et al. (2017). Phyto-rhizoremediation of polychlorinated biphenyl contaminated soils: An outlook on plant-microbe beneficial interactions. *Science of the Total Environment, 575*, 1395–1406. https://doi.org/10.1016/j.scitotenv.2016.09.218.

Vincent, W. F. (1988). *Microbial ecosystems of Antarctica* (p. 290). Cambridge: Cambridge University Press. https://doi.org/10.1017/S0954102089220278.

Vincent, W. F. (2002). Cyanobacterial dominance in the polar regions. In B. A. Whitton & M. Potts (Eds.), *The ecology of cyanobacteria: Their diversity in time and space* (pp. 321–340). Berlin: Springer.

Vishniac, H. S. (1993). The microbiology of Antarctic soils. In E. I. Friedmann (Ed.), *Antarctic microbiology* (pp. 297–341). New York: Wiley-Liss.

Wall, D. H. (2007). Implications of change: Above- and belowground biotic interactions in a low diversity ecosystem. *Transactions of The Royal Society B Biological Sciences, 362*, 2291–2306. https://doi.org/10.1098/rstb.2006.1950.

Wynn-Williams, D. D. (1996). Antarctic microbial diversity: The basis of polar ecosystem processes. *Biodiversity and Conservation, 5*, 1271–1293.

Wyrwas, B., Chrzanowski, Ł., Ławniczak, Ł., et al. (2011). Utilization of Triton X-100 and polyethylene glycols during surfactant-mediated biodegradation of diesel fuel. *Journal of Hazardous Materials, 197*, 97–103. https://doi.org/10.1016/j.jhazmat.2011.09.060.

Yang, S. Z., Jin, H. J., Wei, Z., et al. (2009). Bioremediation of oil spills in cold environments: A review. *Pedosphere, 19*, 371–381.

Yen, K. M., & Gunsalus, I. C. (1982). Plasmid gene organization: Naphthalene/salicylate oxidation. *Proceedings of the National Academy of Sciences, 79*, 874–878.

Yergeau, E. (2013). Fell-field soil microbiology. In D. A. Cowan (Ed.), *Antarctic terrestrial microbiology: Physical and biological properties of Antarctic soils* (pp. 115–129). Heidelberg: Springer. https://doi.org/10.1007/978-3-642-45213-0_7.

Yergeau, E., Newsham, K. K., & Pearce, D. A. (2007). Patterns of bacterial diversity across a range of Antarctic terrestrial habitats. *Environmental Microbiology, 9*, 2670–2682.

Yergeau, E., Bokhorst, S., Kang, S., et al. (2012). Shifts in soil microorganisms in response to warming are consistent across a range of Antarctic environments. *The ISME Journal, 6*, 692702. https://doi.org/10.1038/ismej.2011.124.

Chapter 8
Airborne Microorganisms in Antarctica: Transport, Survival and Establishment

Katie King-Miaow, Kevin Lee, Teruya Maki, Donnabella LaCap-Bugler, and Stephen David James Archer

Abstract Microorganisms are a globally ubiquitous component of the atmosphere, of vital importance to climate, human health and environmental processes. Bioaerosols (which include viable fungi, prokaryotes, pollen and viruses as well as biologically derived remnants) are suspected to have a fundamental role in structuring the composition and function of ecosystems globally. Antarctica presents a tractable opportunity to study the dispersal of airborne microorganisms due to its isolation and its simple, microbially dominated ecosystems. Recent advances in technology have begun to shed light on the poorly understood Antarctic aerosphere, with most research focusing on bacteria. This chapter summarises the current knowledge regarding the movement and behaviour of bioaerosols in the global atmosphere, followed by the role that the air plays as a vector of microbes to Antarctica, and an overview of Antarctic bioaerosol research. Survival mechanisms of microbes in the harsh Antarctic terrestrial and atmospheric environments are outlined, followed by a discussion of the potential effects that aerial input to Antarctic ecosystems may have in the face of climate change. Bioaerosols are found to be highly changeable over space and time, with concentrations and compositions influenced by a myriad of variables, particularly climatic factors such as wind speed and temperature. Although studies of Antarctic bioaerosols have confirmed the extremely low biomass predicted in its atmosphere compared with temperate zones, greater biodiversity has been discovered as technology has improved. Multiple lines of evidence indicate that bioaerosols have been globally transported over great distances.

K. King-Miaow · K. Lee · D. LaCap-Bugler
Institute for Applied Ecology New Zealand, Auckland University of Technology, Auckland, New Zealand

T. Maki
Department of Chemical Engineering, Kanazawa University, Kanazawa, Japan

S. D. J. Archer (✉)
Institute for Applied Ecology New Zealand, Auckland University of Technology, Auckland, New Zealand

Yale-NUS College, National University of Singapore, Singapore, Singapore
e-mail: sarcher@aut.ac.nz

S. Castro-Sowinski (ed.), *The Ecological Role of Micro-organisms in the Antarctic Environment*, Springer Polar Sciences,
https://doi.org/10.1007/978-3-030-02786-5_8

While many microbes are believed to survive in the atmosphere as spores, some species may remain metabolically active and could contribute to certain atmospheric processes. The evidence of continual bioaerosol deposition and theorised significance to current ecosystem structuring suggests that as the Antarctic climate changes, deposited microorganisms could drive rapid community shifts. This chapter identifies numerous knowledge gaps in the field, including the variability, environmental drivers, source (where) and extent (how much) of Antarctic airborne microorganisms. Given the predicted importance of airborne transportation to Antarctic ecosystems, it is essential to substantially increase research effort to gain a more comprehensive view of the extreme Antarctic aerosphere.

Keywords Airborne microorganisms · Aerosphere particles · Bioaerosol particles · Microbial survival mechanisms · Propagule bank

8.1 Introduction

Microorganisms (including fungi, bacteria and viruses) have been shown to be numerous and ubiquitous in the atmosphere and are an important component of bioaerosols, which include all particles of biological origin (Burrows et al. 2009b; Reche et al. 2018; Pearce et al. 2016). Bioaerosols within the natural environment are poorly understood, due to a lack of standardised methodology and little data, resulting in abundant conjectures (Burrows et al. 2009b; Pearce et al. 2016). Recent technological innovations have enabled DNA recovery and interrogation of airborne microorganisms from a wide range of environments, which allows researchers to address fundamental questions regarding the importance of this biosphere.

The Antarctic aerial biome represents one of the most challenging environments on Earth. Near surface temperatures drop to as low as −93.2 °C, and summer temperatures reach up to 15 °C (Mackintosh 2001). Ultraviolet (UV) radiation is high, winds frequently exceed 100 km/h and humidity is very low (Pearce et al. 2009). As a result, the aerosol biomass is anticipated to be one of the lowest and is the least understood in the world (Burrows et al. 2009b). To achieve intercontinental transport and reach Antarctica, it is thought that microbes need to access the upper levels of the atmosphere, where wind speeds are consistently high. The harsh Antarctic environment shares characteristics with the upper levels of the global atmosphere (Herbold et al. 2014), suggesting that intercontinental airborne transport could select for survival after deposition in Antarctica (Pearce et al. 2009). There is strong evidence that bioaerosols use the atmosphere as a long-distance transport vector to Antarctica, for example, South American pollen is frequently found in Antarctica (Wynn-Williams 1991; Vincent 2000; Smith 1991). Continental Antarctica provides an excellent natural laboratory to study global aerial transport processes as it is isolated by the Southern Ocean and prevailing wind and water currents (Pearce et al. 2016), its reduced human and other animal vectoring means the atmosphere becomes virtually the sole transport mechanism available for microbes (Pearce et al. 2016) and understanding transport processes is a tractable problem due to Antarctica's simple microbially dominated ecosystems (Bottos et al. 2014). The Antarctic continent is the most rapidly warming region on Earth, with the Antarctic Peninsula experiencing a temperature increase of 3 °C over the last 50 years (Pearce et al. 2009). In a warming world, shifts in microbial communities may result in extinction of unique endemic species. Microbes are known to be first responders to change. Therefore, understanding how Antarctic communities may shift, informs what can be expected globally in the future (Bottos et al. 2014). Despite technological developments, study of bioaerosols is challenged by their stochasticity in distribution, variability and low biomass (Womack et al. 2010), which is accentuated in Antarctica (Pearce et al. 2016). As a result, very little is known about Antarctic bioaerosols and methodology development remains a barrier to widespread investigation (Pearce et al. 2016).

Most researchers believe that the majority of passively transported bioaerosols are in a dormant form (Womack et al. 2010; Pearce et al. 2016); however, there is growing evidence that some may continue to metabolise while in transit. These organisms may represent atmospheric residents (Womack et al. 2010) and could substantially alter the chemical constituents of the atmosphere. Most bacteria in the atmosphere are thought to act as cloud condensation nuclei (Burrows et al. 2009b), and ice nucleation activity could increase cloud formation and precipitation that could fundamentally affect global weather patterns (Behzad et al. 2015; Sattler et al. 2001; Burrows et al. 2009b; Pearce et al. 2016). Many important plant and animal diseases are also aerially transmitted, such as foot and mouth disease and legionnaires disease (Nguyen et al. 2006), highlighting the need to understand atmospheric microbial transfer. It was once thought that "everything is everywhere and the environment selects" (put by Dutch microbiologist Lourens G. M. Bass Becking (O'Malley 2008)); however more recent research has revealed bioaerosols are highly variable over space and time (Bowers et al. 2011a, 2013; Woo et al. 2013; Fierer et al. 2008). Many microbes appear to display biogeography (varying species distribution over space), despite their ease of dispersal (Womack et al. 2010; Bahl et al. 2011; Pointing et al. 2015; O'Malley 2008). A microbe's propensity for aerial transport is thought to be a key influencer of its ability to disperse (Sokol et al. 2013; Sommaruga and Casamayor 2009; Pearce et al. 2016). Research suggests the presence of two biogeographic zones in Antarctica, the Peninsula and the rest of the continent, which is theorised to be driven by differing microbial exchange into and across Antarctica between the zones (Chong et al. 2015).

This chapter explores what is known about the aerial microbiome in Antarctica and some of the questions that remain. The global atmosphere is defined and described, followed by the process of bioaerosol launching and elucidation of aerosol behaviour once in suspension. Mechanisms of long-range aerial dispersal are discussed, followed by the processes that remove particles from suspension. What is known regarding variation of bioaerosols over space and time is summarised. Bioaerosol sampling challenges and solutions to difficulties are listed. The importance of the atmosphere as a vector to Antarctica, as well as the status of Antarctic bioaerosol research, is described. The mechanisms that microbes use to survive in the atmosphere are then reviewed and the existence of aerial residents is explored.

The chapter finishes with effects of aerial vectoring on Antarctic ecosystem structuring and how microbial communities may respond to continued climate change. The chapter content represents a comprehensive review of the field of bioaerosol study, with a focus on the small body of work that has been performed in Antarctica. Given the hypothesised importance of aerial vectoring to Antarctic ecosystems, understanding the aerial contribution is crucial to understanding microorganisms in the Antarctic environment.

8.2 The Aerosphere and Bioaerosol Particles

8.2.1 Particle Movement in the Atmosphere

The atmosphere consists of a layer of gases which surrounds the Earth, containing particles of various sizes and types, such as water droplets, inorganic particles like mineral dust and particles of biological origin. Particles suspended in air are called aerosols; those of biological origin are termed bioaerosols. Most aerosol particle monitoring is focused on human health and as a result uses two health relevant size classifications, $PM_{2.5}$ and PM_{10}, which refer to the aerodynamic diameter of the particle (<2.5 μm and <10 μm). Bioaerosol particles vary widely in size, from pollen up to 1000 μm in diameter, bacteria which range between 0.25 and 8 μm in diameter, to viruses at <0.3 μm in diameter, and fragments of biological material thereof. Air is in constant motion, driven by differences in atmospheric pressure, which cause its constituents to be highly variable over time and space. The concentration of aerosol particles in the atmosphere is influenced by a complex set of variables not comprehensively understood, including local biological sources and changes in meteorological conditions (Burrows et al. 2009b).

The vertical structure of the atmosphere is divided into predictable layers which are driven by temperature (Fig. 8.1), although the altitude at which layers transition varies (Pearce et al. 2009). At greater altitude wind speeds typically increase and conditions become less favourable for survival (Pearce et al. 2009). The characteristics of the atmosphere determine both the transport range (Archer and Caldeira 2009) and the viability of bioaerosols (Womack et al. 2010). The troposphere stretches from the ground to about 10 km and is where the majority of atmospheric mass, including bioaerosols, is located. The troposphere also contains the atmospheric boundary layer, the region of the atmosphere that transitions from turbulent air flow from the Earth's surface to a calmer, laminar flow layer. The boundary layer mediates exchange of particles between the Earth and the atmosphere and its altitude varies depending on atmospheric conditions and terrestrial topography (Rotach et al. 2015). At between around 10 km and 50 km, altitude sits the stratosphere, where airflow is fast, predictable and horizontal. The troposphere and the stratosphere are of most interest for the purposes of bioaerosol research, as most microbial isolates are from these levels (Burrows et al. 2009b), although some studies have found culturable organisms in the mesosphere, at up to 77 km altitude (Imshenetsky et al. 1978).

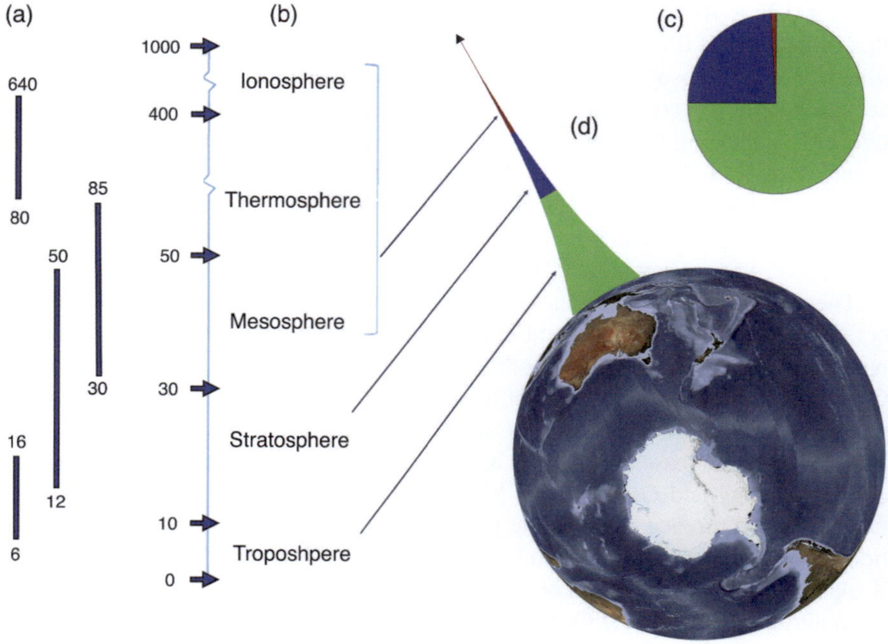

Fig. 8.1 (**a**) Altitude ranges of each atmospheric layer (in km), (**b**) names of atmospheric layers, (**c**) adjusted logarithmic scale by height and density with cumulative area as in (**d**), (**d**) speculative proportions of bioaerosols found in each layer (Pearce et al. 2009), artwork by Chris King

The process of aerosolisation of particles is known as launching (Pepper et al. 2015) and can be either active or passive. Active launching includes forcible ejection of biological material as seen with fungal spores or the process of sneezing. Passive launching results from abiotic processes acting on a reservoir of particles, such as wind blowing over soil or plants, waves breaking on a beach or the bursting of bubbles in water (Burrows et al. 2009b). Rates of passive emission vary based on density of particles on a surface and local meteorological conditions. Turbulence from the Earth's surface creates chaotic air movement in the atmospheric boundary layer, including vertical movement, which can propel particles to higher atmospheric regions. Turbulence disrupts the flow of particles when air is forced around an object or subjected to excessive shear. The Reynolds number (velocity × dimension/viscosity) can estimate the amount of turbulence and is based on wind velocity, viscosity of the air and dimensions of the interfering surface (Pepper et al. 2015). Any result over 2000 is deemed to be turbulent air flow (Pepper et al. 2015). The higher this number, the more movement of particles will occur in a given time, and the higher the extent of aerosolisation. As events that propel particles to high altitudes are rarer, the majority of particles are thought to remain in the atmospheric boundary layer and as a result have short transport ranges. The minority of particles that do escape the boundary layer, to the tropopause, or stratosphere where the airflow is much faster and more uniform, can be rapidly transported on a global scale. Reche et al. (2018) estimated downwards viral flux above the atmospheric boundary layer of 0.26×10^9 to $>7 \times 10^9$ m^{-2} viral particles per day. They found downwards flux of bacteria of 0.3×10^7 to $>8 \times 10^7$ m^{-2} per day. So, although this represents a minority of bioaerosols, significant numbers still escape the boundary layer. There are few studies of vertical distribution of bioaerosols and most of these are qualitative, only indicating the presence or absence of microbes (Maki et al. 2008, 2017; Wainwright et al. 2003). Much of the current research supports reductions in concentration with altitude (Fulton and Mitchell 1966; Fulton 1966a, b); however some studies do not show a clear relationship between concentration and altitude (Matsuki et al. 2003; Andreeva et al. 2002).

There are various events which can propel particles to different atmospheric layers (Fig. 8.2). Storms are a good example of frequent processes which transport bioaerosols upwards. DeLeon-Rodriguez et al. (2013) found hurricanes in the Caribbean created large amounts of bioaerosols that were launched into the tropopause. Dust storms in desert areas are also known to launch large quantities of aerosols, with increases in number and diversity of microbes evident during dust events (Kellogg and Griffin 2006; Maki et al. 2011). Volcanic eruptions are less frequent in occurrence but have the potential to propel substantial numbers of particles to the stratosphere and some cases even the mesosphere (Diallo et al. 2017). Krakatoa erupted in 1883 and observers at the time estimated that ash was ejected to an altitude of 80 km (O'Connell 2006). The effective transportation of thermophilic organisms over large distances through volcanic eruption is supported by the detection of similar organisms in widely dispersed geothermal sites (Herbold et al. 2014). Other impact events such as meteorites landing are rarer still but could feasibly transport particles throughout the atmosphere. Survival at these high altitudes has been shown to be possible, with some experiments showing bacterial survival alongside spacecraft lift-off and re-entry (Rettberg et al. 2002). There is also evidence that *Bacillus subtilis*, a spore-forming, gram-positive bacterium that is frequently found in bioaerosols, can survive in space for at least 6 years (Horneck 1993).

Although large-scale movements are driven primarily by air currents, small-scale movements of airborne particles depend on Brownian motion (the random movements in a fluid from collisions with fast moving molecules and other suspended particles). Particles diffuse from a source down a concentration gradient, with trajectory and speed of diffusion influenced by air currents and gravity (Pepper et al. 2015). Particles with high mass require more force to change direction (force = mass × acceleration) meaning that they are more likely to impact on a surface and be removed from the air (Pepper et al. 2015).

Particles from deserts are small in size, have long residence times and are likely to be significant contributors to global transportation of bioaerosols (Fig. 8.3) (Kellogg and Griffin 2006; Hara and Zhang 2012). The transport of dust and associated organisms has been supported by similarities of dust in different areas. Caribbean and African isolates show very high ribosomal RNA gene sequence similarity, suggesting a common source of bioaerosols (Kellogg et al. 2004; Kellogg and Griffin 2006). Reche et al. (2018) found Saharan particles in Spanish air and particles in Florida in summer are often found to be of African origin (Prospero

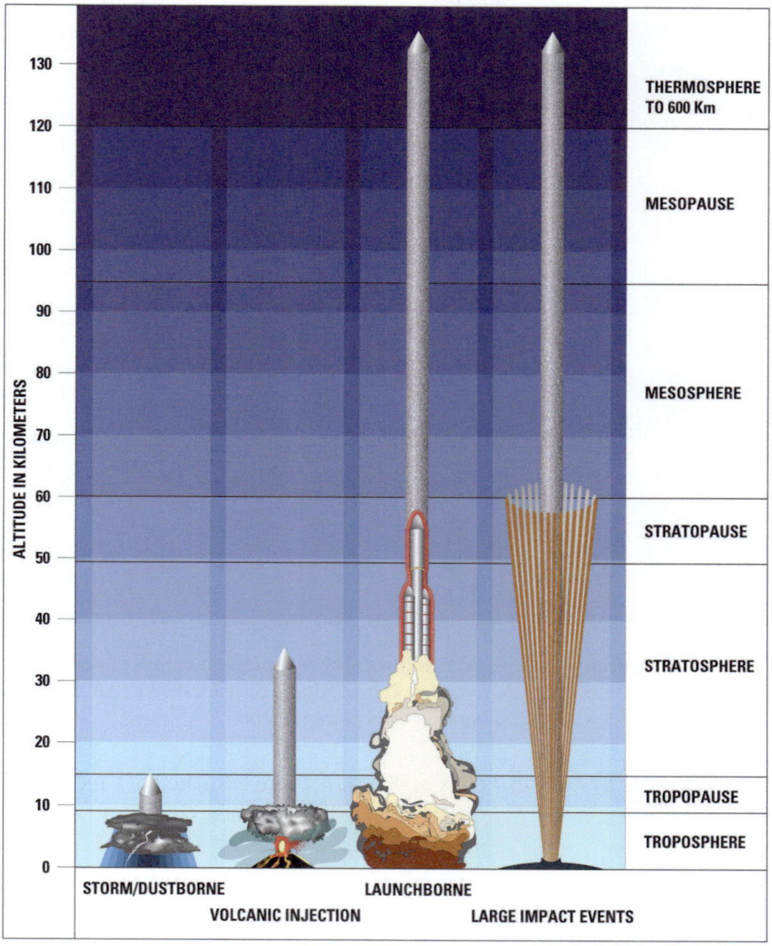

Fig. 8.2 Mechanisms of microbial movement into the atmosphere (Griffin 2004), artwork by Bevy Boyton. Storms can launch aerosols into the tropopause and volcanoes to the stratosphere, and rockets or meteorites can propel particles as far as space

1999). Maki et al. (2011) studied Asian kosa dust and found that after dust events in China, there were significantly more culturable bacteria in snows in Japan, which had been contaminated by the dust.

Air cells are created by differential heating of the Earth between the equator and the poles and the Earth's rotation. Predictable air cells above the boundary layer facilitate movement in the atmosphere horizontally in the prevailing wind direction of the cell (Fig. 8.4). For long-distance transport, bioaerosols need to get above the atmospheric boundary layer, to these air cells, where wind speeds have been estimated to be up to 90 km/h (Miller et al. 2011). Bacterial residence times (the time that a bacterial cell, on average, is expected to remain aloft in the atmosphere) are modelled to be between 3 and 7 days, although this may differ for other bioaero-

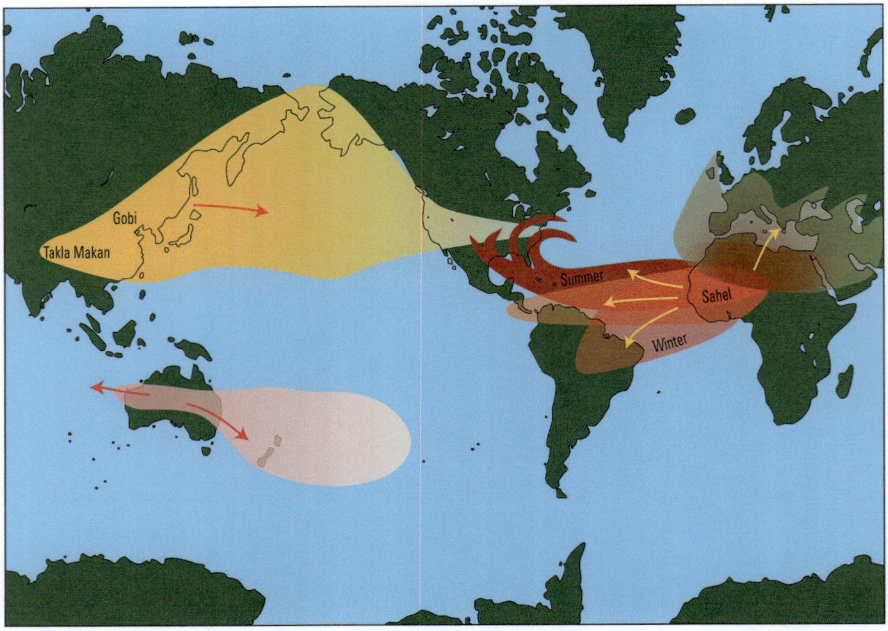

Fig. 8.3 Storms have been shown to be capable of transporting large numbers of particles (Kellogg and Griffin 2006). The three main sources of worldwide desert dust are shown (Australia, the Sahel in Africa and China) along with their typical seasonal movements

sols (Burrows et al. 2009b). In 7 days at 90 km/h, there is sufficient time for a microorganism to be transported over 15,000 km (the distance from the South Pole to London is 15,710 km).

Particles are removed from aerial suspension through particle deposition. Gravitational settling is the gravitational pull of any particle heavier than air downwards and is the most common cause of deposition. Therefore, bioaerosols with lower mass will tend to stay aloft longer and have longer residence times and consequently greater transport range. Gravitational settling can be described by Stokes's law, which considers gravity, particle density, particle diameter and air viscosity.

$$\mathrm{V} = \left[D^2 \times \left(\mathrm{p_p} - \mathrm{p_l}\right) \times \mathrm{g}\right] / 18\mathrm{p}$$

V = velocity of fall (m/s)
g = gravity (9.8 m/s^2)
D = diameter of particle (m)
p_p = density of particle (kg/m^3)
p_l = density of dispersion medium (kg/m^3)
p = viscosity of dispersion medium (kg/m*s)

Particles can be deposited on a surface through impaction (when they collide with the surface) also known as dry deposition. The particle loses kinetic energy and

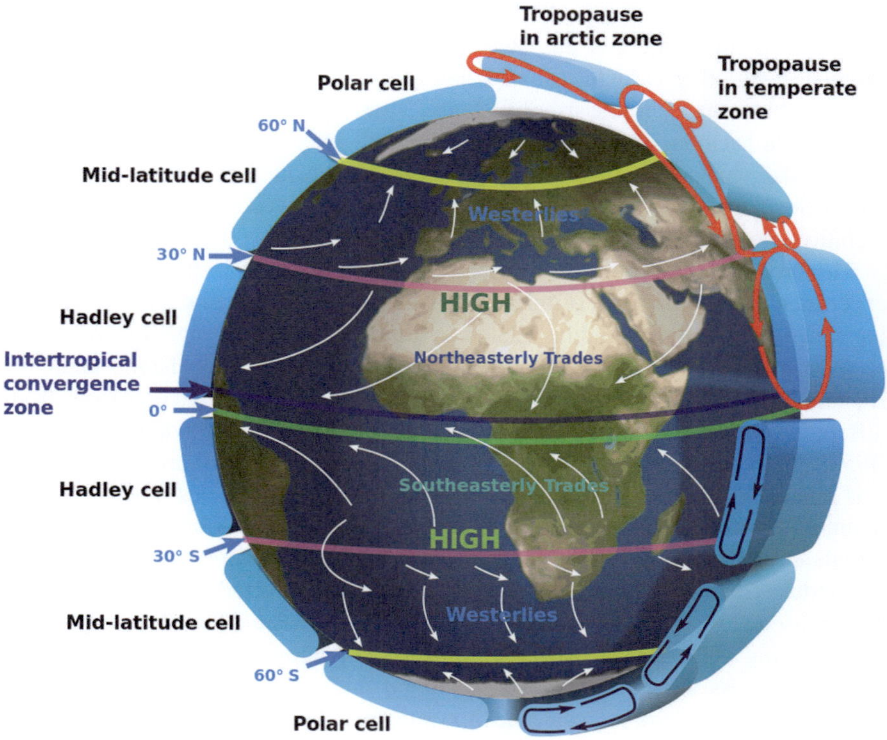

Fig. 8.4 Diagram of the major air cells and wind currents in the troposphere (Kaidor 2013). These winds influence trajectory of airborne transport of microbes

can come to rest on the surface, or deflect and return to the air flow with reduced kinetic energy, increasing the chance of settling. Probability of impaction on a surface depends on velocity, particle diameter and the size of the surface. Wet deposition is mediated by rain or other precipitation. As rain or snow falls, it collides and combines with aerosol particles creating a particle of greater mass that can settle faster. Efficiency of wet deposition depends on the spread area of the particle plume with diffused plumes experiencing stronger impaction. Electrostatic deposition can also occur when particles with opposing charges are attracted to one another, creating particles with greater mass. Bacteria are often negatively charged and therefore have a tendency to become attracted to positively charged particles in the atmosphere or surfaces, which increases deposition (Pepper et al. 2015).

8.2.2 *Bioaerosol Particles: Temporal and Spatial Variation*

Bioaerosols constitute a significant and variable portion of aerosol particles, up to 28% of total aerosol volume (Matthias-Maser et al. 2000). Bioaerosols are thought to frequently exist in assemblages with other inorganic or organic particles (Burrows et al. 2009b). A bacterium is around 1 μm in diameter, whereas typical bacterial associated particles are 2–4 μm, suggesting that multiple cells clump together or associate with other particles (Shaffer and Lighthart 1997). Huffman et al. (2010) observed bioaerosol particle sizes of 1.5 μm, 3 μm, 5 μm and 13 μm. 1.5 μm is likely to be single bacteria, 3 μm is likely to be multiple bacterial cells or fungal spores, 5 μm is likely to be fungal spores and the 13 μm is likely to be pollen (Huffman et al. 2010). Several authors have proposed particle association could improve bioaerosol viability (see Sect. 8.4 onwards).

The broad and rapid dispersal of particles from a source in the air means that experimental determination of microbial residence times remains infeasible, leaving much of our understanding related to atmospheric modelling. Burrows et al. (2009a) modelled the aerosol concentration of bacteria in the global atmosphere for six simulated years, using global meteorological models, estimated emissions from different ecosystems, estimated residence times and estimated deposition rates (Fig. 8.5). In the simulation, a mean microbial residence time of around one week was calculated. There was significant variation, with some particles staying aloft for months. Meteorological variables significantly affect residence times, for instance, turbulence in the atmosphere will increase residence times, while precipitation in an area reduces residence times. Dry regions have longer expected residence times due to increased vertical movement in the boundary layer, related to greater turbulence from heating and reduced wet deposition. As the number of bioaerosol studies increases across a range of spatial, temporal and environmental variables, our understanding of microbial residence time influences will increase. Physical characteristics of particles, such as size or surface structure, also affect residence time. Particles with a smaller diameter or lower mass tend to have greater residence times,

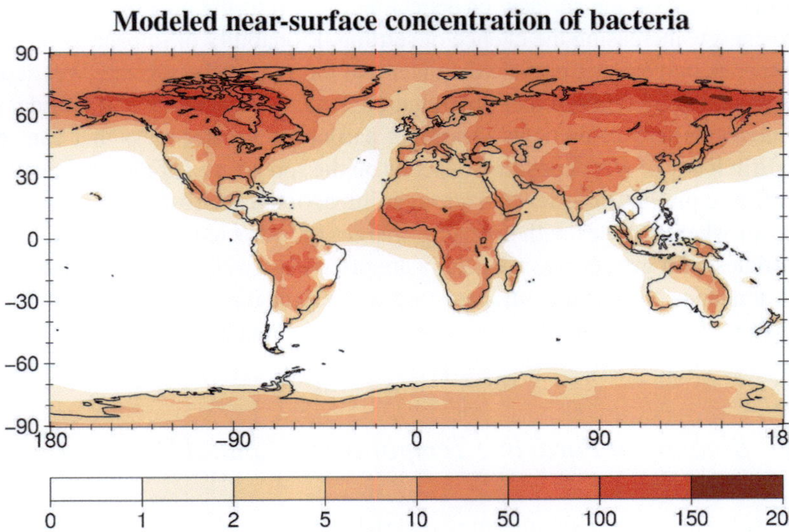

Fig. 8.5 Heat map of the simulated concentration of 1 μm bacteria in the troposphere according to modelling by Burrows et al. (2009a)

due to reduced deposition. It is thought that particles under 1 μm in diameter (which include viruses and some free-living bacteria) fall within the "scavenger gap" and have the longest residence times (Burrows et al. 2009a). This assumption is one of the bases for Burrows's simulation. The surface structure of pollen can alter its aerodynamic properties and aid its dispersal by wind (Niklas 1985). Given similar selective pressures apply to other types of bioaerosol, they may use similar mechanisms to enhance dispersal. Some bacteria, such as *Pseudomonas* spp., have cell surface ice nucleation proteins (Pearce et al. 2009) and are believed to act as cloud condensation nuclei (Bauer et al. 2002). It is thought that organisms which act as cloud condensation nuclei have increased wet deposition in the atmosphere and therefore have reduced residence times (Burrows et al. 2009a).

Reliable estimates of bioaerosol concentrations from all ecosystems are not available and variable methodologies make comparison difficult. Estimates based on available information show that there is significant variation in bioaerosol concentration by ecosystem. Generally higher concentrations are observed over more productive ecosystems (grassland and crops), likely due to larger microbial source populations from which microbes can be aerosolized (Harrison et al. 2005; Tong and Lighthart 2000). Cities showed high concentrations due to frequent vehicle movements aerosolising microorganisms (Shaffer and Lighthart 1997). In the USA, it was found that cities harbour a distinct aerial microbiome to rural areas and that bioaerosols in cities tended to be more homogenous than in rural areas (Barberán et al. 2015). Desert areas have low estimated concentrations of bioaerosols by mass due to low source biomass. However, due to the ease of aerosolisation of particles from desert surfaces and long residence times of desert particles, desert dust is still thought to play a significant role in global microbial dispersal (Shaffer and Lighthart 1997; Maki et al. 2017; Kellogg and Griffin 2006; Bowers et al. 2011b).

Burrows's model shows that after 3 years, bacteria reach an effective distribution equilibrium and are present in all locations. This suggests that short- and long-range atmospheric transportation are highly effective at dispersing microorganisms globally (Fig. 8.5). Later studies of bioaerosols support the high concentration heterogeneity predicted by the model (Bowers et al. 2011b, 2013; Barberán et al. 2014; Grantham et al. 2015). The application of the model to Antarctic bioaerosols is discussed in Sect. 8.3. Bioaerosols have been found to be highly variable over various time scales with stochasticity in concentration and community structure over a short timeframe frequently more pronounced than seasonal changes (Burrows et al. 2009b; Fierer et al. 2008). Some cultivation-based studies have shown that seasonal variation is very likely to be driven by changes in meteorological conditions, occurring predictably over the course of the year. The highest average concentration of bioaerosols is thought to be at times of maximum productivity, generally in the summer (Tong and Lighthart 2000; Lighthart and Stetzenbach 1994). However, this could be confounded by significant seasonal variations in culturability of microbes (Burrows et al. 2009b). In a molecular-based study, Woo et al. (2013) found highest microbial loading during summer and a community composition shift over the 12-month period, thought to be driven by changes in air mass origin from terrestrial to marine. This variation extends to observed diurnal patterns linked to solar heating as net upwards flux was found to be highest in the warmest part of the day (Tong and Lighthart 2000; Shaffer and Lighthart 1997; Chen et al. 2001). Higher temperatures increase turbulence, wind speeds, vertical mixing, residence times and division rates, increasing the microbial load and aerosolisation.

8.3 The Antarctic Aerosphere

Bacterial concentrations over Antarctica were estimated to be very low from Burrows's simulation and this has since been verified (Pearce et al. 2016; Bottos et al. 2014). Polar areas are thought to have low concentrations of bioaerosols, due to a lack of microbial activity in those areas and the physical stability of frozen surfaces. Pearce et al. (2009) found a large variation of microbial concentrations in different polar and ice ecosystems, although all showed evidence of life. Since there is a lack of data from ice environments, Burrows's simulation used average background estimates of bioaerosol concentration as a maximum. The other relevant ecosystem for consideration of Antarctic bioaerosols is the sea, since it surrounds Antarctica. Seas are estimated to have low bacterial concentrations of 1×10^4 m^{-3} based on cultivation studies (Harrison et al. 2005; Bauer et al. 2002; Kellogg and Griffin 2006) although this could be an underestimate given marine bacteria are on average less culturable than terrestrial bacteria (Parks et al. 2017). In nutrient-rich marine regions, such as those found around Antarctica (Fripiat et al. 2017), concentrations of bacteria are often much higher likely resulting in higher emissions (Cho and Azam 1990). Bioaerosol sampling in Antarctica is insufficient to reveal the level of marine aerial input, although a couple of studies show limited marine taxa despite proximity to water (Bottos et al. 2014; Pearce et al. 2010). In Burrows's simulation, Antarctica was decoupled from the rest of the world, due to its isolation by the Southern Ocean. However, residence times of particles in Antarctica were estimated to be high, despite low emissions and most particles that circulated in Antarctica were expected to originate there. Some of these assumptions have since been supported by other authors, based on later Antarctic bioaerosol research (Bottos et al. 2014; Crawford et al. 2017; Pearce et al. 2010). Estimates of the bioaerosol fraction of Antarctic aerosols indicate that they constitute only a small proportion of total aerial particles (under around 2%) but also that they can vary significantly (Crawford et al. 2017). This low biological particle fraction in Antarctica is likely driven by low ecosystem productivity and large amounts of dust and other aerosols in the atmosphere.

Westerlies move air from Africa and South America towards Antarctica (Fig. 8.4). Frequent discoveries of temperate pollen (Wynn-Williams 1991) in Antarctica indicate that these winds could facilitate intercontinental transport. Within the Antarctic continent, there are also characteristic air movement patterns such as the circumpolar vortex, which is a ring of low pressure systems that creates circular air movement around the continent and provides a barrier to entry to the continent (Fig. 8.6). Powerful Antarctic katabatic winds, created as cold high density air flows downwards towards the sea under the force of gravity (Parish and Cassano 2003) which, combined with local weather cells, are thought to distribute microorganisms effectively around the continent (Nkem et al. 2006; Pearce et al. 2009). Deuterated methane was released from a plane at 5.5 km altitude in the maritime Antarctic and within a week was detected all around Antarctica, indicating rapid dispersal on a continental scale (Mroz et al. 1989). For the purposes of bioaerosol movements, the Antarctic Peninsula can be thought to be somewhat isolated from the rest of the continent, as the Peninsula sits outside of these typical wind patterns and forms a distinct biogeographic zone (Chong et al. 2015).

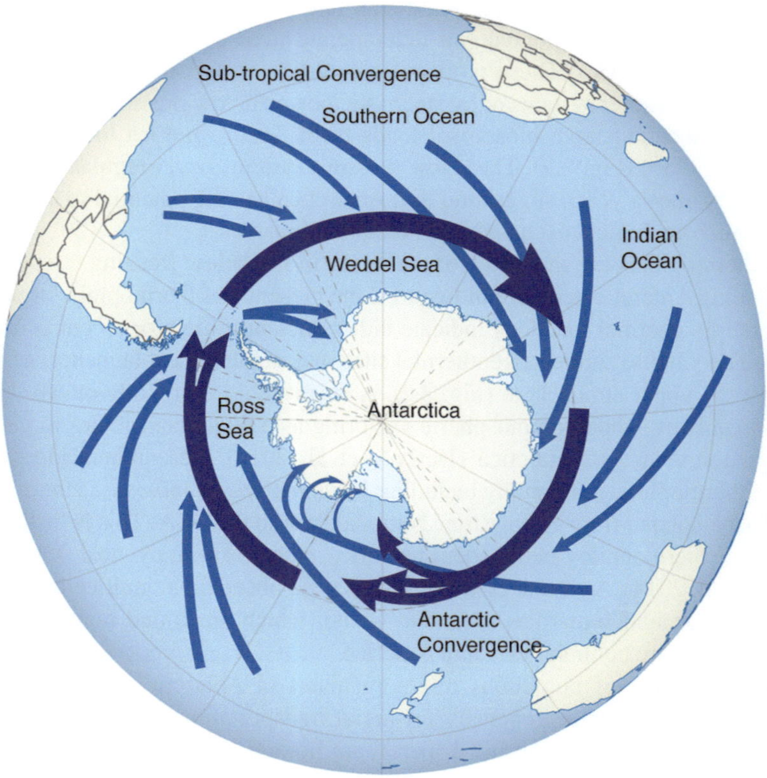

Fig. 8.6 Predominant wind movements over Antarctica during summer at 4 km altitude adapted by Chris King from (Wynn-Williams 1991). Broken arrows are frequent cyclone tracks, solid ones are more occasional tracks. The circumpolar vortex is shown

8.3.1 Importance of Aerial Transport of Microorganisms to and Within Antarctica

There are three lines of evidence that support the atmosphere as a long-range vector to Antarctica based on findings in studies of ecosystems throughout Antarctica: firstly, exotic propagules found in Antarctica, secondly isolated thermophiles at remote geothermal sites and thirdly the presence of globally ubiquitous microorganisms. Exotic propagules have consistently been found in remote locations by numerous researchers (Wynn-Williams 1991; Marshall 1996a, 1997; Smith 1991). These include lichen spores and pollen granules from Patagonia that have been found at King George Island (Wynn-Williams 1991) and exotic species in Antarctica in ice sheets up to 400,000 years old (Vincent 2000), as well as plant pollen and fungal spores at Halley Bay (Gregory 1961), Signy Island and South Georgia Island (Gregory 1961; Smith 1991). The lack of alternative transport pathways strongly suggests these propagules arrived via the air and have travelled thousands of kilometres from their source populations in temperate areas of South America (Vincent 2000; Marshall 1996a). These propagules could be transported continually from high-altitude weather systems or by stochastic powerful low pressure systems capable of translocating large bioaerosol loads from South America to the Antarctic (Marshall 1996b) (Fig. 8.7). These low pressure systems occur regularly (approximately 1.5 times a year), suggesting that Antarctica has the potential to experience regular microbiological exchange with other land masses.

As thermophiles are incapable of surviving in the below freezing temperatures surrounding Antarctic geothermal features, the presence of thermophilic organisms at remote geothermal locations indicate either ancient refugia or recent aerial dispersal. The volatile nature of geothermal sites makes them ideal launchers of material into the upper atmosphere (Fig. 8.2). A recent molecular study on Mt. Erebus microorganisms supports atmospheric vectoring as a likely contributor to geothermal environments in Antarctica (Herbold et al. 2014). Researchers identified a diverse thermophilic community including *Mastigocladus laminosus*, a thermophile previously isolated from Yellowstone National Park in the USA. This is further supported by thermophiles being identified in a bioaerosol study near Mt. Erebus (Bottos et al. 2014). A similar distribution has been observed in multicellular organisms that surround fumaroles (volcanic vents), which are found nowhere else in Antarctica but are common in South America.

The abundance of ubiquitous microorganisms in Antarctic air and terrestrial samples also suggests a propensity for aerial transportation and establishment in Antarctica where other means of dispersal are limited. The McMurdo Dry Valleys have been well studied as one of the largest ice-free areas in Antarctica, where comparatively high microbial biomass is found (Pearce et al. 2016). The valleys are located on the coast, to the west of the McMurdo Sound and ice shelf. Bioaerosol samples taken in the McMurdo Dry Valleys showed the most prolific bacterial taxa are consistent with bioaerosols from other continents (Bottos et al. 2014). Microorganisms of exotic origin have also been detected in a range of terrestrial and

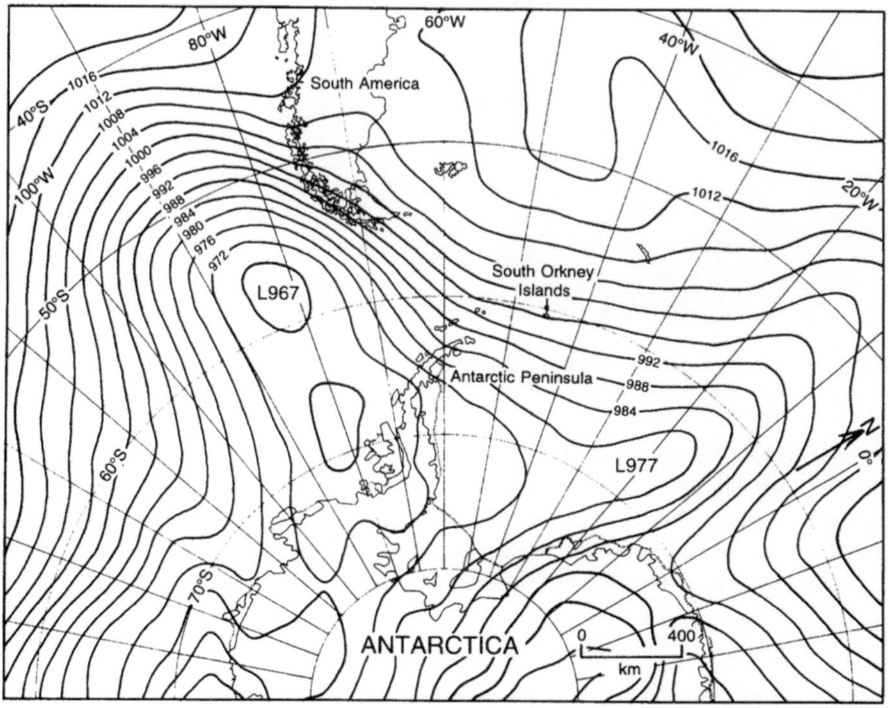

Fig. 8.7 A synoptic chart of 11/11/1993 showing a large low pressure system between South America and the South Orkney Islands (Marshall 1996b)

aquatic environments in Antarctica (Herbold et al. 2014; Archer et al. 2014; Sokol et al. 2013; Lee et al. 2012). Interestingly, many Antarctic studies noted a high sequence similarity between samples within Antarctica, suggesting organisms which reach Antarctica can effectively distribute themselves around the continent (Vincent 2000; Bottos et al. 2014; Pearce et al. 2016).

There is evidence that most organisms detected in Antarctica originated there (Bottos et al. 2014; Crawford et al. 2017; Pearce et al. 2010), which can be expected due to Antarctica's extreme isolation. This is consistent with the decoupling of the Antarctic from the rest of the world predicted by modelling (Burrows et al. 2009a). Therefore, it seems long-range aerial dispersal to Antarctica, although present, is rare and acts with more common short-range transport to fundamentally affect Antarctic microbial communities. Local wind has been directly shown to be an important factor for local dispersal of cyanobacteria and small eukaryotic organisms (Wood et al. 2008; Nkem et al. 2006). Dry Valleys with lakes had more cyanobacteria in the nearby soils than valleys without lakes (Wood et al. 2008) (Fig. 8.8), and wind-borne dispersal of faecal coliforms and avian-associated bacteria has been detected downwind of research stations and bird colonies (Hughes 2003; Kobayashi et al. 2016; Pearce et al. 2010). Nkem et al. (2006) observed frequent short-range wind dispersal of rotifers and tardigrades (small multicellular invertebrates) in the

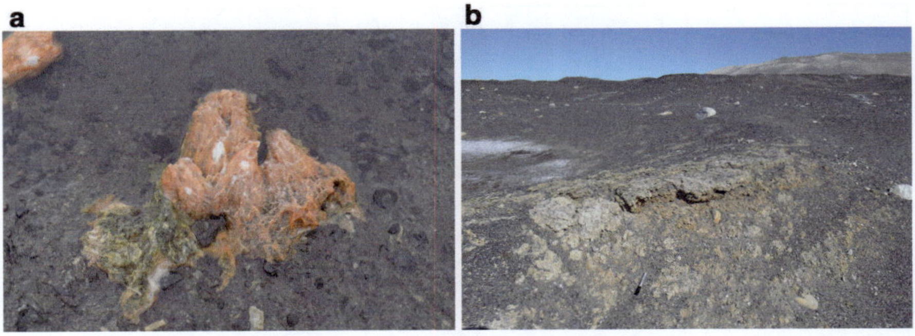

Fig. 8.8 Cyanobacterial mat in an Antarctic lake (**a**). These mats can dry out when water levels drop. When dry they break up and are easily transported by the wind (**b**)

McMurdo Dry Valleys. Additionally, large volumes of dust are relocated (typically under 30 km but up to 120 km) onto sea ice in the McMurdo Sound, from the Dry Valleys (Atkins and Dunbar 2009). Given microbes are known to frequently associate with dust and other particles (Burrows et al. 2009b), it is reasonable to assume substantial numbers of microbes would also be transferred.

8.3.2 Antarctic Bioaerosol Sampling: Challenges and Solutions

The extremely low biomass of the air presents a significant challenge to the collection and analysis of an unbiased representative bioaerosol community. Most air sampling to date either uses passive gravitational settling or active pumping, with impaction or liquid impingement to capture aerosol particles. Although early work on bioaerosols was a critical basis for later study, it drastically underestimated microbial biodiversity (Burrows et al. 2009b; Pearce et al. 2009). Early description of microorganisms in air was based on cultivation or microscopic identification; however, a large majority of microorganisms (70–99%) are uncultivatable (Burrows et al. 2009b), and microscopic identification cannot differentiate organisms with similar morphologies (Haig et al. 2016). Recent culture-independent community analysis based on nucleic acid (DNA or RNA) sequencing has revealed a plethora of diversity not apparent from microscopy or culture. Molecular analysis can provide information on total community's identity (amplicon sequencing) whole genomes (metagenomics) or gene transcription (transcriptomics) of bioaerosols (Behzad et al. 2015; Yoo et al. 2016).

Aerial sampling challenges are exacerbated in Antarctica where biomass sources in all systems are typically far lower and logistical constraints for studies are greater (Bottos et al. 2014), resulting in a marked lack of data. The current best practice for Antarctic bioaerosol sampling is to pump air through a 0.2 μm polycarbonate filter (Pearce et al. 2016). These filters are easy to run and can be left for long periods of

Fig. 8.9 Bottos is pictured setting up the air pump that was used to take the sample leading to the published paper "Airborne Bacterial Populations Above Desert Soils of the McMurdo Dry Valleys, Antarctica"

time to collect samples; however they have low flow rates and can take from 24 h to 2 months to collect sufficient biomass for analysis (as with Bottos et al. (2014) Fig. 8.9). The long sample duration on polycarbonate filters has been shown to disproportionately degrade the DNA of gram-negative bacteria in under 24 h due to desiccation stress (Luhung et al. 2015), indicating serious sample bias in existing Antarctic air sampling methodology. Samplers with liquid collection mediums and higher flow rates have been developed which aim to reduce sample time and bias but are limited in their use in sub-zero temperatures (Dybwad et al. 2014). Comparative testing reveals different samplers work best for capturing different types of bioaerosol, so careful choice of sampler for the desired application is crucial (Dybwad et al. 2014; Haig et al. 2016).

8.3.3 Bioaerosol Research in Antarctica

Between 1994 and 2014, 12 studies were conducted (both published and unpublished) on bioaerosols in Antarctica (Pearce et al. 2016) (Table 8.1, Fig. 8.10). All studies have shown that there was a much lower bioaerosol concentration during the time of sampling than would be expected in warmer climates. Most aerobiological studies to date have identified limited biodiversity; however as techniques and resolution have advanced, more recent molecular studies (Bottos et al. 2014; Pearce

Table 8.1 Summary of aerobiological studies undertaken in Antarctica 1994–2017 available on Google Scholar – search terms "Antarctic Aerobiology," "Antarctica Bioaerosol"

Reference	Study area	Sampling method	Analysis method	Summary of findings
Kobayashi et al. (2016)	Hukuro Cove, Langhovde	Sampled for 1 h near Adelie penguins. On to 0.45 μm filter. One sample upwind and one downwind of the colony	Molecular 16S rRNA sequencing. Air mass back trajectory	19.4 times more *Bacillus* downwind of penguins from penguin faeces
Bottos et al. (2014)	McMurdo dry valleys – Miers Valley	Air filtered on to 0.2 μm polycarbonate filters, total sample air volume 75,000 L at each of two locations. 1 m elevation. 55 day sample time. Summer	Molecular 16S rRNA sequencing. Air mass back trajectory	Aerosols dominated by firmicutes suggesting volcanic activity. Most abundant taxa common to aerosols from other continents, representing a distinct widely dispersed bioaerosol community. Minimal marine input. Air masses originated from Antarctic plateau. Some taxa in common with Halley station work. Input likely from within Antarctica rather than intercontinental
Pearce et al. (2010)	Halley V Research Station	Two-week sampling in summer and winter with HiVol sampler with 0.2 mm polycarbonate filter	Molecular 16S rRNA sequencing. Air mass back trajectory	Low diversity, many sequence replicates and sequences from uncultivated organisms. No significant patterns detected between summer and winter. Few marine sequences irrespective of the distance to water. 1/3 sequences similar to those found in human studies indicating possible contamination of local environment from research station
Hughes et al. (2004)	Halley V Research Station	Two-week sampling in summer and winter with HiVol sampler using 0.2 μm polycarbonate filter	Molecular 16S rRNA sequencing. Air mass back trajectory	Low diversity, many sequence replicates and sequences from uncultivated organisms. No significant differences detected between summer and winter. Few marine sequences irrespective of the distance to water. 1/3 sequences similar to those found in human studies indicating possible contamination of local environment from research station

(continued)

Table 8.1 (continued)

Reference	Study area	Sampling method	Analysis method	Summary of findings
Hughes (2003)	Rothera Research Station	Exposed agar plates	Plate counts	Faecal coliform bacteria detected 75 m downwind of the sewage outfall. within 1 h of deposition UV and desiccation kills most bacteria
Marshall (1997)	Signy Island, South Orkney Islands	Rotorod samplers. Three sites. 1 m and 0.15 m above ground level. Four rotorods at two separate 24 h periods each week at all three sites for 14 months	Microscopy	Low concentrations of fungal spores in the air compared to the rest of the world. Concentrations increased in summer. *Chlamydospores* and *Cladosporium* spp. were the most and second most abundant spores, respectively. Evidence of long-distance transport of spores
Marshall (1996a)	Signy Island, South Orkney Islands	Rotorod samplers. Three sites. 1 m and 0.15 m elevation. Four rotorods for two separate 24 h periods each week at all three sites for 2 years	Microscopy	Lichen soredia most abundant bioaerosols. Dominance of soredia over ascospores decreases with more mature fell field sites. No correlation with temperature, humidity or wind speed. 1 m elevation not significantly different to ground level. Soredia peak in numbers after winter snow melt, demonstrating they are produced at sub-zero temperatures

et al. 2010; Hughes 2003) have shown an increasingly diverse and distinctive bioaerosol community, compared to local soils. Unfortunately, the limited studies conducted, samples collected and information gained from the samples have resulted in persistent knowledge gaps. Additionally, the lack of standardised techniques and restricted spatial and temporal coverage mean that it is difficult to make any broad inferences on the significance or extent of bioaerosol transportation to Antarctic ecosystems (Pearce et al. 2016). The studies in Table 8.1 and other terrestrial-based works in Antarctica strongly suggest that the atmosphere is crucial to both inter- and intracontinental Antarctic transport and a distinct microbial community may reside in Antarctic air. Future studies will be able to utilise current knowledge and techniques to conduct bioaerosol sampling in Antarctica that is higher resolution, less biased, inclusive of fungi and viruses that may be more abundant in bioaerosols than bacteria (Reche et al. 2018) and determine how these organisms survive and where they originate (Sect. 8.4).

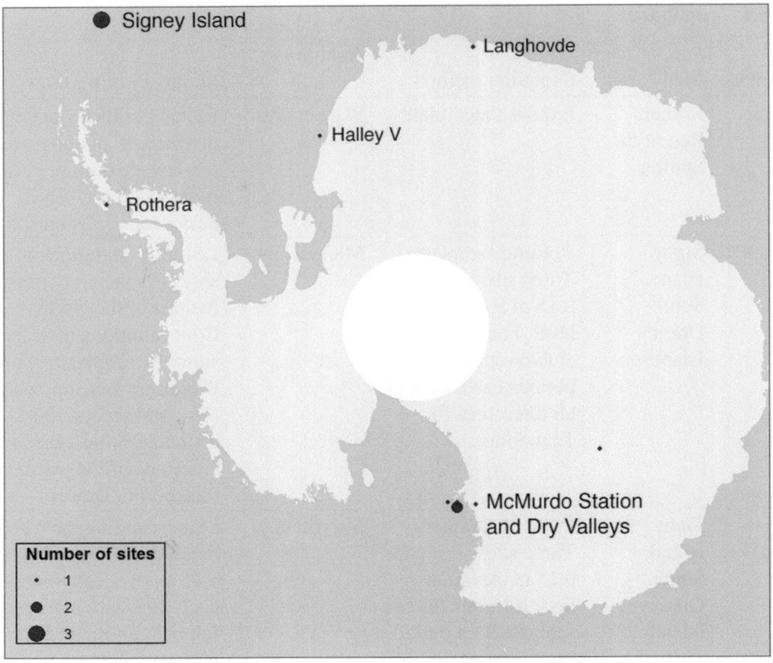

Fig. 8.10 Aerobiological study sites in Antarctica published 1994–2014 (Adapted from Pearce et al. 2016). Wider circles indicate site no per study. Most data comes either from the Peninsula or the McMurdo Dry Valleys

8.4 Microbial Survival Mechanisms in the Global Atmosphere and Antarctica

Microorganisms are highly resilient to environmental stressors, as shown by their successful colonisation of almost every niche explored by humans to date (Nuwer 2014). The global atmosphere and terrestrial Antarctica are both extreme environments which share many survival challenges including high ultraviolet (UV) irradiance, low temperature, low humidity and low nutrients (Womack et al. 2010). The extreme selective pressures experienced by microorganisms in the atmosphere would result in microbes that survive transport to Antarctica being well suited to colonise Antarctic ecosystems (Pearce et al. 2009). Dormancy (through spore formation) is thought to be the principle bioaerosol survival mechanism (Pearce et al. 2016; Bottos et al. 2014; Womack et al. 2010).

Spore Formation To survive transportation in the air, microorganisms can expend energy to counteract each stressor individually and remain metabolically active, or they can become dormant through spore formation, becoming resistant to all stressors with a single strategy. Species which become inactive therefore use the atmosphere as a vector only and aren't considered to be residents of the aerial habitat.

Spore formation is thought to be the most common mechanism for airborne survival, given that the majority of culturable isolates from high altitudes are spore-forming bacteria and fungi (Griffin 2004; Smith et al. 2011). Several authors have suggested that the air is dominated by Firmicutes (Pearce et al. 2009; Bottos et al. 2014), a bacterial phylum where spore formers are common. Bacterial spores are small cells with a highly reduced cytoplasm and a tough outer coating, allowing them to remain dormant for millions of years and rapidly reactivate when conditions are appropriate for growth (Cano and Borucki 1995). Their DNA is bound with various proteins to protect from UV, heat, cold, desiccation and any other stressors the organism might be likely to encounter (Lennon and Jones 2011). Suspension of cellular metabolism protects against nutrient starvation and the toughened outer coating may protect from desiccation and UV (Lennon and Jones 2011). Single-cell eukaryotes and fungi also form spores or cysts which operate on a similar principle. Small multicellular eukaryotes, like rotifers and tardigrades, undergo anhydrobiosis (desiccation of their bodies and significant reduction or suspension of their metabolism). Anhydrobiosis similarly facilitates survival in disadvantageous conditions and aerial disposal in Antarctica (Nkem et al. 2006). However, multicellular organisms have a much more limited transport range due to their larger body sizes.

UV Tolerance UV has been found to be the limiting factor determining survival of aerial microbes (Smith et al. 2011) deactivating foreign microorganisms in Antarctic samples in under an hour (Hughes 2003). UV intensity rises with altitude in the aerosphere and is elevated in Antarctica, due to its thin atmosphere and the localised depletion of the ozone layer (Pearce et al. 2009). UV radiation is highly damaging to most molecules in cells but mediates its lethal impacts through DNA damage, which interrupts all cellular functions. UV causes harm to DNA in various ways, the most significant being cyclobutane pyrimidine dimers, where adjacent bases bind together. This erroneous binding causes destruction of normal base pairing resulting in mutation and distorts the DNA double helix, preventing genes from being transcribed (Sinha and Häder 2002). As UV damage has such serious consequences, all organisms have developed defences against it. Pigmentation or carotenoids are commonly employed by microorganisms to absorb UV radiation before it can cause damage within a cell (Sinha and Häder 2002). Isolates of culturable bacteria and fungi from the stratosphere often show pigmentation, indicating it could be an important factor for their survival (Womack et al. 2010; Smith et al. 2011). Bioaerosols generally exist as aggregates of particles, often including organic and inorganic matter bound together. Several authors have suggested that this might protect against UV radiation, as could persistence within clouds (Womack et al. 2010; Pearce et al. 2009; Burrows et al. 2009b). Microorganisms are also commonly seen in challenging environments growing in biofilms, which are layered communities of microbes, and likely provide protection from UV in a similar fashion (Pointing et al. 2015). DNA repair mechanisms, which correct UV induced damage, are ubiquitous in living organisms. They include use of enzymes such as photolyases, which harvest energy from light to repair damaged DNA. Excision repair is another very common pathway, which works by cutting out the damaged portion of DNA and resynthesizing it using the complementary strand as a template (Sinha and Häder 2002). Some microorganisms such as *Deinococcus* have highly efficient DNA repair mechanisms, which confer extreme resistance to UV radiation (Pepper et al. 2015).

Cold Resistance Most organisms have growth optima well above temperatures commonly experienced in Antarctica and the aerosphere. With increased altitude temperature drops to below zero towards the tropopause, to around −60 °C, before recovering to approximately zero in the stratosphere (NASA 1962). Antarctic winter temperatures become as low as −93.2 °C and summer temperatures reach up to 15 °C, with a summer mean of around zero (Pointing et al. 2015). Below freezing temperatures cause ice crystal formation on cell surfaces and slowing of metabolic processes, which can either kill cells or severely limit their growth rates (Pepper et al. 2015). Psychrophilic (cold loving) microorganisms have been shown to grow successfully down to −18 °C (Rothschild and Mancinelli 2001) and various adaptions allow these organisms to remain metabolically active. Ratios of unsaturated to saturated fatty acids in cell membranes can be increased to counteract reductions in membrane fluidity at lower temperatures. Some organisms have developed enzymes which have optimal activity at lower temperatures, and some employ antifreeze proteins, which help prevent crystallisation (Laybourn-Parry 2002). Cold-tolerant organisms can be isolated from bacterial communities in temperate environments (Wilson and Walker 2010), indicating psychrotolerant organisms likely originated outside Antarctica, transported from warmer clines. Although there has been no evidence of airborne microorganisms employing these mechanisms to tolerate cold environments, with increasing molecular studies of bioaerosols, similar survival mechanisms are likely to be detected.

Low Humidity Tolerance While microorganisms differ in their responses to changes in humidity, very low humidity seems universally intolerable (Pepper et al. 2015). Both Antarctic and high atmospheric relative humidity levels have been shown to be low, with desiccation tolerant organisms frequently found in both environments (Luhung et al. 2015; Pearce et al. 2009). Low humidity causes damage to the lipid bilayers in the cell membrane and can change the membrane from a crystalline structure to a gel structure. This affects configuration of cell surface proteins, interrupting their function and deactivating the cell (Pepper et al. 2015). Viral survival is also thought to vary in a way which is dependent on humidity, with encapsulated viruses showing better resilience to lower humidity (Mohr 2007). As with protection from UV, persistence within a cloud and association with other particles are thought to help protect against low humidity (Womack et al. 2010; Pearce et al. 2009). Biofilms with extracellular matrixes adhering the organisms are also thought to reduce water loss (Pointing et al. 2015). Gram-positive cells tend to be more resilient to desiccation (Mohr 2007), thought to be due to a number of factors such as a thicker peptidoglycan layer, ability to accumulate magnesium and linkages to radiation tolerance (probably due to strong DNA repair capability conferring protection against desiccation as well as UV) (Makarova et al. 2001). Accordingly, the majority of bacterial bioaerosols found to date are gram positive (Griffin 2004; Smith et al. 2011; Bottos et al. 2014). However, this observation could be confounded by sample bias noted against gram-negative bacteria (Luhung et al. 2015) or the fact that gram-positive bacteria are also frequently spore formers.

Oligotrophy Resistance Another shared characteristic and challenge to survival, in the aerosphere and Antarctica generally, is low nutrient concentrations. The atmosphere is generally assumed to be lacking in the nutrients microorganisms require; however, key nutrients for microbial survival such as carbon, sulphate and nitrate can be found in cloud water, at similar levels to lake water (Pearce et al. 2009). Bauer et al. (2002) observed carbonaceous material constituted up to 20% of total aerosol mass in atmospheric aerosol samples. In addition to metabolising nutrients in the atmosphere, photosynthetic microbes that can independently fix carbon have been found in air samples, such as cyanobacteria and *Chloroflexi* (Brodie et al. 2007). Evidence shows that aerosolised bacteria can multiply and metabolise organic compounds generally present in clouds, some even at super-cooled temperatures (Dimmick and Wolochow 1979; Dimmick et al. 1975; Sattler et al. 2001). Vaïtilingom and Deguillaume (2013) and Amato et al. (2007) showed microorganisms are capable of degrading formaldehyde and carboxylic acids, carbon sources often present in cloud water. Current bioaerosol work is insufficient to determine the extent of metabolism and division in the atmosphere and what portion of the bioaerosol community typically remains metabolically active. It is supposed that there is unlikely to be significant reproduction of bacteria within clouds (Burrows et al. 2009a), given that it is estimated that most bioaerosols only spend a tiny fraction of their time suspended within cloud droplets (Lelieveld and Heintzenberg 1992). The ability to remain metabolically active is the key differentiator between dormant microbes using the atmosphere as a vector and active microbes using the air at a habitat.

8.5 Impact of Aerial Dispersal on Antarctic Microbial Populations

Long-range aerial transportation of microorganisms has long been suspected to play a significant role in structuring the Antarctic biological community (Horowitz et al. 1972). We have suggested that the air is likely to be the dominant vector for microbes to Antarctica (Sect. 8.3) and that microorganisms transported via the atmosphere are likely to be preselected to colonise Antarctica, with sporulation the likely main survival mechanism (Sect. 8.4). Previous research has shown that most Antarctic environments have a relatively low turnover rate (Cary et al. 2010) so aerially deposited microorganisms are unlikely to be consumed by the resident community. This implies that a large pool of atmospherically vectored temperate organisms is present in Antarctica, which will come out of dormancy when conditions become favourable under continued global warming (Kussell et al. 2005). Atmospherically vectored microbes, therefore, are likely to have the potential to substantially impact Antarctic ecosystems, and understanding the potential impacts from this shifting microbial landscape to the continent is particularly pertinent.

8.5.1 The Propagule Bank

The propagule bank describes a reservoir of dormant microorganisms awaiting favourable conditions to reactivate (Wynn-Williams 1991). It represents the adaptive potential of an ecosystem, allowing microbial communities to rapidly respond to change. Up to 80% of microbial cells in soil are estimated to be in a dormant state (Lennon and Jones 2011), with no published equivalent figure for the atmosphere. The presence of many dormant organisms represents a significant risk to the existing microbial community, which could change drastically, resulting in biodiversity loss (Lennon and Jones 2011). In Antarctica, aerially deposited microbes are thought to comprise a large proportion of the propagule bank. An example of this is found at Lake Vostok (Fig. 8.11), located below the central East Antarctic ice sheet and named for the Russian research station on its surface. Here, viable microorganisms were recovered at different depths. The lake itself is sealed under the ice sheet and remains liquid due to the enormous pressure from the weight of ice above (Vincent 2000). Ice cores collected from above the lake contained a continuous chronology of microbial deposition since the lake was covered around 400,000 years ago (Sinha and Krishnan 2013). From the ice cores, viable propagules were found, with culturable yeasts, fungi and bacteria present up to around 3000 years old. At around 10,000 years old, microbial communities became dominated by spore formers (Vincent 2000). At around 3.6 km depth, ice was derived from the underlying lake water and has been shown to contain viable bacteria (Karl et al. 1999). In 1998 (around the time Fig. 8.11 was published), drilling was temporarily halted due to concerns of contaminating the pristine ecosystem. In 2012 drilling was completed and sampling revealed unique microbial life; however concerns remained regarding contamination from the drilling process (Bulat 2016). It is speculated that as Antarctic ice melts, a multitude of ancient organisms could reactivate from dormancy (Fox-Skelly 2017) and alter microbial community composition.

To model changes in ecosystems from the revival of dormant species, several cloche experiments have been performed in Antarctica (Convey and Wynn-Williams 2002; Wynn-Williams 1996; Smith 1991, 1994; Kennedy 1994). A cloche is a cover placed on the ground designed to warm the ground beneath and sometimes to reduce UV exposure or increase humidity. Although this changes multiple variables simultaneously in an unnatural manner, these experiments provide insight into the potential effects of climate change. The experiments showed similar rapid increase in biodiversity and abundance of nematode worms, cyanobacteria, bryophytes and microarthropods (Convey and Wynn-Williams 2002; Wynn-Williams 1996; Smith

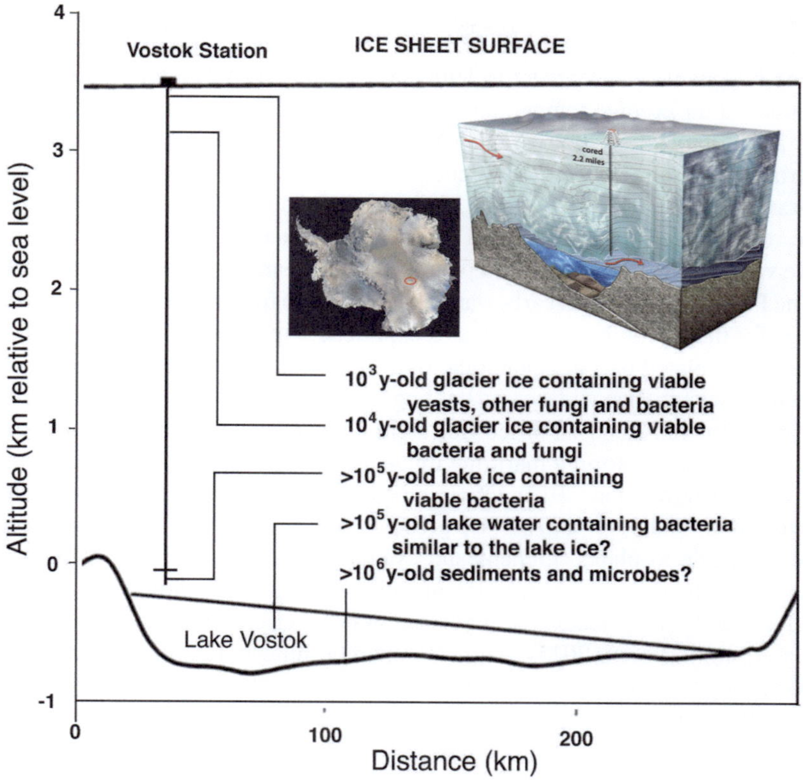

Fig. 8.11 Different types of microorganisms present in cross section of ice layers above Lake Vostok (adapted from Vincent by Chris King (2000))

1991, 1994; Kennedy 1994). The increase in abundance of particular species can be driven by increased reproduction of existing populations. However, the increased diversity can only be driven by activation and reproduction of dormant organisms in the propagule bank. The rapid increase in abundance and diversity in these cloche experiments demonstrated that substantial changes can occur very rapidly and furthermore display both the latent potential and the speed at which the propagule bank can respond to even slight increases in average temperature. The outcome of these complex cascading ecological interactions could have rapid and fundamental effects on the Antarctic ecosystems.

8.5.2 Future Changes to Antarctica

It is highly likely that climate change will induce substantial changes to Antarctic microbial communities, fuelled by the propagule bank and aerial input (Cowan et al. 2011; Convey and Wynn-Williams 2002). However, current understanding in the rate and direction of change in microbial communities is limited due to lack of studies in this field (Kennicutt et al. 2014). The warming of Antarctica is providing a more hospitable environment to temperate species. The Antarctic Peninsula has experienced an average temperature increase of 3 °C in the last 50 years (Turner et al. 2005) which has led to an increase in free water availability, an extension of the growing season (Convey 2006), and precipitation is more frequently observed as rain rather than snow (Pearce et al. 2009). These changing conditions are believed to be responsible for the increased range and abundance of the only two known native vascular plants in Antarctica over the last 25 years (Fowbert and Smith 1994).

The Peninsula acts as an early warning indicator to forecast impacts from climate change for the rest of the continent, as it experiences more temperate conditions and is the least isolated part of Antarctica. In the near future, the Antarctic continent is expected to experience widespread cooling, followed by warming as the ozone hole recovers. Generally, coastal or ice-free areas are expected to warm first, as they are most exposed to a warming maritime influence and less affected by glacial cooling (Fig. 8.12) (Chown et al. 2012). Observed changes and cloche experiments indicate increase in ground cover and surface greening will occur, as the conditions become more amenable to plants. Lichens and cyanobacteria will likely grow more optimally, given many are psychrotolerant rather than psychrophilic. These changes could create positive feedback effects on atmospheric carbon levels and climate change (Pointing et al. 2015). A green Antarctic reflects less light, and increased carbon fixing from higher photosynthetic rates may be more than offset by increased release of carbon in soils, through greater decomposition (Pointing et al. 2015).

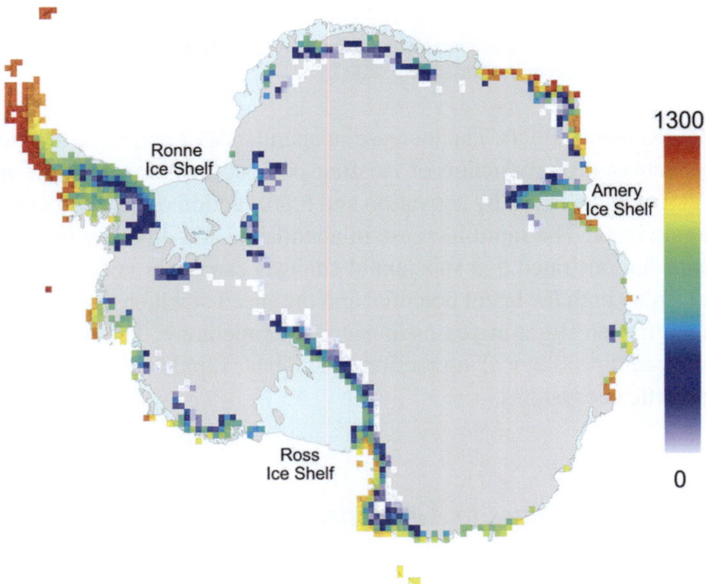

Fig. 8.12 Projected increase in annual cumulative degree days, between 2007 and 2100 (Chown et al. 2012)

Antarctica's future changes continue to make it increasingly hospitable to temperate organisms; therefore dormant bioaerosols originating in nearby regions are likely to activate, as are other temperate microbes previously deposited in soil and water via the atmosphere and other vectors.

8.6 Study of Antarctic Bioaerosols: Filling the Gaps

This chapter has outlined the evidence for and mechanisms of airborne transportation of microorganisms to Antarctica. The limited studies to date have confirmed speculation that the biomass in Antarctic air is extremely low. Although most microorganisms likely originate from local species pools, at least some originate from outside the Antarctic continent, although tangible evidence is lacking. Given the sensitivity of the terrestrial biome to non-local organisms, addressing key knowledge gaps of airborne transportation is essential to understand future responses to climate change. In the future we hope that technological advancements will allow studies to shed more light onto the extent, sources (local or intercontinental), survival mechanisms and significance of airborne microorganisms to Antarctic ecosystem structuring. With the technological advancements made in the past few years, we have an unprecedented opportunity to resolve these core biological questions. However, to understand the underlying physical processes discussed in this chapter that drive this transportation, interdisciplinary studies must be conducted.

References

Amato, P., Demeer, F., Melaouhi, A., Fontanella, S., Martin-Biesse, A. S., Sancelme, M., Laj, P., & Delort, A. M. (2007). A fate for organic acids, formaldehyde and methanol in cloud water: Their biotransformation by micro-organisms. *Atmospheric Chemistry and Physics Discussions, 7*(2), 5253–5276. https://doi.org/10.5194/acpd-7-5253-2007.

Andreeva, I. S., Borodulin, A. I., Buryak, G. A., Zhukov, V. A., Zykov, S. V., Marchenko, Y. V., Marchenko, V. V., Olkin, S. E., Petrishchenko, V. A., Pyankov, O. V., Reznikova, I. K., Repin, V. E., Sa-Atov, A. S., Sergeev, A. N., Raputa, V., Penenko, V. V., Tsvetova, E. A., Arshinov, M. Y., Belan, B. D., Panchenko, M. V., Ankilov, A. N., Baklanov, A. M., Vlasenko, A. L., Koutsenogil, K. P., Makarov, V. I., & Churkina, T. V. (2002). Biogenic component of atmospheric aerosol in the South of West Siberia. *Chemistry for Sustainable Development, 10*, 523–537. https://doi.org/10.1134/S1024856010010148.

Archer, C. L., & Caldeira, K. (2009). Global assessment of high-altitude wind power. *Energies, 2*, 307–319. https://doi.org/10.3390/en20200307.

Archer, S. D., McDonald, I. R., Herbold, C. W., & Cary, S. C. (2014). Characterisation of bacterioplankton communities in the meltwater ponds of Bratina Island, Victoria Land, Antarctica. *FEMS Microbiology Ecology, 89*(2), 451–464. https://doi.org/10.1111/1574-6941.12358.

Atkins, C. B., & Dunbar, G. B. (2009). Aeolian sediment flux from sea ice into Southern McMurdo Sound, Antarctica. *Global and Planetary Change, 69*(3), 133–141. https://doi.org/10.1016/j.gloplacha.2009.04.006.

Bahl, J., Lau, M. C. Y., Smith, G. J. D., Vijaykrishna, D., Cary, C. S., Lacap, D. C., Lee, C. K., Papke, T. R., Warren-Rhodes, K. A., Wong, F. K. Y., McKay, C. P., & Pointing, S. B. (2011). Ancient origins determine global biogeography of hot and cold desert cyanobacteria. *Nature Communications, 2*, 163. https://doi.org/10.1038/ncomms1167.

Barberán, A., Henley, J., Fierer, N., & Casamayor, E. O. (2014). Structure, inter-annual recurrence, and global-scale connectivity of airborne microbial communities. *Science of the Total Environment, 487*, 187–195.

Barberán, A., Ladau, J., Leff, J. W., Pollard, K. S., Menninger, H. L., Dunn, R. R., & Fierer, N. (2015). Continental-scale distributions of dust-associated bacteria and fungi. *Proceedings of the National Academy of Sciences, 112*(18), 5756–5761. https://doi.org/10.1073/pnas.1420815112.

Bauer, H., Kasper-Giebl, A., Löflund, M., Giebl, H., Hitzenberger, R., Zibuschka, F., & Puxbaum, H. (2002). The contribution of bacteria and fungal spores to the organic carbon content of cloud water, precipitation and aerosols. *Atmospheric Research, 64*(1), 109–119. https://doi.org/10.1016/S0169-8095(02)00084-4.

Behzad, H., Gojobori, T., & Mineta, K. (2015). Challenges and opportunities of airborne metagenomics. *Genome Biology and Evolution, 7*(5), 1216–1226. https://doi.org/10.1093/gbe/evv064.

Bottos, E. M., Woo, A. C., Zawar-Reza, P., Pointing, S. B., & Cary, S. C. (2014). Airborne bacterial populations above desert soils of the McMurdo Dry Valleys, Antarctica. *Microbial Ecology, 67*(1), 120–128. https://doi.org/10.1007/s00248-013-0296-y.

Bowers, R. M., McLetchie, S., Knight, R., & Journal, F.-N. (2011a). Spatial variability in airborne bacterial communities across land-use types and their relationship to the bacterial communities of potential source environments. *The ISME Journal, 5*, 601–612. https://doi.org/10.1038/ismej.2010.167.

Bowers, R. M., Sullivan, A. P., & Costello, E. K. (2011b). Sources of bacteria in outdoor air across cities in the midwestern United States. *Applied and Environmental Microbiology, 77*(18), 6350–6356. https://doi.org/10.1128/AEM.05498-11.

Bowers, R. M., Clements, N., Emerson, J. B., Wiedinmyer, C., Hannigan, M. P., & Fierer, N. (2013). Seasonal variability in bacterial and fungal diversity of the near-surface atmosphere. *Environmental Science & Technology, 47*(21), 12097–12106. https://doi.org/10.1021/es402970s.

Brodie, E. L., DeSantis, T. Z., & Parker, J. P. M. (2007). Urban aerosols harbor diverse and dynamic bacterial populations. *Proceedings of the National Academy of Sciences, 104*(1), 299–304. https://doi.org/10.1073/pnas.0608255104.

Bulat, S. A. R. (2016). Microbiology of the subglacial Lake Vostok: First results of borehole-frozen lake water analysis and prospects for searching for lake inhabitants. *Philosophical Transactions of the Royal Society A, 374*, 20140292. https://doi.org/10.1098/rsta.2014.0292.

Burrows, S. M., Butler, T., Jöckel, P., Tost, H., Kerkweg, A., Poschl, U., & Lawrence, M. G. (2009a). Bacteria in the global atmosphere–part 2: Modeling of emissions and transport between different ecosystems. *Atmospheric Chemistry and Physics, 9*, 9281–9297. https://doi.org/10.5194/acp-9-9281-2009.

Burrows, S. M., Elbert, W., Lawrence, M. G., & Poschl, U. (2009b). Bacteria in the global atmosphere – part 1: Review and synthesis of literature data for different ecosystems. *Atmospheric Chemistry and Physics, 9*, 9263–9280. https://doi.org/10.5194/acp-9-9263-2009.

Cano, R. J., & Borucki, M. K. (1995). Revival and identification of bacterial spores in 25- to 40-million-year-old Dominican amber. *Science, 268*(5213), 1060–1064. https://doi.org/10.1126/science.7538699.

Cary, C. S., McDonald, I. R., Barrett, J. E., & Cowan, D. A. (2010). On the rocks: The microbiology of Antarctic Dry Valley soils. *Nature Reviews Microbiology, 8*(2), 129–138. https://doi.org/10.1038/nrmicro2281.

Chen, M., Jin, L., Sun, Z., Lu, J., Wang, Q., & Hu, Q. (2001). Concentration and flux of bioaerosol and environmental factors. *Progress in Natural Science, 11*(9), 686–687.

Cho, B. C., & Azam, F. (1990). Biogeochemical significance of bacterial biomass in the ocean's euphotic zone. *Marine Ecology Progress Series, 63*(2/3), 253–259. https://doi.org/10.3354/meps063253.

Chong, C. W., Pearce, D. A., & Convey, P. (2015). Emerging spatial patterns in Antarctic prokaryotes. *Frontiers in Microbiology, 6*, 1058. https://doi.org/10.3389/fmicb.2015.01058.

Chown, S. L., Huiskes, A. H. L., Gremmen, N. J. M., Lee, J. E., Terauds, A., Crosbie, K., Frenot, Y., Hughes, K. A., Imura, S., Kiefer, K., Lebouvier, M., Raymond, B., Tsujimoto, M., Ware, C., de Vijver, B., & Bergstrom, D. (2012). Continent-wide risk assessment for the establishment of nonindigenous species in Antarctica. *Proceedings of the National Academy of Sciences, 109*(13), 4938–4943. https://doi.org/10.1073/pnas.1119787109.

Convey, P. (2006). *Antarctic climate change and its influence on terrestrial ecosystems*. Dordrecht: Springer.

Convey, P., & Wynn-Williams, D. D. (2002). Antarctic soil nematode response to artificial climate amelioration. *European Journal of Soil Biology, 38*, 255–259. https://doi.org/10.1016/S1164-5563(02)01155-X.

Cowan, D. A., Chown, S. L., Convey, P., Tuffin, M., Hughes, K., Pointing, S., & Vincent, W. F. (2011). Non-indigenous microorganisms in the Antarctic: Assessing the risks. *Trends in Microbiology, 19*(11), 540–548. https://doi.org/10.1016/j.tim.2011.07.008.

Crawford, I., Gallagher, M. W., Bower, K. N., Choularton, T. W., Flynn, M. J., Ruske, S., Listowski, C., Brough, N., Lachlan-Cope, T., Fleming, Z. L., Foot, V. E., & Stanley, W. R. (2017). Real-time detection of airborne fluorescent bioparticles in Antarctica. *Atmospheric Chemistry and Physics, 17*(23), 14291–14307. https://doi.org/10.5194/acp-17-14291-2017.

DeLeon-Rodriguez, N., Lathem, T. L., Rodriguez-R, L. M., Barazesh, J. M., Anderson, B. E., Beyersdorf, A. J., Ziemba, L. D., Bergin, M., Nenes, A., & Konstantinidis, K. T. (2013). Microbiome of the upper troposphere: Species composition and prevalence, effects of tropical storms, and atmospheric implications. *Proceedings of the National Academy of Sciences, 110*(7), 2575–2580. https://doi.org/10.1073/pnas.1212089110.

Diallo, M., Ploeger, F., Konopka, P., Birner, T., Müller, R., Riese, M., Garny, H., Legras, B., Ray, E., Berthet, G., & Jegou, F. (2017). Significant contributions of volcanic aerosols to decadal changes in the stratospheric circulation. *Geophysical Research Letters, 44*(20), 10,780–10,791. https://doi.org/10.1002/2017GL074662.

Dimmick, R. L., & Wolochow, H. (1979). Evidence for more than one division of bacteria within airborne particles. *Applied and Environmental Microbiology, 38*(4), 642–643.

Dimmick, R. L., Straat, P. A., Wolochow, H., & Levin, G. V. (1975). Evidence for metabolic activity of airborne bacteria. *Journal of Aerosol Science, 6*, 387–393.

Dybwad, M., Skogan, G., & Blatny, J. (2014). Comparative testing and evaluation of nine different air samplers: End-to-end sampling efficiencies as specific performance measurements for bioaerosol applications. *Aerosol Science and Technology, 48*(3), 282–295. https://doi.org/10.1080/02786826.2013.871501.

Fierer, N., Liu, Z., & Rodríguez-Hernández, M. (2008). Short-term temporal variability in airborne bacterial and fungal populations. *Applied and Environmental Microbiology, 74*(1), 200–207. https://doi.org/10.1128/AEM.01467-07.

Fowbert, J. A., & Smith, R. I. L. (1994). Rapid population increases in native vascular plants in the argentine islands, Antarctic Peninsula. *Arctic and Alpine Research, 26*(3), 290–296. https://doi.org/10.1080/00040851.1994.12003068.

Fox-Skelly J (2017). *There are diseases in the ice and they are waking up*. http://www.bbc.com/earth/story/20170504-there-are-diseases-hidden-in-ice-and-they-are-waking-up. Accessed 1/7/18 2018.

Fripiat, F., Meiners, K. M., Vancoppenolle, M., Papadimitriou, S., Thomas, D. N., Ackley, S. F., Arrigo, K. R., Carnat, G., Cozzi, S., & Delille, B. (2017). Macro-nutrient concentrations in Antarctic pack ice: Overall patterns and overlooked processes. *Elementa: Science of the Anthropocene, 5*, 13. https://doi.org/10.1525/elementa.217.

Fulton, J. D. (1966a). Microorganisms of the upper atmosphere. V. Relationship between frontal activity and the micropopulation at altitude. *Applied Microbiology, 14*(2), 245–250.
Fulton, J. D. (1966b). Microorganisms of the upper atmosphere: III. Relationship between altitude and micropopulation. *Applied Microbiology, 14*(2), 237–240.
Fulton, J. D., & Mitchell, R. B. (1966). Microorganisms of the upper atmosphere. II. Microorganisms in two types of air masses at 690 meters over a city. *Applied Microbiology, 14*(2), 232–236.
Grantham, N. S., Reich, B. J., Pacifici, K., Laber, E. B., Menninger, H. L., Henley, J. B., Barberán, A., Leff, J. W., Fierer, N., & Dunn, R. R. (2015). Fungi identify the geographic origin of dust samples. *PLoS One, 10*(4), e0122605. https://doi.org/10.1371/journal.pone.0122605.
Gregory PH (1961). The microbiology of the atmosphere.
Griffin, D. W. (2004). Terrestrial microorganisms at an altitude of 20,000 m in Earth's atmosphere. *Aerobiologia, 20*, 135–140. https://doi.org/10.1023/B:AERO.0000032948.84077.12.
Haig, C. W., Mackay, W. G., Walker, J. T., & Williams, C. (2016). Bioaerosol sampling: Sampling mechanisms, bioefficiency and field studies. *Journal of Hospital Infection, 93*(3), 242–255. https://doi.org/10.1016/j.jhin.2016.03.017.
Hara, K., & Zhang, D. (2012). Bacterial abundance and viability in long-range transported dust. *Atmospheric Environment, 47*, 20–25. https://doi.org/10.1016/j.atmosenv.2011.11.050.
Harrison, R. M., Jones, A. M., Biggins, P. D. E., Pomeroy, N., Cox, C. S., Kidd, S. P., Hobman, J. L., Brown, N. L., & Beswick, A. (2005). Climate factors influencing bacterial count in background air samples. *International Journal of Biometeorology, 49*(3), 167–178. https://doi.org/10.1007/s00484-004-0225-3.
Herbold, C. W., Lee, C. K., McDonald, I. R., & Cary, C. S. (2014). Evidence of global-scale aeolian dispersal and endemism in isolated geothermal microbial communities of Antarctica. *Nature Communications, 5*, 3875. https://doi.org/10.1038/ncomms4875.
Horneck, G. (1993). Responses of *Bacillus subtilis* spores to space environment: Results from experiments in space. *Origins of Life and Evolution of the Biosphere : The Journal of the International Society for the Study of the Origin of Life, 23*(1), 37–52. https://doi.org/10.1007/BF01581989.
Horowitz, N., Cameron, R. E., & Hubbard, J. S. (1972). Microbiology of the dry valleys of Antarctica. *Science, 176*(4032), 242–245.
Huffman, J. A., Treutlein, B., & Poschl, U. (2010). Fluorescent biological aerosol particle concentrations and size distributions measured with an ultraviolet aerodynamic particle sizer (UV-APS) in Central Europe. *Atmospheric Chemistry and Physics, 10*, 3215–3233. https://doi.org/10.5194/acp-10-3215-2010.
Hughes, K. A. (2003). Aerial dispersal and survival of sewage-derived faecal coliforms in Antarctica. *Atmospheric Environment, 37*(22), 3147–3155. https://doi.org/10.1016/S1352-2310(03)00207-3.
Hughes, K. A., McCartney, H. A., Lachlan-Cope, T. A., & Pearce, D. A. (2004). A preliminary study of airborne microbial biodiversity over Peninsular Antarctica. *Cellular and Molecular Biology (Noisy-le-Grand, France), 50*(5), 537–542.
Imshenetsky, A. A., Lysenko, S. V., & Kazakov, G. A. (1978). Upper boundary of the biosphere. *Applied and Environmental Microbiology, 35*(1), 1–5.
Kaidor (2013). Earth global circulation. Inkscape. Wikipedia.
Karl, D. M., Bird, D. F., Björkman, K., Houlihan, T., Shackelford, R., & Tupas, L. (1999). Microorganisms in the accreted ice of Lake Vostok, Antarctica. *Science, 286*(5447), 2144–2147. https://doi.org/10.1126/science.286.5447.2144.
Kellogg, C. A., & Griffin, D. W. (2006). Aerobiology and the global transport of desert dust. *Trends in Ecology & Evolution, 21*(11), 638–644. https://doi.org/10.1016/j.tree.2006.07.004.
Kellogg, C. A., Griffin, D. W., Garrison, V. H., Peak, K. K., Royall, N., Smith, R. R., & Shinn, E. A. (2004). Characterization of aerosolized bacteria and fungi from desert dust events in Mali, West Africa. *Aerobiologia, 20*(2), 99–110. https://doi.org/10.1023/B:AERO.0000032947.88335.bb.

Kennedy, A. D. (1994). Simulated climate change: A field manipulation study of polar microarthropod community response to global warming. *Ecography, 17*(2), 131–140. https://doi.org/10.1111/j.1600-0587.1994.tb00085.x.

Kennicutt, M. C., Chown, S. L., Cassano, J. J., Liggett, D., Massom, R., Peck, L. S., Rintoul, S. R., Storey, J. W. V., Vaughan, D. G., Wilson, T. J., & Sutherland, W. J. (2014). Polar research: Six priorities for Antarctic science. *Nature News, 512*(7512), 23. https://doi.org/10.1038/512023a.

Kobayashi, F., Maki, T., Kakikawa, M., Noda, T., Mitamura, H., Takahashi, A., Imura, S., & Iwasaka, Y. (2016). Atmospheric bioaerosols originating from Adélie penguins (*Pygoscelis adeliae*): Ecological observations of airborne bacteria at Hukuro Cove, Langhovde, Antarctica. *Polar Science, 10*, 71. https://doi.org/10.1016/j.polar.2015.12.002.

Kussell, E., Kishony, R., Balaban, N. Q., & Leibler, S. (2005). Bacterial persistence: A model of survival in changing environments. *Genetics, 169*, 1807–1814. https://doi.org/10.1534/genetics.104.035352.

Laybourn-Parry, J. (2002). Survival mechanisms in Antarctic lakes. *Philosophical Transactions of the Royal Society B: Biological Sciences, 357*, 863–869. https://doi.org/10.1098/rstb.2002.1075.

Lee, C. K., Barbier, B. A. A., Bottos, E. M., McDonald, I. R., & Cary, S. C. (2012). The Inter-Valley soil comparative survey: The ecology of Dry Valley edaphic microbial communities. *The ISME Journal, 6*(5), 1046–1057. https://doi.org/10.1038/ismej.2011.170.

Lelieveld, J., & Heintzenberg, J. (1992). Sulfate cooling effect on climate through in-cloud oxidation of anthropogenic SO2. *Science, 258*(5079), 117–120. https://doi.org/10.1126/science.258.5079.117.

Lennon, J. T., & Jones, S. E. (2011). Microbial seed banks: The ecological and evolutionary implications of dormancy. *Nature Reviews Microbiology, 9*(2), 119–130. https://doi.org/10.1038/nrmicro2504.

Lighthart, B., & Stetzenbach, L. D. (1994). Distribution of microbial bioaerosol. In B. Lighthart & A. J. Mohr (Eds.), *Atmospheric microbial aerosols: Theory and applications* (pp. 68–98). Boston: Springer. https://doi.org/10.1007/978-1-4684-6438-2_4.

Luhung, I., Wu, Y., Ng, C., Miller, D., Cao, B., & Chang, V. (2015). Protocol improvements for low concentration DNA-based bioaerosol sampling and analysis. *PLoS One, 10*(11), e0141158. https://doi.org/10.1371/journal.pone.0141158.

Mackintosh L (2001). How cold is the Antarctic? NIWA. https://www.niwa.co.nz/education-and-training/schools/resources/climate/antarctic. Accessed 1/7/2018 2018.

Makarova, K. S., Aravind, L., Wolf, Y. I., Tatusov, R. L., Minton, K. W., Koonin, E. V., & Daly, M. J. (2001). Genome of the extremely radiation-resistant bacterium Deinococcus radiodurans viewed from the perspective of comparative genomics. *Microbiology and Molecular Biology Reviews, 65*(1), 44–79. https://doi.org/10.1128/MMBR.65.1.44-79.2001.

Maki, T., Susuki, S., Kobayashi, F., Kakikawa, M., Yamada, M., Higashi, T., Chen, B., Shi, G., Hong, C., & Tobo, Y. (2008). Phylogenetic diversity and vertical distribution of a halobacterial community in the atmosphere of an Asian dust (KOSA) source region, Dunhuang City. *Air Quality, Atmosphere & Health, 1*(2), 81–89. https://doi.org/10.1007/s11869-008-0016-9.

Maki, T., Aoki, K., Kobayashi, F., Kakikawa, M., Tobo, Y., Matsuki, A., Hasegawa, H., & Iwasaka, Y. (2011). Characterization of halotolerant and oligotrophic bacterial communities in Asian desert dust (KOSA) bioaerosol accumulated in layers of snow on Mount Tateyama Central Japan. *Aerobiologia, 27*, 277–290. https://doi.org/10.1007/s10453-011-9196-0.

Maki, T., Hara, K., Iwata, A., Lee, K. C., Kawai, K., Kai, K., Kobayashi, F., Pointing, S. B., Archer, S., Hasegawa, H., & Iwasaka, Y. (2017). Variations in airborne bacterial communities at high altitudes over the Noto Peninsula (Japan) in response to Asian dust events. *Atmospheric Chemistry and Physics, 17*(19), 11877–11897. https://doi.org/10.5194/acp-17-11877-2017.

Marshall, W. A. (1996a). Aerial dispersal of lichen soredia in the maritime Antarctic. *New Phytologist, 134*(3), 523–530. https://doi.org/10.1111/j.1469-8137.1996.tb04370.x.

Marshall, W. A. (1996b). Biological particles over Antarctica. *Nature, 383*(6602), 680–680. https://doi.org/10.1038/383680a0.

Marshall, W. A. (1997). Seasonality in antarctic airborne fungal spores. *Applied and Environmental Microbiology, 63*(6), 2240–2245.

Matsuki, A., Iwasaka, Y., & Osada, K. (2003). Seasonal dependence of the long-range transport and vertical distribution of free tropospheric aerosols over East Asia: On the basis of aircraft and lidar measurements and isentropic trajectory analysis. *Journal of Geophysical Research, 108*(D23), 8663. https://doi.org/10.1029/2002JD003266.

Matthias-Maser, S., Obolkin, V., Khodzer, T., & Jaenicke, R. (2000). Seasonal variation of primary biological aerosol particles in the remote continental region of Lake Baikal/Siberia. *Atmospheric Environment, 34*(22), 3805–3811. https://doi.org/10.1016/S1352-2310(00)00139-4.

Miller, L. M., Gans, F., & Kleidon, A. (2011). Jet stream wind power as a renewable energy resource: Little power, big impacts. *Earth System Dynamics, 2*(2), 201–212. https://doi.org/10.5194/esd-2-201-2011.

Mohr. (2007). *Manual of environmental microbiology* (3rd ed.). Washington, DC: ASM Press. https://doi.org/10.1128/9781555815882.

Mroz, E., Mohammed, A., Cappis, J. H., Guthals, P. R., Mason, A. S., & Rokop, D. J. (1989). Antarctic atmospheric tracer experiments. *Journal of Geophysical Research: Atmospheres, 94*(D6), 8577–8583. https://doi.org/10.1029/JD094iD06p08577.

NASA (1962). *US standard atmosphere 1962*. NASA.

Nguyen, T. M. N., Ilef, D., Jarraud, S., Rouil, L., & Desenclos, J.-C. (2006). A community-wide outbreak of legionnaires disease linked to industrial cooling towers—how far can contaminated aerosols spread? *Journal of Infectious Diseases, 193*, 102–111. https://doi.org/10.1093/oxfordjournals.aje.a112917.

Niklas, K. J. (1985). The aerodynamics of wind pollination. *The Botanical Review, 51*(3), 328. https://doi.org/10.1007/bf02861079.

Nkem, J. N., Wall, D. H., Virginia, R. A., Barrett, J. E., Broos, E. J., Porazinska, D. L., & Adams, B. J. (2006). Wind dispersal of soil invertebrates in the McMurdo Dry Valleys, Antarctica. *Polar Biology, 29*(4), 346–352. https://doi.org/10.1007/s00300-005-0061-x.

Nuwer R (2014). *The last place on Earth without life*. http://www.bbc.com/future/story/20140303-last-place-on-earth-without-life. Accessed 1/7/2018 2018.

O'Connell S (2006) *How Krakatoa made the biggest bang*. https://www.independent.co.uk/news/science/how-krakatoa-made-the-biggest-bang-5336165.html. Accessed 1/7/2018 2018.

O'Malley, M. A. (2008). 'Everything is everywhere: But the environment selects': Ubiquitous distribution and ecological determinism in microbial biogeography. *Studies in History and Philosophy of Biological and Biomedical Sciences, 39*(3), 314–325. https://doi.org/10.1016/j.shpsc.2008.06.005.

Parish, T. R., & Cassano, J. J. (2003). The role of katabatic winds on the Antarctic surface wind regime. *Monthly Weather Review, 131*(2), 317–333. https://doi.org/10.1175/1520-0493(2003)131<0317:Trokwo>2.0.Co;2.

Parks, D. H., Rinke, C., Chuvochina, M., Chaumeil, P.-A., Woodcroft, B. J., Evans, P. N., Hugenholtz, P., & Tyson, G. W. (2017). Recovery of nearly 8,000 metagenome-assembled genomes substantially expands the tree of life. *Nature Microbiology, 2*(11), 1533–1542. https://doi.org/10.1038/s41564-017-0012-7.

Pearce, D. A., Bridge, P. D., Hughes, K. A., Sattler, B., Psenner, R., & Russell, N. J. (2009). Microorganisms in the atmosphere over Antarctica. *FEMS Microbiology Ecology, 69*(2), 143–157. https://doi.org/10.1111/j.1574-6941.2009.00706.x.

Pearce, D. A., Hughes, K. A., Lachlan-Cope, T., Harangozo, S. A., & Jones, A. E. (2010). Biodiversity of air-borne microorganisms at Halley Station, Antarctica. *Extremophiles : Life Under Extreme Conditions, 14*(2), 145–159. https://doi.org/10.1007/s00792-009-0293-8.

Pearce, D. A., Alekhina, I. A., Terauds, A., Wilmotte, A., Quesada, A., Edwards, A., Dommergue, A., Sattler, B., Adams, B. J., Magalhães, C., Chu, W.-L., Lau, M. C. Y., Cary, C., Smith, D. J., Wall, D. H., Eguren, G., Matcher, G., Bradley, J. A., de Vera, J.-P., Elster, J., Hughes, K. A., Cuthbertson, L., Benning, L. G., Gunde-Cimerman, N., Convey, P., Hong, S., Pointing, S. B.,

Pellizari, V. H., & Vincent, W. F. (2016). Aerobiology over Antarctica – a new initiative for atmospheric ecology. *Frontiers in Microbiology, 7*, 16. https://doi.org/10.3389/fmicb.2016.00016.
Pepper, I., Gerber, C., Gentry, T. (Eds.) (2015). *Environmental microbiology* (3rd edn). https://doi.org/10.1016/B978-0-12-394626-3.00005-3.
Pointing, S. B., Büdel, B., Convey, P., Gillman, L. N., Körner, C., Leuzinger, S., & Vincent, W. F. (2015). Biogeography of photoautotrophs in the high polar biome. *Frontiers in Plant Science, 6*, 692. https://doi.org/10.3389/fpls.2015.00692.
Prospero, J. M. (1999). Assessing the impact of advected African dust on air quality and health in the eastern United States. *Human and Ecological Risk Assessment: An International Journal, 5*(3), 471–479. https://doi.org/10.1080/10807039.1999.10518872.
Reche, I., D'Orta, G., Mladenov, N., Winget, D. M., & Suttle, C. A. (2018). Deposition rates of viruses and bacteria above the atmospheric boundary layer. *The ISME Journal, 12*(4), 1154–1162. https://doi.org/10.1038/s41396-017-0042-4.
Rettberg, P., Eschweiler, U., Strauch, K., Reitz, G., Horneck, G., Wänke, H., Brack, A., & Barbier, B. (2002). Survival of microorganisms in space protected by meteorite material: Results of the experiment EXOBIOLOGIE of the PERSEUS mission. *Advances in Space Research, 30*(6), 1539–1545. https://doi.org/10.1016/S0273-1177(02)00369-1.
Rotach, M. W., Gohm, A., Lang, M. N., Leukauf, D., Stiperski, I., & Wagner, J. S. (2015). On the vertical exchange of heat, mass, and momentum over complex, mountainous terrain. *Frontiers in Earth Science, 3*, 76. https://doi.org/10.3389/feart.2015.00076.
Rothschild, L. J., & Mancinelli, R. L. (2001). Life in extreme environments. *Nature, 409*, 1092–1101. https://doi.org/10.1038/35059215.
Sattler, B., Puxbaum, H., & Psenner, R. (2001). Bacterial growth in supercooled cloud droplets. *Geophysical Research Letters, 28*(2), 239–242. https://doi.org/10.1029/2000GL011684.
Shaffer, B. T., & Lighthart, B. (1997). Survey of culturable airborne bacteria at four diverse locations in Oregon: Urban, rural, forest, and coastal. *Microbial Ecology, 34*, 167–177. https://doi.org/10.1007/s002489900046.
Sinha, R. P., & Häder, D.-P. (2002). UV-induced DNA damage and repair: A review. *Photochemical & Photobiological Sciences, 1*(4), 225–236. https://doi.org/10.1039/B201230H.
Sinha, R. K., & Krishnan, K. P. (2013). Novel opportunity for understanding origin and evolution of life: Perspectives on the exploration of subglacial environment of Lake Vostok, Antarctica. *Annals of Microbiology, 63*(2), 409–415. https://doi.org/10.1007/s13213-012-0525-5.
Smith, L. R. I. (1991). Exotic sporomorpha as indicators of potential immigrant colonists in Antarctica. *Grana, 30*(2), 313–324. https://doi.org/10.1080/00173139109431986.
Smith, R. I. L. (1994). Vascular plants as bioindicators of regional warming in Antarctica. *Oecologia, 99*(3), 322–328. https://doi.org/10.1007/bf00627745.
Smith, D. J., Griffin, D. W., McPeters, R. D., Ward, P. D., & Schuerger, A. C. (2011). Microbial survival in the stratosphere and implications for global dispersal. *Aerobiologia, 27*(4), 319–332. https://doi.org/10.1007/s10453-011-9203-5.
Sokol, E. R., Herbold, C. W., Lee, C. K., Cary, S. C., & Barrett, J. E. (2013). Local and regional influences over soil microbial metacommunities in the Transantarctic Mountains. *Ecosphere, 4*(11), 136. https://doi.org/10.1890/ES13-00136.1.
Sommaruga, R., & Casamayor, E. O. (2009). Bacterial 'cosmopolitanism' and importance of local environmental factors for community composition in remote high-altitude lakes. *Freshwater Biology, 54*, 944–1005. https://doi.org/10.1111/j.1365-2427.2008.02146.x.
Tong, Y., & Lighthart, B. (2000). The annual bacterial particle concentration and size distribution in the ambient atmosphere in a rural area of the Willamette Valley, Oregon. *Aerosol Science & Technology, 32*(5), 393–403. https://doi.org/10.1080/027868200303533.
Turner, J., Colwell, S. R., Marshall, G. J., Lachlan-Cope, T. A., Carleton, A. M., Jones, P. D., Lagun, V., Reid, P. A., & Iagovkina, S. (2005). Antarctic climate change during the last 50 years. *International Journal of Climatology, 25*(3), 279–294. https://doi.org/10.1002/joc.1130.

Vaïtilingom, M., & Deguillaume, L. (2013). Potential impact of microbial activity on the oxidant capacity and organic carbon budget in clouds. *Proceedings of the National Academy of Sciences, 110*(2), 59–564. https://doi.org/10.1073/pnas.1205743110.

Vincent, W. F. (2000). Evolutionary origins of Antarctic microbiota: Invasion, selection and endemism. *Antarctic Science, 12*(3), 374–385. https://doi.org/10.1017/S0954102000000420.

Wainwright, M., Wickramasinghe, N. C., Narlikar, J. V., & Rajaratnam, P. (2003). Microorganisms cultured from stratospheric air samples obtained at 41 km. *FEMS Microbiology Letters, 218*, 161–165. https://doi.org/10.1111/j.1574-6968.2003.tb11513.x.

Wilson, S. L., & Walker, V. K. (2010). Selection of low temperature resistance in bacteria and potential applications. *Environmental Technology, 31*(8–9), 943–956. https://doi.org/10.1080/09593331003782417.

Womack, A. M., Bohannan, B. J. M., & Green, J. L. (2010). Biodiversity and biogeography of the atmosphere. *Philosophical Transactions of the Royal Society B: Biological Sciences, 365*(1558), 3645–3653. https://doi.org/10.1098/rstb.2010.0283.

Woo, A. C., Brar, M. S., Chan, Y., Lau, M., Leung, F., Scott, J. A., Vrijmoed, L., Zawar-Reza, P., & Pointing, S. B. (2013). Temporal variation in airborne microbial populations and microbially-derived allergens in a tropical urban landscape. *Atmospheric Environment, 74*, 291–300. https://doi.org/10.1016/j.atmosenv.2013.03.047.

Wood, S. A., Rueckert, A., Cowan, D. A., & Cary, C. S. (2008). Sources of edaphic cyanobacterial diversity in the Dry Valleys of Eastern Antarctica. *The ISME Journal, 2*(3), 308–320. https://doi.org/10.1038/ismej.2007.104.

Wynn-Williams, D. D. (1991). Aerobiology and colonization in Antarctica — the BIOTAS Programme. *Grana, 30*(2), 380–393. https://doi.org/10.1080/00173139109431994.

Wynn-Williams, D. D. (1996). Antarctic microbial diversity: The basis of polar ecosystem processes. *Biodiversity and Conservation, 5*(11), 1271–1293. https://doi.org/10.1007/BF00051979.

Yoo, K., Lee, T., Choi, E., Yang, J., Shukla, S., Hwang, S.-I., & Park, J. (2016). Approach of molecular methods for the detection and monitoring of microbial communities in bioaerosols: A review. *Journal of Environmental Sciences*. https://doi.org/10.1016/j.jes.2016.07.002.

Chapter 9
Molecular Biology of RNA Viruses Isolated in Antarctica

Juan Cristina

Abstract RNA viruses exist as collections of closely related viral genomes, termed quasispecies, subjected to a continuous process of genetic variation, competition among the variants generated, and selection of the most fitted distributions in a given environment. Large population sizes and high evolutionary rates make RNA viruses unique. Detailed studies on the molecular biology and evolution of RNA viruses are extremely important for our understanding of the role of these viruses in Antarctica, as well as to gain insight into their emergence and spread. Antarctic ecosystems are dominated by microorganisms, where RNA viruses play an important role as an active, persistent, and important component of the aquatic microbial community, both at Antarctic sea and freshwater environments. RNA virus infection of Antarctic wildlife is more significant than previously anticipated, where at least members of five different RNA virus families have been identified. RNA plant viruses have been also observed. Recent studies reveal the introduction, persistence, and evolution of emerging RNA viruses of extreme importance for human and/or animal health in Antarctica. To determine the role of Antarctica in the global emergence of these viruses is extremely important for the development of appropriate strategies to control infection caused by these viruses.

Keywords RNA viruses · Emergence and spread of virus · Paramyxoviridae · Orthomyxoviridae · Birnaviridae · Flaviviridae · Togaviridae

J. Cristina (✉)
Laboratorio de Virología Molecular, CIN, Facultad de Ciencias, Universidad de la República, Montevideo, Uruguay
e-mail: cristina@cin.edu.uy

S. Castro-Sowinski (ed.), *The Ecological Role of Micro-organisms in the Antarctic Environment*, Springer Polar Sciences,
https://doi.org/10.1007/978-3-030-02786-5_9

9.1 Introduction to RNA Virus Evolution

Viruses whose genomes are constituted by RNA are normally termed RNA viruses. These viruses are characterized by high mutation rates and unique dynamics that confer RNA viruses a great capability of evolution. This represents one of the major obstacles for the control and prevention of emerging RNA viral diseases, such as *Enterovirus*, Dengue virus, Influenza virus, Hepatitis C virus, Ebola virus, West Nile virus, Zika virus, etc.

More than 50 years ago, studies performed in enterovirus coxsackie A9 virus populations provided evidence of variants displaying a variety of phenotypes (Eggers and Tamm 1965). But it was in 1978, when Domingo and colleagues, studying *QB* phage, established that this phage cannot be described as a defined unique structure, but rather as a weighted average of a large number of different individual sequences (Domingo et al. 1978).

These important findings were highlighted by the work of Manfred Eigen, who formulated the first mathematical framework linking Darwinian natural selection to self-organizing primitive entities (which presumably were RNA molecules), assuming erroneous replication of the template molecules (Eigen 1971). This framework was later extended by Eigen and Schuster to become known as the *quasispecies theory* (Eigen and Schuster 1977). This theory describes the behavior of self-replicating macromolecules that, due to high error frequency and lack of proofreading mechanisms, replicate into dynamic distributions of genetically closely related molecules surrounding a master sequence (Eigen 1992). Due to the fact that RNA-dependent RNA polymerases of these viruses lack of proofreading activity and they are prone to error, this conceptual framework was suitably employed to describe RNA virus population. Then, in virology, the term refers to the mutant spectrum, a "cloud' of nonidentical, yet closely related genetic variants (Domingo 2006). As an important consequence, the theory predicts that genetic variation, competition or cooperation, and selection do not act on a single genome, but rather on the larger mutant swarm spectrum (Perales et al. 2010). So, RNA viruses generate highly heterogeneous populations and exploit all known mechanisms of genetic variation to ensure their survival. They share a powerful potential for adaptation and rapid evolution involving two key factors: high mutation rates and low replication fidelity.

The mutation rates of RNA viruses are thousandfold higher than those observed in DNA-based organisms (Sanjuan 2012) and can range from 10^{-2} to 10^{-5} substitutions per nucleotide site per replication cycle (Vignuzzi et al. 2006).

Due to these facts, it is extremely important to study the genetic variability of RNA viruses in all regions of the world, including Antarctica, in order to gain insight into their great capability of adaptation and evolution in Antarctic environments and to study their possible emergence and spread in Antarctica and beyond. The purpose of this review is to highlight the most important aspects of RNA viruses isolated in Antarctica, their adaptation, and evolution.

9.2 RNA Viruses as Major Contributors to Antarctic Sea

First reports on RNA viruses in the oceans of the world revealed the isolation of RNA viruses that infect aquatic animals of economic importance, infecting invertebrates, fishes, seabirds, and marine mammals (Lang et al. 2009). RNA viruses that infect microorganisms are now recognized as an active, persistent, and important component of the aquatic microbial community (Culley 2018). Recent studies, using metagenomics, have expanded our knowledge of RNA viruses in different environments (Krishnamurthy and Wang 2017).

A recent study, carried out at Palmer Station on the Western Antarctic Peninsula (WAP), revealed that although the majority of the sequences obtained were unknown, a group of RNA viruses related to the order *Picornavirales* were dominant throughout the southern hemisphere summer (Miranda et al. 2016). Moreover, precipitous declines in chlorophyll were concomitant with increases in the total mass of viruslike RNA, a relationship that suggests that the most likely hosts of these viruses were diatoms (Miranda et al. 2016). Then, RNA viruses in particular play an important role in the summer phytoplankton bloom in Antarctic waters. The typical bloom in the WAP is known to be initiated by diatoms, which reach their highest concentrations in early to mid-summer (Ducklow et al. 2007). The diatom populations then decline with a transition to dominance by photosynthetic flagellates, which can generate a second peak in chlorophyll (Moline et al. 2004). These studies suggest that RNA viruses contributed up to 65% of the total virioplankton (8–65%) (Miranda et al. 2016), in agreement with previous studies on the relative contribution of RNA viruses (Steward et al. 2013).

Very recent studies suggest that future warming within the WAP will cause changes in sea ice that will influence viruses and their microbial hosts through changes in the timing, magnitude, and composition of the phytoplankton bloom (Evans et al. 2017). This is also in agreement with results suggesting that temperate virus-host interactions are critical to predicting changes in microbial dynamics brought on by warming in polar marine systems (Brum et al. 2017).

9.3 RNA Viruses in Antarctic Freshwater Environments

A major problem to understand the biology of microorganisms inhabiting the Antarctic environments is the logistical constraint of conducting fieldwork all over the year (Foreman et al. 2010). However, microorganisms persist throughout the year under severe environmental conditions. Previous studies on Antarctic lakes and ponds have focused primarily on plankton community structure (Glatz et al. 2006). However, ice is increasingly being recognized as a suitable habitat for microorganisms (Priscu et al. 1998), and complex microbial biota have been found in a wide range of icy systems including glaciers (Zhang et al. 2001) (see Fig. 9.1). These studies provide evidence that the microbial community overwinters in the ice

Fig. 9.1 Collins Glacier at King George Island, Antarctic Peninsula. The figure shows part of the Collins Glacier at Fildes Peninsula, King George Island. (The photo was taken by the author on May 1, 2011)

column and returns to a highly active metabolic state when spring melt is initiated (Foreman et al. 2010). Again, Antarctic ecosystems are dominated by microorganisms, and viruses play particularly important roles in the food webs (Cavicchioli and Erdmann 2015).

In 2015, Lopez-Bueno et al. examined RNA viruses in the Limnopolar Lake, a freshwater lake on Livingston Island north of the Antarctic Peninsula, describing RNA viruses for the first time in an Antarctic freshwater environment (Lopez-Bueno et al. 2015). Positive-sense single-stranded picorna-like viruses were the dominant type of RNA virus, and four distinct types were identified: Antarctic picorna-like virus (APLV) 1–4. Importantly, the extent of the quasispecies was determined in this study (Lopez-Bueno et al. 2015). APLV1 was the most abundant RNA virus in the lake and had the least amount of genome variation, compared to APLV4 and APLV2 and APLV3 which had a higher complexity of quasispecies. The authors proposed that the RNA viruses with a stable, low complexity population of quasispecies represented viruses that were well adapted to their hosts. In contrast, the presence of high quasispecies complexity was indicative of infection occurring in a variety of hosts and/or input into the lake from diverse sources (e.g., cyanobacterial mats) which each harbored distinct, low complexity quasispecies. Moreover, this study, carried out on a freshwater Antarctic lake, highlights again that RNA viruses play a role in Antarctic aquatic systems (Cavicchioli and Erdmann 2015). Recent metagenomics studies revealed that the community of RNA viruses in Antarctic lakes and soil ecosystems is dominated by the order *Picornavirales*. Their quasispecies composition suggests that its high genetic variability may correlate with viral adaptation to new environmental conditions (Rastrojo and Alcami 2018).

9.4 RNA Viruses Associated with Antarctic wildlife

Inhabiting Antarctica and the pack ice surrounding it is a unique assemblage of seals: crabeater (*Lobodon carcinophaga*), leopard (*Hydrurga leptonyx*), Ross (*Ommatophoca rossii*), and Weddell seal (*Leptonychotes weddellii*). Southern elephant seals (*Mirounga leonina*) occupy subantarctic islands for breeding and then migrate south to Antarctica. Antarctic fur seals (*Arctocephalus gazella*) breed on peri-Antarctic islands, as well as islands of the northern Antarctic Peninsula (Siniff et al. 2008) (see Fig. 9.2). Confined to pack ice-affected waters south of the Southern boundary of the Antarctic Circumpolar Current are the Adélie (*Pygoscelis adeliae*) and emperor (*Aptenodytes forsteri*) penguins (see Fig. 9.3); north of that boundary are three other penguin species: gentoo (*P. papua*), chinstrap (*P. antarctica*), and macaroni (*Eudyptes chrysolophus*) (Williams 1995). While mostly distributed to the north, these three penguin species populations breed on the islands of the northwestern Antarctic Peninsula, overlapping with Adélie penguins. Colonies of king penguins (*A. patagonicus*), royal penguins (*E. schlegeli*) and rockhopper penguins (*E. chrysocome*) are found on subantarctic islands and do not inhabit the coastal

Fig. 9.2 Antarctic fur seals (*Arctocephalus gazella*). The figure shows Antarctic fur seals at Drake Passage of Fildes Peninsula in King George Island. (Photo taken by the author on November 6, 2013)

Fig. 9.3 Adélie penguins (*Pygoscelis adeliae*). The figure shows Adélie penguins at Drake Passage of Fildes Peninsula, King George Island. (Photo taken by the author on April 30th, 2011)

Antarctic continent. Flying birds inhabiting Antarctica and subantarctic regions include skuas, petrels, terns, gulls, and albatross (see Fig. 9.4). Also endemic into the sea ice zone, the snow (*Pagodroma nivea*) and Antarctic petrels (*Thalassoica antarctica*) can be found (Brooke 2004; Murphy 1936).

Introduction of pathogen microorganisms, particularly emerging viral diseases endemic in other regions of the world, in these animal populations may have detrimental effects. On the other hand, viruses circulating at an isolated zoogeographical area like Antarctica may spread to other latitudes, having detrimental effects as well, for animal wildlife and for the human population. Understanding the potential effects of these facts requires a detailed characterization of viruses circulating in Antarctica (Smeele et al. 2018), as this will provide insight into the health of Antarctic wildlife.

Historically, there have been a few reported cases of mass mortality observed in Antarctic wildlife, and the disease-causing agent was not clearly determined. In 1971, several hundred gentoo penguin chicks at a Signy Island colony, South Orkneys, were found dead (MacDonald and Conroy 1971), and no isolation of the disease agent was possible in this opportunity (Barbosa and Palacios 2009). In 1972, about 65% of Adélie penguin chicks were reported dead in a colony near Mawson Station (Kerry et al. 2009). Again, the cause of this disease was not determined. The only mass mortality reported in seals was the death of over 1500 crabeater seals in a colony around Crown Prince Gustav Channel in the Antarctic Peninsula in 1955. Although the nature of the disease was not established, a viral infection was suggested (Laws and Taylor 1957).

Fig. 9.4 Antarctic skuas and gulls. The figure shows Kelp gulls (*Larus dominicanus*) and Brown skuas (*Stercorarius Antarctica*) near Collins Glacier at King George Island. (Photo taken by the author on May 1, 2011)

Early work, based on serology assays, was particularly focused in detecting pathogens that were capable of posing a risk for animal health (Morgan et al. 1981; Alexander et al. 1989; Stenvers et al. 1992). Today, advancements in high-throughput sequencing (HTS) approaches are beginning to have an impact on our knowledge of Antarctic animal virology, as several viral families have been detected and characterized (Smeele et al. 2018). Importantly, many of these viral families are RNA viruses, like *Paramyxoviridae*, *Orthomyxoviridae*, *Birnaviridae*, *Flaviviridae*, and *Togaviridae* (see Table 9.1). Moreover, very recent studies revealed also the isolation of a picorna-like virus from the Antarctic fur seal (*Arctocephalus gazella*) (Krumbholz et al. 2017). Since these families have unique and different characteristics, they will be treated separately in the following paragraphs.

9.5 Paramyxoviruses Infecting Antarctic Wildlife

The family *Paramyxoviridae* is a family of enveloped, non-segmented negative-sense RNA viruses with genomes of ~15 kb (Lamb and Kolakofsky 2001). Paramyxoviruses are responsible for considerable disease burden in human and wildlife populations: measles, mumps, and respiratory syncytial virus continue to affect the health of children worldwide, while canine distemper virus causes serious morbidity and mortality in a wide range of mammalian species (Lamb and

Table 9.1 RNA viruses infecting Antarctic wildlife

Family	Virus	Host	Location of isolation	Date of isolation	Reference
Paramyxoviridae	Avian paramyxovirus	Adélie penguins (*Pygoscelis adeliae*)	Wilkes base	–	Morgan and Westbury (1981)
			Peterson Island	–	
			Midgley Island	–	
			Shirley Island	–	
			Cameron Island	–	
			Casey Station	–	Morgan and Westbury (1988)
			Vestfold Hills	1981	Austin and Webster (1993)
			Ross Island	1978	
		Royal penguins (*Eudyptes chrysolophus*)	Macquarie Island	–	Morgan et al. (1981)
		King penguins (*Aptenodytes patagonicus*)	Macquarie Island	–	Morgan et al. (1981)
		South polar skuas (*Stercorarius maccormicki*)	Ross Island	1978	Austin and Webster (1993)
				1986	
	Canine distemper virus	Leopard seal (*Hydrurga leptonyx*)	Antarctic Peninsula	1989	Bengtson et al. (1991)
		Crabeater seal (*Lobodon carcinophaga*)	Antarctic Peninsula	1989	Bengtson et al. (1991)
	Newcastle disease virus	Adélie penguins (*Pygoscelis adeliae*)	King George Island	2006	Thomazelli et al. (2010)
	Phocine distemper virus	Weddell seals (*Leptonychotes weddellii*)	Vestfold Hills	1999	McFarlane (2009)
	Antarctic penguin virus A-C	Gentoo penguins (*Pygoscelis papua*)	Kopaitic Island	2014–2016	Neira et al. (2017)

Orthomyxoviridae	Influenza A virus	Adélie penguins (*Pygoscelis adeliae*)	Peterson Island	–	Morgan et al. (1981)
			Ross Island	1978	Austin and Webster (1993)
			Hope Bay	1998, 2001	Baumeister et al. (2004)
				2002	
			Rada Covadonga	2013	Hurt et al. (2014)
			Antarctic Peninsula		
			King George Island		
		South polar skua (*Stercorarius maccormicki*)	Ross Island	1978, 1986	Austin and Webster (1993)
			Potter Peninsula	1998, 2001	Baumeister et al. (2004)
			Hope Bay	2002	
				1998, 2001	
				2002	
		Chinstrap penguins (*Pygoscelis antarctica*)	Potter Peninsula	1998, 2001	Baumeister et al. (2004)
				2002	
			Antarctic Peninsula	2015	Hurt et al. (2016)
			Aitcho Island		
					Barriga et al. (2016)
		Gentoo penguins (*Pygoscelis papua*)	Potter Peninsula	1998, 2001	Baumeister et al. (2004)
			Aitcho Island	2002	Barriga et al. (2016)
				2015	
		Giant petrel (*Macronectes giganteus*)	Harmony Peninsula	1998, 2001	Baumeister et al. (2004)
				2002	
Birnaviridae	Infectious bursal	Adélie penguins (*Pygoscelis adeliae*)	Mawson Station	1995, 1996	Gardner et al. (1997)
	Disease virus		Mawson coast	1996, 2002	Watts et al. (2009)
			Davis Coast	1996, 2002	
			Terra Nova Bay	1996, 2002	

(continued)

Table 9.1 (continued)

Family	Virus	Host	Location of isolation	Date of isolation	Reference
		Emperor penguin (*Aptenodytes forsteri*)	Mawson Station	1995, 1996	Gardner et al. (1997)
			Auster Rookery	1996–2001	Watts et al. (2009)
			Amanda Bay rookery	1996–2001	
			Cape Washington rookery	1996–2001	
		King penguin (*Aptenodytes patagonicus*)	Davis Station	1999	Miller et al. (1008)
			Vestfold	1999, 2002	
			Davis Station	1999, 2002	Watts et al. (2009)
Flaviviridae	Not determined	South polar skua (*Stercorarius maccormicki*)	Davis Station	1999	Miller et al. (2008)
Togaviridae	Southern elephant seal virus	Southern elephant seals (*Mirounga leonina*)	Macquarie Island	–	Forrester et al. (2012)
					La Linn et al. (2001)

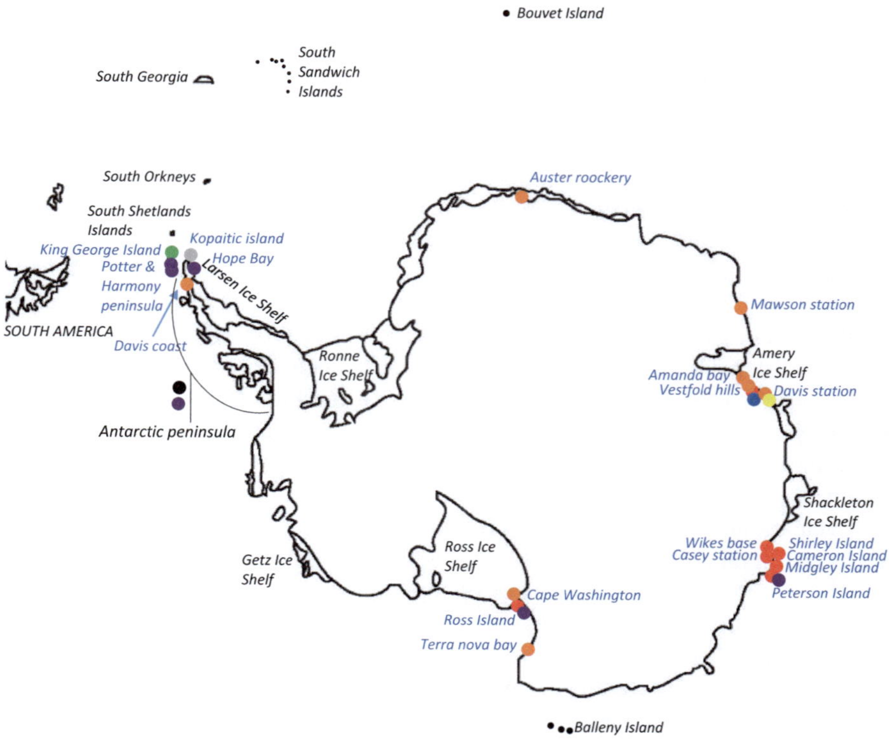

Fig. 9.5 Schematic representation of the distribution of RNA viruses isolated in Antarctica. Colored circles denote the site of isolation, and the location of isolation is shown in blue next to the circle. The locations of isolation for avian paramyxovirus (red), Newcastle disease virus (green), phocine distemper virus (blue), Antarctic penguin virus (gray), influenza A virus (violet), infectious bursal disease virus (orange), Flaviviruses (yellow), and canine distemper virus (black) are shown

Kolakofsky 2001; Pomeroy et al. 2008; Collins and Melero 2011). A major disease burden has also been associated with rinderpest in cattle, Newcastle disease virus (NDV) in poultry, and phocine distemper virus in seals (Lamb and Kolakofsky 2001; Kim and Samal 2016). Most of the research done on paramyxoviruses affecting wildlife in Antarctica has been based on serological studies to detect antibodies against paramyxoviruses in serum samples.

A high prevalence of antibodies to NDV has been previously reported in south polar skuas (*Stercorarius maccormicki*) (Miller et al. 2010), and partial genome sequences of NDV infecting Adélie penguins (*Pygoscelis adeliae*) in King George Island has been obtained (Thomazelli et al. 2010) (see Table 9.1 and Fig. 9.5). Recent phylogenetic studies, carried out on these NDV nucleotide sequences, assigned them to genotype I of class II (Soñora et al. 2015), in agreement with previous antigenic studies (Thomazelli et al. 2010; Alexander et al. 1989). Viruses of this genotype have been associated with outbreaks in Australia occurred in 1998 and 2000 (Gould et al. 2001). The presence of NDV strains in Antarctica, where other avian species live, highlights the importance of NDV strain characterization in all regions of the world (Soñora et al. 2015). Genotypes V, VI, and VII of class II are currently circulating worldwide in chickens (Xiao et al. 2013). The role of Antarctica in maintaining other NDV genotypes, not circulating in this moment, reinforces the relevance of in-depth NDV surveillance studies that have been carried out in Antarctica (Soñora et al. 2015).

Since sled dogs (*Canis familiaris*) were used during the early Antarctic expeditions, concern about canine distemper virus (CDV) infection among Antarctic pinnipeds gave rise to monitor this virus in seal populations. Antibodies to CDV have been detected in leopard (*Hydrurga leptonyx*) and crabeater seals (*Lobodon carcinophaga*) around the Antarctic Peninsula (Bengtson et al. 1991), and antibodies to phocine distemper virus (PDV) were found in Weddell seals from Vestfold Hills, East Antarctica (McFarlane 2009) (see Table 9.1 and Fig. 9.5). Unfortunately, no genetic sequences are available from CDV isolated from Antarctic seals. This would shed light on the genetic relations among CDV circulating in canines and pinnipeds.

Avian paramyxoviruses have been found in different species of penguins and skuas in several different locations in Antarctica (see Table 9.1 and Fig. 9.1).

Recently, the genomic sequences of three novel avian paramyxoviruses have been obtained from strains isolated from gentoo penguins (*Pygoscelis papua*) at Kopaitic Island, in the northern tip of the Antarctic Peninsula (see Table 9.1). These viruses were named Antarctic penguin virus A to C (APVA-C) (Neira et al. 2017). These findings suggest that Antarctica has a higher degree of paramyxovirus diversity than previously anticipated (Neira et al. 2017).

9.6 Orthomyxoviruses Infecting Antarctic Wildlife

Influenza A virus (IAV) is a member of the family *Orthomyxoviridae* and contains eight segments of a single-stranded RNA genome with negative polarity (Neumann et al. 2004). IAV continues to be a major challenge to public health worldwide, causing annual influenza epidemics in humans, emerging epizootics in domestic animals, and occasionally an unpredictable emergence of human pandemic viruses with antigenically novel hemagglutinin (HA) subtypes (Wright et al. 2007). Eighteen distinct HA subtypes, characterized by high genetic and antigenic variability, have been identified so far, 16 of which (subtypes 1–16) are found in multiple avian host species (Munster et al. 2007; Dugan et al. 2008). Only 4 out of 16 of these subtypes have been incorporated in mammals (H1, H2, and H3 in humans, H1 and H3 in swine, H7 and H3 in equines, and H3 in canines; Taubenberger and Kash 2010). The genetic pool of avian IAV contains 16 HA and 6 neuraminidase (NA) subtypes and constitutes a major threat for the generation of novel viruses that can potentially spread to domestic animals and human populations.

Wild aquatic birds such as dabbling ducks, gulls, and other shorebirds are considered the natural reservoir for avian influenza viruses (AIVs) (Hurt et al. 2014). Wild migratory birds play a key role in the spread of AIVs on a local, regional, and intercontinental scale via broadly established flyways (Hill et al. 2012; Olsen et al. 2006). Although major migratory flyways may not extend to Antarctica, birds such as the Arctic tern (*Sterna paradisaea*) (Egevang et al. 2010) and south polar skua (*Stercorarius maccormicki*) (Kopp et al. 2011) conduct transhemispheric migrations to Antarctica and therefore have the potential to facilitate movement of avian influenza viruses into and out of Antarctica (Hurt et al. 2014).

Earlier studies on AIV in Antarctica reported the presence of Influenza A virus antibodies in penguins and other birds (Baumeister et al. 2004; Wallensten et al. 2006; Miller et al. 2008) (see Table 9.1). These studies have also revealed that skuas (*Stercorarius maccormicki*) and giant petrels (*Macronectes giganteus*), who travel long distances both within Antarctica and between continents, have a high frequency of AIV antibodies, ranging from 1.0% to 11.1% for skuas and up to 51.5–58.8% for giant petrels (Baumeister et al. 2004; Miller et al. 2008; Austin and Webster 1993). These birds have very close interactions with penguins, often nesting alongside their colonies and attacking penguin chicks. This may provide a route for AIV transmission (Hurt et al. 2014). Interestingly, recent studies revealed the presence of H11N2 AIV in Adélie penguins (*Pygoscelis adeliae*) (Hurt et al. 2014). H11 subtype has not been reported so far for birds from South America (although studies on AIV in South America are scarce); they have been detected in migratory birds in North America (being the second most frequent HA-detected subtype) (Krauss et al. 2004), as well as in Australia (Hurt et al. 2006) and New Zealand (Langstaff et al. 2009), and in ducks in other regions of the world (Munster et al. 2007). While no symptoms were noticed in the H11N2-infected penguins (Hurt et al. 2014), given the fact that highly pathogenic AIVs have been detected in South America (Senne 2007) and Australia (Selleck et al. 2003), the possibility that such viruses could be transferred to Antarctica by migratory birds cannot be ruled out, posing a significant threat to Antarctic wildlife populations. Importantly, as an AIV surveillance in penguins and other birds continues in Antarctica, detections of additional AIVs have been recently reported (Hurt et al. 2016), revealing the presence of a novel H5N5 reassortant AIV in chinstrap penguins (*Pygoscelis antarctica*), with genes derived from both American and Eurasian genetic lineages (Hurt et al. 2016; Barriga et al. 2016). Again, these finding raise an important concern, given the recent emergence of highly pathogenic AIVs belonging to H5 genotype in North America in 2004 (Bevins et al. 2016), although the analysis of the HA cleavage site of H5 strains isolated in the Antarctic peninsula revealed the presence of low pathogenicity AIV (LPAIV) strains, containing cleavage motif PQRETRGLF (Senne et al. 1996). Nevertheless, many of these viruses were detected in apparently healthy wild waterfowl, which may facilitate the spread of the virus (Hurt et al. 2016), posing a potential impact for the Antarctic Peninsula.

Early studies studying the presence of antibodies against IAV in pinnipeds did not find positives results, although several studies have been carried out at crabeater and Weddell seals (Austin and Webster 1993; McFalrane 2009).

9.7 Birnaviruses Infecting Antarctic Wildlife

Infectious bursal disease virus (IBDV) is a member of the family *Birnaviridae*. IBDV is a double-stranded RNA virus that has a bi-segmented genome, being each segment ~2.3 to 3 kb in length (Delmas et al. 2011). This virus causes a highly contagious disease in chickens and an important burden to the poultry industry worldwide.

Neutralization assays against IBDV revealed high titers of antibodies in Adélie penguins (*Pygoscelis adeliae*) at colonies around Mawson and Davis Stations and at Terra Nova Bay in East Antarctica (see Table 9.1). At the same locations, a similar situation was observed for emperor penguins (*Aptenodytes forsteri*) at Auster, Amanda y Cape Washington colonies (Gardner et al. 1997; Watts et al. 2009) (see Fig. 9.1). Antibodies against IBDV have also been found in south polar skuas living around Vestfold Hills and Davis Station (Miller et al. 2008; Watts et al. 2009). High titers of IBDV antibodies have also been observed beyond populations of penguins and south polar skuas living in East Antarctica, suggesting that this virus may be normally circulating among Antarctic birds (Watts et al. 2009) (see Fig. 9.1). Unfortunately, IBDV has not been isolated so far from Antarctic birds, despite several studies detecting antibodies against the virus.

9.8 Flaviviruses Infecting Antarctic Wildlife

Flaviviruses are enveloped, positive-sense single-stranded RNA viruses with a genome of approximately 9.4–13 kb in length (Zhang et al. 2017). Most flaviviruses are zoonotic, so infections by these viruses may spread between animals and humans (Junglen et al. 2017). Many flaviviruses are associated with very important human diseases, like Dengue virus (DENV), Yellow fever virus (YFV), West Nile virus (WNV), or Zika virus (ZIKV) (Klema et al. 2015; Moratorio et al. 2013; Ramirez et al. 2010; Fajardo et al. 2016).

Although no genomic information is currently available for flaviviruses circulating in Antarctic wildlife, antibodies to flavivirus have been detected in the serum of south polar skuas (*Stercorarius maccormicki*) at Davis Station, East Antarctica (see Table 9.1 and Fig. 9.1) (Miller et al. 2008). Although there are reports indicating the isolation of flaviviruses from seabirds ticks (*Ixodes uriae*) infecting king penguins (*Aptenodytes patagonicus*), these isolations took place in penguins living in Macquarie Island but not so far from penguins at the Antarctic continent (Major et al. 2009). More studies will be needed in order to identify the flavivirus species infecting Antarctic animals.

9.9 Togaviruses Infecting Antarctic Wildlife

Togaviruses are enveloped positive-sense RNA viruses with a genome length of about 11–12 kb in length. This family has two genera: *Rubivirus*, represented by Rubella virus, an important human pathogen, and *Alphavirus*, a genus to which all animal togaviruses belong (Power et al. 2017). Alphaviruses life cycle requires an arthropod vector for transmission to their vertebrate host. The first evidence of alphaviruses infecting marine mammals was shown in southern elephant seals (*Mirounga leonina*) living in Macquarie Island (see Table 9.1). A seal alphavirus was isolated from the blood-sucking louse (*Lepidophthirus macrorhini*), which is widespread among these seals. Nevertheless, a high seroprevalence of antibodies against the southern elephant seal alphavirus suggest its transmission from lice to seals (La Linn et al. 2001). The full genome of this alphavirus (southern elephant seal virus) was recently obtained (Forrester et al. 2012).

9.10 RNA Plant Viruses in Antarctica

Studies have been performed in order to observe the presence of plant viruses in the Argentina Islands, which are a group of islands in the Wilhelm Archipelago of the Antarctic Peninsula, using enzyme-linked immunosorbent assay (ELISA) to test for the presence of common plant viruses (Polischuk et al. 2007). Interestingly, samples from moss genera *Barbilophozia* and *Polytrichum* were found to contain antigens of viruses from the genus *Tobamovirus* (e.g., Tobacco mosaic virus and Cucumber green mottle mosaic virus). These viruses are single-stranded positive-sense RNA viruses that belong to the family *Virgaviridae* (Adams et al. 2017). This viruses can infect many plants, including tobacco, potato, tomato, and squash. Interestingly, in the same study, samples of the monocotyledon Antarctic hair grass *Deschampsia antarctica* were positive for viruses typically infecting dicotyledons: Cucumber green mottle mosaic virus (CGMMV), Cucumber mosaic virus (CMV), and Tomato spotted wilt virus (TSWV) (Polischuk et al. 2007). While CGMMV belongs to the family *Virgaviridae* (Adams et al. 2017), CMV belongs to the family *Bromoviridae*, another family of single-stranded positive-sense RNA viruses that infect a great variety of plants. On the other hand, TSWV belongs to the family *Tospoviridae*, a family of single-stranded negative-sense RNA viruses that also infect a large number of plant families (Scholthof et al. 2011). These studies demonstrate the presence of highly diverse plant RNA viruses in Antarctica. More studies will be needed in order to determine virus presence, virus specificity, and host susceptibility in plants from Antarctic ecosystems.

9.11 Remarks and Future Directions of RNA Virus Research in Antarctica

RNA viruses play an important active role as a component of the Antarctic aquatic microbial community, both at sea and freshwater environments. RNA virus infection of Antarctic wildlife is more significant than previously anticipated, where at least members of five different RNA virus families have been identified, *Paramyxoviridae*, *Orthomyxoviridae*, *Birnaviridae*, *Flaviviridae*, and *Togaviridae* (see Table 9.1), and probably many more will be identified, as the recent picorna-like virus isolated from the Antarctic fur seal (*Arctocephalus gazella*) (Krumbholz et al. 2017). Moreover, three other RNA virus families infecting plants, *Virgaviridae*, *Bromoviridae*, and *Tospoviridae*, were also observed in Antarctic moss and grass. This highlights the great genomic plasticity of RNA viruses to adapt and evolve to ensure its survival.

From the last years on, viral metagenomics will provide a dramatic increase of viral discoveries through HTS. Recent studies, using HTS approach, have identified ~1400 novel RNA viruses, from which over 200 correspond to invertebrate species (Shi et al. 2016). This approach has been used in virology studies in Antarctica to study soil (Adriaenssens et al. 2017), lakes (Aguirre de Cárcer et al. 2016; Lopez-Bueno et al. 2015), and marine viral ecology (Brum et al. 2017; Miranda et al. 2016). Nevertheless, despite the development of HTS, relatively little is known about RNA viruses infecting Antarctic animals by comparison, and vast geographical areas have not been studied in detail for virus isolation (see Fig. 9.1). This approach will permit a fast advance in studying these viruses in the Antarctic context, particularly for the study of emerging RNA viruses of extreme importance for human and/or animal health in Antarctica, like Influenza A virus, Newcastle disease virus, canine distemper virus, or infectious bursal disease virus. To determine the role of Antarctica in the global emergence of these viruses is extremely important for the development of appropriate strategies to control infection caused by these viruses.

Acknowledgments JC acknowledge the support from Comisión Sectorial de Investigación Cientifica (CSIC), through CSIC Grupos program, and PEDECIBA, Universidad de la República. I also acknowledge the support from Agencia Nacional de Investigación e Innovación (ANII) and Sistema Nacional de Investigadores, Uruguay.

References

Adams, M. J., Adkins, S., Bragard, C., Gilmer, D., & Li, D. (2017). ICTV virus taxonomy profile: Virgaviridae. *The Journal of General Virology, 98*, 1999–2000.

Adriaenssens, E. M., Kramer, R., Van Goethem, M. W., Makhalanyane, T. P., Hogg, I., & Cowan, D. A. (2017). Environmental drivers of viral community composition in Antarctic soils identified by viromics. *Microbiome, 5*, e83.

Aguirre de Cárcer, D., López-Bueno, A., Alonso-Lobo, J. M., Quesada, A., et al. (2016). Metagenomic analysis of lacustrine viral diversity along a latitudinal transect of the Antarctic Peninsula. *FEMS Microbiology Ecology, 92*, fiw074. https://doi.org/10.1093/femsec/fiw074.

Alexander, D. J., Manvell, R. J., Collins, M. S., Brockman, S. J., Westbury, H. A., et al. (1989). Characterization of paramyxoviruses isolated from penguins in Antarctica and sub-Antarctica during 1976–1979. *Archives of Virology, 109*, 135–143.

Austin, F. J., & Webster, R. G. (1993). Evidence of ortho- and paramyxoviruses in fauna from Antarctica. *Journal of Wildlife Diseases, 29*, 568–571.

Barbosa, A., & Palacios, M. J. (2009). Health of Antarctic birds: A review of their parasites, pathogens and diseases. *Polar Biology, 32*, 1095–1115.

Barriga, G. P., Boric-Bargetto, D., San Martin, M. C., Neira, V., van Bakel, H., et al. (2016). Avian influenza virus H5 strain with North American and Eurasian lineage genes in an Antarctic penguin. *Emerging Infectious Diseases, 22*, 2221–2223.

Baumeister, E., Leotta, G., Pontoriero, A., Campos, A., Montalti, D., et al. (2004). Serological evidences of influenza A virus infection in Antarctica migratory birds. *International Congress Series, 1263*, 737–740.

Bengtson, J. L., Boveng, P., Franzen, U., Have, P., Heide-Jorgensen, M. P., et al. (1991). Antibodies to canine distemper virus in antarctic seals. *Marine Mammal Science, 7*, 85–87.

Bevins, S. N., Dusek, R. J., White, C. L., Gidlewski, T., Bodenstein, B., et al. (2016). Widespread detection of highly pathogenic H5 influenza viruses in wild birds from the Pacific flyway of the United States. *Scientific Reports, 6*, 28980.

Brooke, M. (2004). *Albatrosses and petrels across the world*. Oxford: Oxford University Press.

Brum, J. R., Hurwitz, B. L., Schofield, O., Ducklow, H. W., & Sullivan, M. B. (2017). Seasonal time bombs: Dominant temperate viruses affect southern ocean microbial dynamics. *The ISME Journal, 10*, 437–449.

Cavicchioli, R., & Erdmann, S. (2015). The discovery of Antarctic RNA viruses: A new game changer. *Molecular Ecology, 24*, 4809–4811.

Collins, P. L., & Melero, J. A. (2011). Progress in understanding and controlling respiratory syncytial virus: Still crazy after all these years. *Virus Research, 162*, 80–99.

Culley, A. (2018). New insight into the RNA aquatic virosphere via viromics. *Virus Research, 244*, 84–89.

Delmas, B., Mundt, E., Vakharia, V. N., & Wu, J. L. (2011). Family Birnaviridae. In A. M. Q. King, E. Lefkowitz, M. J. Adams, & E. B. Carstens (Eds.), *Virus taxonomy: Ninth report of the international committee on taxonomy of viruses* (pp. 496–507). San Diego: Elsevier.

Domingo, E. (2006). *Quasispecies: Concept and implications for virology*. Heidelberg: Springer.

Domingo, E., Sabo, D., Taniguchi, T., & Weissmann, C. (1978). Nucleotide sequence heterogeneity of an RNA phage population. *Cell, 13*, 735–744.

Ducklow, H. W., Baker, K., Martinson, D. G., Quetin, L. B., et al. (2007). Marine pelagic ecosystems: The West Antarctic Peninsula. *Philosophical Transactions of the Royal Society B, 362*, 67–94.

Dugan, V. G., Chen, R., Spiro, D. J., Sengamalay, N., Zaborsky, J., et al. (2008). The evolutionary genetics and emergence of avian influenza viruses in wild birds. *PLoS Pathogens, 4*, e1000076.

Egevang, C., Stenhouse, I. J., Phillips, R. A., Petersen, A., Fox, J. W., et al. (2010). Tracking of Arctic terns *Sterna paradisaea* reveals longest animal migration. *Proceedings of the National Academy of Sciences of the United States of America, 107*, 10182–10189.

Eggers, H. J., & Tamm, I. (1965). Coxsackie A9 virus: Mutation from drug dependence to drug independence. *Science, 148*, 97–98.

Eigen, M. (1971). Self organization of matter and the evolution of biological macromolecules. *Naturwissenschaften, 58*, 465–523.

Eigen, M. (1992). *Steps towards life: A perspective on evolution*. Oxford: Oxford University Press.

Eigen, M., & Schuster, P. (1977). The hypercycle. A principle of natural self-organization. Part A: Emergence of the hypercycle. *Naturwissenschaften, 64*, 541–565.

Evans, C., Brandsma, J., Pond, D. W., & Venables, H. J. (2017). Drivers of interannual variability in virioplankton abundance at the coastal western Antarctic peninsula and the potential effects of climate change. *Environmental Microbiology, 19*, 740–755.

Fajardo, A., Cristina, J., & Moreno, P. (2016). Emergence and spreading potential of Zika virus. *Frontiers in Microbiology, 7*, 1667.

Foreman, C. M., Dieser, M., Greenwood, M., Cory, R. M., Laybourn-Parry, J., et al. (2010). When a habitat freezes solid: Microorganisms over-winter within the ice column of a coastal Antarctic lake. *FEMS Microbiology Ecology, 76*, 401–412.

Forrester, N. L., Palacios, G., Tesh, R. B., Savji, N., Guzman, H., et al. (2012). Genome-scale phylogeny of the alphavirus genus suggests a marine origin. *Journal of Virology, 86*, 2729–2738.

Gardner, H., Kerry, K., Riddle, M., Brouwer, S., & Gleeson, L. (1997). Poultry virus infection in Antarctic penguins. *Nature, 387*, 245.

Glatz, R. E., Lepp, P. W., Ward, B. B., & Francis, C. A. (2006). Planktonic microbial community composition across steep physical/chemical gradients in permanently ice-covered Lake Bonney, Antarctica. *Geobiology, 4*, 53–67.

Gould, A. R., Kattenbelt, J. A., Selleck, P., Hansson, E., Della-Porta, A., et al. (2001). Virulent Newcastle disease in Australia: Molecular epidemiological analysis of viruses isolated prior to and during the outbreaks of 1998–2000. *Virus Research, 77*, 51–60.

Hill, N. J., Takekawa, J. Y., Cardona, C. J., Meixell, B. W., Ackerman, J. T., et al. (2012). Cross-seasonal patterns of avian influenza virus in breeding and wintering migratory birds: A flyway perspective. *Vector Borne and Zoonotic Diseases, 12*, 243–253.

Hurt, A. C., Hansbro, P. M., Selleck, P., Olsen, B., Minton, et al. (2006). Isolation of avian influenza viruses from two different transhemispheric migratory shorebird species in Australia. *Archives of Virology, 151*, 2301–2309.

Hurt, A. C., Vijaykrishna, D., Butler, J., Baas, C., Maurer-Stroh, S., et al. (2014). Detection of evolutionarily distinct avian influenza a viruses in Antarctica. *MBio, 5*(3), e01098–e01040.

Hurt, A. C., Su, Y. C., Aban, M., Peck, H., Lau, H., et al. (2016). Evidence for the introduction, reassortment, and persistence of diverse influenza a viruses in Antarctica. *Journal of Virology, 90*, 9674–9682.

Junglen, S., Korries, M., Grasse, W., Wieseler, J., & Kopp, A. (2017). Host range restriction of insect-specific flaviviruses occurs at several levels of the viral life cycle. *mSphere, 2*, e00375–e00316. https://doi.org/10.1128/mSphere.00375-16.

Kerry, K., Irvine, L., Beggs, A., & Watts, J. (2009). An unusual mortality event among Adélie penguins in the vicinity of Mawson Station, Antarctica. In *Health of Antarctic wildlife* (pp. 107–112). Berlin: Springer.

Kim, S. H., & Samal, S. K. (2016). Newcastle disease virus as a vaccine vector for development of human and veterinary vaccines. *Viruses, 8*, 183. https://doi.org/10.3390/v8070183.

Klema, V. J., Padmanabhan, R., & Choi, K. H. (2015). Flaviviral replication complex: Coordination between RNA synthesis and 5′-RNA capping. *Viruses, 7*, 4640–4656.

Kopp, M., Peter, H. U., Mustafa, O., Lisovski, S., Ritz, M. S., et al. (2011). South polar skuas from a single breeding population overwinter in different oceans though show similar migration patterns. *Marine Ecology Progress Series, 435*, 263–267.

Krauss, S., Walker, D., Pryor, S. P., Niles, L., Chenghong, L., et al. (2004). Influenza A viruses of migrating wild aquatic birds in North America. *Vector Borne and Zoonotic Diseases, 4*, 177–189.

Krishnamurthy, S. R., & Wang, D. (2017). Origins and challenges of viral dark matter. *Virus Research, 239*, 136–142.

Krumbholz, A., Groth, M., Esefeld, J., Peter, H. U., & Zell, R. (2017). Genome sequence of a novel Picorna-like RNA virus from feces of the Antarctic fur seal (*Arctocephalus gazella*). *Genome Announcement, 5*(36), e01001–e01017. https://doi.org/10.1128/genomeA.01001-17.

La Linn, M., Gardner, J., Warrilow, D., Darnell, G. A., McMahon, C. R., et al. (2001). Arbovirus of marine mammals: A new alphavirus isolated from the elephant seal louse, *Lepidophthirus macrorhini*. *Journal of Virology, 75*, 4103–4109.

Lamb, R. A., & Kolakofsky, D. (2001). Paramyxoviridae: The viruses and their replication. In B. N. Fields, D. M. Knipe, & P. M. Howley (Eds.), *Virology* (pp. 1305–1340). Philadelphia: Lippincott Williams and Wilkins.

Lang, A. S., Rise, M. L., Culley, A. I., & Steward, G. F. (2009). RNA viruses in the sea. *FEMS Microbiology Reviews, 33*, 295–323.

Langstaff, I. G., McKenzie, J. S., Stanislawek, W. L., Reed, C. E., et al. (2009). Surveillance for highly pathogenic avian influenza in migratory shorebirds at the terminus of the East Asian-Australasian flyway. *New Zealand Veterinary Journal, 57*, 160–165.

Laws, R. M., & Taylor, R. (1957). A mass dying of crabeater seals, *Lobodon carcinophagus* (gray). *Journal of Zoology, 129*, 315–324.

López-Bueno, A., Rastrojo, A., Peiró, R., Arenas, M., & Alcamí, A. (2015). Ecological connectivity shapes quasispecies structure of RNA viruses in an Antarctic lake. *Molecular Ecology, 24*, 4812–4825.

MacDonald, D. M., & Conroy, J. W. H. (1971). Virus disease resembling puffinosis in the gentoo penguin (*Pygoscelis papua*). *British Antarctica Survey Bulletin, 60*, 80–83.

Major, L., Linn, M. L., Slade, R. W., Schroder, W. A., Hyatt, A. D., et al. (2009). Ticks associated with Macquarie island penguins carry arboviruses from four genera. *PLoS One, 4*, e4375.

McFarlane, R. (2009). Health assessment and diseases of the Weddell seal, Leptonochotes Weddelli, in Vestfold Hills, East Antarctica. In *Health of Antarctic wildlife* (pp. 139–166). Berlin: Springer.

Miller, G. D., Watts, J. M., & Shellam, G. R. (2008). Viral antibodies in south polar skuas around Davis Station, Antarctica. *Antarctic Science, 20*, 455–461.

Miller, P. J., Afonso, C. L., Spackman, E., Scott, M. A., Pedersen, J. C., et al. (2010). Evidence for a new avian paramyxovirus serotype 10 detected in rockhopper penguins from the Falkland Islands. *Journal of Virology, 84*, 11496–11504.

Miranda, J. A., Culley, A. I., Schvarcz, C. R., & Steward, G. F. (2016). RNA viruses as major contributors to Antarctic virioplankton. *Environmental Microbiology, 18*, 3714–3727.

Moline, M. A., Claustre, H., Frazer, T. K., Schofield, O., & Vernet, M. (2004). Alteration of the food web along the Antarctic Peninsula in response to a regional warming trend. *Global Change Biology, 10*, 1973–1980.

Moratorio, G., Iriarte, A., Moreno, P., Musto, H., & Cristina, J. (2013). A detailed comparative analysis on the overall codon usage patterns in West Nile virus. *Infection, Genetics and Evolution, 14*, 396–400.

Morgan, I. R., & Westbury, H. A. (1981). Virological studies of Adélie Penguins (*Pygoscelis adeliae*) in Antarctica. *Avian Diseases, 25*, 1019–1026.

Morgan, I. R., & Westbury, H. A. (1988). Studies of viruses in penguins in the Vestfold Hills. *Hydrobiologia, 165*, 263–269.

Morgan, I. R., Westbury, H. A., Caple, I. W., & Campbell, J. (1981). A survey of virus infection in sub-antarctic penguins on Macquarie Island, Southern Ocean. *Australian Veterinary Journal, 57*, 333–335.

Munster, V. J., Baas, C., Lexmond, P., Waldenström, J., Wallensten, A., et al. (2007). Spatial, temporal, and species variation in prevalence of influenza A viruses in wild migratory birds. *PLoS Pathogens, 3*, e61.

Murphy, R. C. (1936). *The oceanic birds of South America*. New York: Macmillan.

Neira, V., Tapia, R., Verdugo, C., Barriga, G., Mor, S., et al. (2017). Novel avulaviruses in penguins, Antarctica. *Emerging Infectious Diseases, 23*, 1212–1121.

Neumann, G., Brownlee, G. G., Fodor, E., & Kawaoka, Y. (2004). Orthomyxovirus replication, transcription, and polyadenylation. *Current Topics in Microbiology and Immunology, 283*, 121–143.

Olsen, B., Munster, V. J., Wallensten, A., Waldenström, J., Osterhaus, A. D., et al. (2006). Global patterns of influenza A virus in wild birds. *Science, 312*, 384–388.

Perales C, Lorenzo-Redondo R, López-Galíndez C, Martínez MA, Domingo E (2010) Mutant spectra in virus behavior. *Future Virology* 5. https://doi.org/10.2217/fvl.10.61

Polischuk, V., Budzanivska, I., Shevchenko, T., & Oliynik, S. (2007). Evidence for plant viruses in the region of Argentina Islands, Antarctica. *FEMS Microbiology Ecology, 59*, 409–417.

Pomeroy, L. W., Bjørnstad, O. N., & Holmes, E. C. (2008). The evolutionary and epidemiological dynamics of the Paramyxoviridae. *Journal of Molecular Evolution, 66*, 98–106.

Power, A., Huang, H., Roehrig, J., Strauss, E., & Weaver, S. (2017). Togaviridae. In J. N. Maclachlan, E. J. Dubovi, S. W. Barthold, D. F. Swayne, & J. R. Winton (Eds.), *Fenner's veterinary virology* (pp. 511–524). Boston: Academic.

Priscu, J. C., Fritsen, C. H., Adams, E. E., Giovannoni, S. J., Paerl, H. W., et al. (1998). Perennial Antarctic lake ice: An oasis for life in a polar desert. *Science, 280*, 2095–2098.

Ramirez, A., Fajardo, A., Moros, Z., Gerder, M., Caraballo, G., et al. (2010). Evolution of dengue virus type 3 genotype III in Venezuela: Diversification, rates and population dynamics. *Virology Journal, 7*, 329.

Rastrojo, A., & Alcamí, A. (2018). Viruses in polar lake and soil ecosystems. *Advances in Virus Research, 101*, 39–54.

Sanjuán, R. (2012). From molecular genetics to phylodynamics: Evolutionary relevance of mutation rates across viruses. *PLoS Pathogens, 8*, e1002685.

Scholthof, K. B., Adkins, S., Czosnek, H., Palukaitis, P., Jacquot, E., et al. (2011). Top 10 plant viruses in molecular plant pathology. *Molecular Plant Pathology, 12*, 938–954.

Selleck, P. W., Arzey, G., Kirkland, P. D., Reece, R. L., Gould, A. R., et al. (2003). An outbreak of highly pathogenic avian influenza in Australia in 1997 caused by an H7N4 virus. *Avian Diseases, 47*, 806–811.

Senne, D. A. (2007). Avian influenza in North and South America, 2002–2005. *Avian Diseases, 51*, 167–173.

Senne, D. A., Panigrahy, B., Kawaoka, Y., Pearson, J. E., Süss, J., et al. (1996). Survey of the hemagglutinin (HA) cleavage site sequence of H5 and H7 avian influenza viruses: Amino acid sequence at the HA cleavage site as a marker of pathogenicity potential. *Avian Diseases, 40*, 425–437.

Shi, M., Lin, X. D., Tian, J. H., Chen, L. J., Chen, X., et al. (2016). Redefining the invertebrate RNA virosphere. *Nature, 540*, 539–543.

Siniff, D. B., Garrott, R. A., Rotella, J. J., Fraser, W. R., & Ainley, D. G. (2008). Opinion: Projecting the effects of environmental change on Antarctic seals. *Antarctic Science, 20*, 425–435.

Smeele, Z. E., Ainley, D. G., & Varsani, A. (2018). Viruses associated with Antarctic wildlife: From serology based detection to identification of genomes using high throughput sequencing. *Virus Research, 243*, 91–105.

Soñora, M., Moreno, P., Echeverría, N., Fischer, S., Comas, V., et al. (2015). An evolutionary insight into Newcastle disease viruses isolated in Antarctica. *Archives of Virology, 160*, 1893–1900.

Stenvers, O., Plötz, J., & Ludwig, H. (1992). Antarctic seals carry antibodies against seal herpesvirus. *Archives of Virology, 123*, 421–424.

Steward, G. F., Culley, A. I., Mueller, J. A., Wood-Charlson, E. M., et al. (2013). Are we missing half of the viruses in the ocean? *The ISME Journal, 7*, 672–679.

Taubenberger, J. K., & Kash, J. C. (2010). Influenza virus evolution, host adaptation, and pandemic formation. *Cell Host and Microbe, 7*, 440–451.

Thomazelli, L. M., Araujo, J., Oliveira, D. B., Sanfilippo, L., Ferreira, C. S., et al. (2010). Newcastle disease virus in penguins from King George Island on the Antarctic region. *Veterinary Microbiology, 146*, 155–160.

Vignuzzi, M., Stone, J. K., Arnold, J. J., Cameron, C. E., & Andino, R. (2006). Quasispecies diversity determines pathogenesis through cooperative interactions in a viral population. *Nature, 439*, 344–348.

Wallensten, A., Munster, V. J., Osterhaus, A. D., Waldenstr, J., & Bonnedahl, J. (2006). Mounting evidence for the presence of influenza A virus in the avifauna of the Antarctic region. *Antarctic Science, 18*, 353–356.

Watts, J. M., Miller, G. D., & Shellam, G. R. (2009). Infectious bursal disease virus and Antarctic birds. In K. R. Kerry & M. Riddle (Eds.), *Health of Antarctic wildlife* (pp. 95–105). Berlin: Springer Berlin Heidelberg.

Williams, T. D. (1995). *The Penguins: Spheniscidae, 2*. Oxford: Oxford University Press.

Wright, P. F., Neumann, G., & Kawaoka, Y. (2007). Orthomyxoviruses. In D. M. Knipe & P. M. Howley (Eds.), *Fields virology* (Vol. 2, 5th ed., pp. 1691–1740). Philadelphia: Lippincott Williams and Wilkins.

Xiao, S., Paldurai, A., Nayak, B., Mirande, A., Collins, P. L., et al. (2013). Complete genome sequence of a highly virulent Newcastle disease virus currently circulating in Mexico. *Genome Announcements*. https://doi.org/10.1128/genomeA.00177-12.

Zhang, X., Yao, T., Ma, X., & Wang, N. (2001). Analysis of the characteristics of microorganisms packed in the ice core of Malan Glacier, Tibet. *Science in China Series D, 44*, 165–170.

Zhang, X., Jia, R., Shen, H., Wang, M., Yin, Z., et al. (2017). Structures and functions of the envelope glycoprotein in Flavivirus infections. *Viruses, 9*, E338. https://doi.org/10.3390/v9110338.

Part III
The Ecological Roles of Microorganisms Inhabiting Specific Antarctic Niches

Chapter 10
The Hidden Life of Antarctic Rocks

Vanesa Amarelle, Valentina Carrasco, and Elena Fabiano

Abstract Microbes are able to colonize almost every part on Earth where liquid water is available, and rocks are not an exception. Moreover, in extremely dry and harsh places, like some found in Antarctica, rocks may represent the main refuge for life. Despite its relevance, our understanding of lithobiontic communities is just at the beginnings. In this chapter we present a brief history of research on Antarctic lithobiontic communities and summarize recent advances in our understanding of this fascinating microbial world. We point up methodological approximations used for its characterization, microbial diversity of lithobionts, and the identification of functional traits that drive lithobiont survival and community assembly. These extreme environmental niches can be considered a barely explored source of microbial life whose function in global processes such as global climate changes remains unclear. Understanding the adaptations that allow lithobionts to successfully compete in their environment is a quest for understanding the fundamentals of life.

Keywords Lithobiontic communities · Rock environment · Carbon input · Nitrogen input · Adaptation

10.1 Introduction

Rocks are suitable niches to support life and may be the main refuge for living organisms in some extremely inhospitable places (Fig. 10.1) (Wierzchos et al. 2018). Microbes able to colonize rocks are termed lithobionts (*bios*, organism; *lithos*, rock) or endoliths (*endo*, inside) (Makhalanyane et al. 2014; Wierzchos et al. 2012; De Los Ríos et al. 2014). Nevertheless the broader term lithobionts is preferable when it is not a specific reference to its location in the rock, as it is described below. In this sense, three main categories are distinguished according to the lithobiontic niche:

V. Amarelle · V. Carrasco · E. Fabiano (✉)
Department of Microbial Biochemistry and Genomics, Biological Research Institute Clemente Estable, Montevideo, Uruguay
e-mail: efabiano@iibce.edu.uy

S. Castro-Sowinski (ed.), *The Ecological Role of Micro-organisms in the Antarctic Environment*, Springer Polar Sciences,
https://doi.org/10.1007/978-3-030-02786-5_10

Fig. 10.1 Rocks are bioreceptive niches. Rocks provide microbes with a safe refuge against harsh life conditions such as low temperatures, water scarcity, strong wind, and harmful UV radiations

epiliths, hypoliths, and endoliths. Epiliths (*epi*, upon) are located on the external surface of the rock in contact with the atmosphere, while hypoliths (*hypo*, under) are found underneath the rock (Fig. 10.2). The term hypoliths is somehow confusing as some authors refer as hypoliths those microbes that inhabit the underside of the rock in contact with the soil, while for others hypoliths are those organisms living under the rocks, regardless of whether they are in the rock (lithobionts) or in the underneath soil/substrate affected by the rock. Additionally, Wierzchos et al. (2012) suggested to use the term "hypoendoliths" to refer to endoliths that colonize the porous spaces on the underside of the rock which is not in contact with the soil. Furthermore, endolithic organisms can be classified as (a) chasmoendoliths (*chasm*, cleft) regarding to those organisms that colonize cracks and fissures in the rocks or (b) cryptoendoliths (*crypto*, hidden) regarding those that inhabit pores and structural cavities within rocks ("rocky tissue"). From a functional perspective, endoliths can be also differentiated as (a) euendoliths (*eu*, good, true) concerning those organisms that actively penetrate the rocky substrate creating new tunnels or cavities and promoting biogenic weathering of rocks or (b) autoendoliths (*auto*, self) which make those that participate in mineral deposition on the rock. This functional classification is quite recent (Marlow et al. 2015), and according to these authors, autoendoliths represent "a distinct category of rock-hosted life." Autoendoliths may contribute directly to the rock formation, through metabolic products that precipitate as minerals, or indirectly

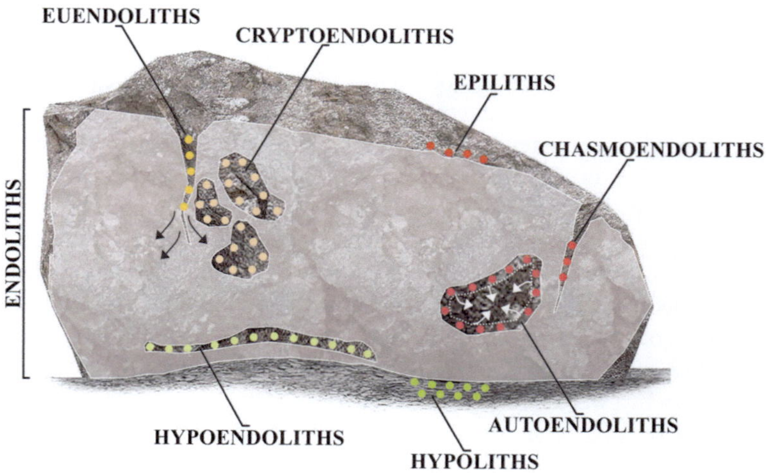

Fig. 10.2 Schematic representation and classification of lithobionts. Epiliths (red dots) are located on the upper surface of the rock, while hypoliths (green dots) are found underneath the rock. Organisms that colonize rock interiors are called endoliths. Among endoliths, cryptoendoliths (orange) are those who inhabit pores, chasmoendoliths (fucsia) are found between cracks or fissures, and hypoendoliths (light green) live within spaces on the underside of the rock that are not in contact with the soil. Autoendoliths (pink) are endoliths that have the ability to contribute to rock formation through mineral deposition, whereas euendoliths (yellow) are those who actively penetrate the rock substrate

by products that promote mineralization. It should be noted that autoendoliths remain as such until they become fully entombed by themselves as their metabolic activity ceases (Marlow et al. 2015).

The categories described above and summarized in Fig. 10.2 are not strict, nor mutually exclusive. Some organisms, such as lichens, can be partially epilithic and partially endolithic. Besides, others can colonize preexisting cavities as cryptoendolithic or chasmoendolithic and produce metabolic compounds that contribute to rock weathering as euendoliths do.

10.2 Antarctic Lithobionts

Rocks represent a primary niche for microbial colonization in Antarctica, where climatic conditions are extremely harsh (Nienow et al. 1988; de los Ríos et al. 2014). Moreover, it has been proposed that they play a relevant role in the ecology and function of extreme environments like those found in Antarctica deserts (Makhalanyane et al. 2014). Although the first report of Antarctic rocks colonization by microbes was published by Friedmann and Ocampo in 1976 (1976), microbial rock colonization was known since 1914, when German geobotanist Ludwig Diels reported endolithic algae inhabiting rocks in the Dolomites Alps (Diels 1914).

Nonetheless, Diels findings and studies on endolithic microbes were not the focus of attention until the publication of Imre Friedmann and his wife Roseli Ocampo (1976) demonstrating the presence of microbes living in rocks of the Ross Desert in the southern Victoria Land of Antarctica, a place where life was not expected to be possible. According to a report of Will Hively to Friedmann published in the *Discover* magazine in 1997 (http://discovermagazine.com/1997/may/lookingforlifein1124, visited: 04/27/2018), Friedmann said: “In this limbo, this gray zone between the limit of adaptability and actual death, there are organisms that live permanently, always hungry, always too cold. Generations and generations live there. They are pushed beyond what any extremophile could possibly like, almost to the absolute limits of existence.” And these facts aroused the attention of NASA in the search for life on Mars. By the 1970s it was presumed that soils from the Ross Desert of Antarctica were abiotic or that microbes that were found there have arrived fortuitously (Horowitz et al. 1972). Nonetheless, Wolf Vishniac and his graduate student Stanley Mainzer employing a simple technique (the “wolf traps”) demonstrated microbial activity in situ in this site (Vishniac and Mainzer 1973). Intriguingly, the first Antarctic rocks studied by Friedmann and Ocampo were collected by Vishniac in the austral summer of 1973 during an Antarctic expedition in which he found his death attempting to retrieve equipment from the crack of a mountain (http://discovermagazine.com/1997/may/lookingforlifein1124, visited: 04/24/2018).

10.3 Continental vs Maritime Antarctic Rock Environment

Rogers (2007) classified terrestrial Antarctica into three zones according to climatic and biotic features: the sub-Antarctic, the maritime Antarctica (that includes the South Shetland Islands as well as part of the western side of the Antarctic Peninsula), and the continental Antarctica. Continental Antarctica includes, besides the continent, some islands near the continent, the eastern side of Antarctic Peninsula, and some part of its western side.

Most of the available knowledge from Antarctic lithobionts comes from rocks collected on continental Antarctica, mainly from the Ross Desert (Table 10.1, Fig. 10.3) (Makhalanyane et al. 2014). The Ross Desert (an unofficial name) comprises 4800 km^2 and constitutes the largest ice-free region in Antarctica. This region is also known as “Dry Valleys” or “McMurdo Dry Valleys” alluding to a row of deep valleys of predominant east-west orientation, immersed in the 2000 m high Transantarctic Mountains. The ground is characterized by polygonal surfaces containing dry permafrost and some rocks of sandstone, granite, and quartz (Pointing et al. 2009). The Dry Valleys are considered within the most hostile regions for life on Earth (Friedmann 1982; Ascaso and Wierzchos 2002; McKay and Friedmann 1985). Higher plants and animals are absent, epilithic rock colonization by lichens is scarce, and negligible microbial biomass could be detected in rock-free permafrost (de los Ríos et al. 2014; Goordial et al. 2017).

Table 10.1 Methodological approximations used for the identification of Antarctic lithobionts

Year	Location[a]	Antarctic zone[b]	Rock type	Lithobiont type	Methodology[c]	Reference
1976	Ross Desert	C	Sandstone	Endoliths	Microscopy (LM, TEM)	Friedmann and Ocampo (1976)
1981	Vestfold Hill, Princess Elizabeth Land	C	Quartz	Chasmoendolithic algae	Microscopy (LM)	Broady (1981a)
1982	Ross Desert	C	Beacon sandstone, granite, granodiorite, marble	Endoliths	Microscopy (LM, TEM)	Friedmann (1982)
1988	Ross Desert	C	Beacon sandstone, granite, marble	Endolithic bacteria	Biochemical characterization of bacterial isolates	Siebert and Hirsch (1988)
1990	King George Island	M	Volcanic andesite	Epiliths	Microscopy (LM,TEM), IR spectroscopy, and X-ray diffraction	Ascaso et al. (1990)
2000	Vestfold Hill and islands close to Davis Station	C	Quartz	Hypoliths	(a) Phenotype of bacterial isolates; (b) 16S rDNA RFLP and sequencing; (c) 16S rDNA clone library analysis	Smith et al. (2000)
2003	Ross Desert	C	Sandstone	Endoliths	(a) Microscopy; (b) 16S rDNA clone library analysis	de la Torre et al. (2003)
2003	Two Step Cliffs, Alexander Island, Antarctic Peninsula	C	Gypsum crust	Endoliths	(a) Microscopy; (b) phenotype and SSU rDNA seq of fungal and bacterial isolates	Hughes and Lawley (2003)
2009	McKelvey Valley, Ross Desert	C	Sandstone, quartz	Endoliths (sandstone), hypoliths (quartz)	(a) T-RFLP of 16SrDNA; (b) 16S rDNA clone library analysis	Pointing et al. (2009)
2013	McKelvey Valley, Ross Desert	C	Sandstone, quartz	Endoliths (sandstone), hypoliths (quartz)	(a) T-RFLP; (b) GeoChip; (c) 16S rDNA clone library analysis	Chan et al. (2013)

(continued)

Table 10.1 (continued)

Year	Location[a]	Antarctic zone[b]	Rock type	Lithobiont type	Methodology[c]	Reference
2013	Miers Valley, Ross Desert	C	Quartz	Hypoliths	(a) T-RFLP; (b) 16S rDNA seq -454-Pyro	Makhalanyane et al. (2013)
2014	Miers Valley, Ross Desert	C	Quartz	Hypoliths	(a) Microscopy (SEM-BSE, LTSEM)	de los Ríos et al. (2014)
2014	Ross Desert	C	Granite, sandstone, quartz, granite, dolerites	Epiliths, endoliths	(a) Microscopy	Zucconi et al. (2014)
2016	Sør Rondane Mountains, Dronning Maud Land	C	Granite, marble	Endoliths, hypoliths	(a) 18S rDNA DGGE; (b) SSU rRNA seq-454-Pyro	Obbels et al. (2016)
2016	Victoria Valley, Ross Desert	C	Quartz, sandstone	Endoliths (sandstone), hypoliths (quartz)	(a) T-RFLP; (b)16S rDNA seq - 454-Pyro	Van Goethem et al. (2016)
2016	Miers Valley, Ross Desert	C	NS	Hypoliths (scraped from stones)	(a) T-RFLP; (b) GeoChip; (c) 16S rDNA-454-Pyro	Wei et al. (2016)
2017	University Valley, Ross Desert	C	Beacon sandstone	Cryptoendoliths	(a) 23S rDNA-454-Pyro; (b) 16S rDNA seq –Illumina; (c) Metagenomics (Illumina)	Goordial et al. (2017)

[a]Locations indicated in the Table are depicted in Fig. 10.3

[b]*C* continental, *M* maritime, Antarctic zone classification was done according to Rogers (2007)

[c]*LM* light microscopy, *TEM* transmission electron microscopy; *SEM-BSE* scanning electron microscopy with backscattered electron imaging, *LTSEM* low-temperature scanning electron microscopy, *SSU rRNA* small subunits ribosomal ribonucleic acid, *DGGE* denaturing gradient gel electrophoresis, *T-RFLP* terminal restriction fragment length polymorphism, *16S rDNA seq –454-Pyro* pyrosequencing of 16S rDNA using 454-Roche technology, *16S rDNA seq –Illumina* pyrosequencing of 16S rDNA using Illumina technology

Maritime Antarctica climatic conditions are less extreme than continental Antarctica, and different forms of living organisms might be present. For instance, during the summer season, some animals such as penguins, seals, sea lions, and

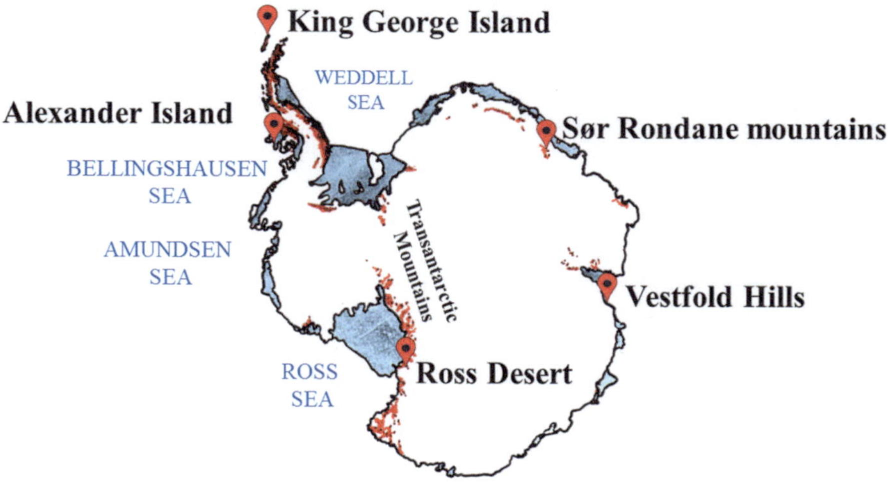

Fig. 10.3 Antarctic map showing the locations depicted in Table 10.1. Transantarctic Mountains are considered a geographic delimitation of West (to their left) and East (to their right) Antarctica. Blue regions correspond to ice shelves

Fig. 10.4 Typical appearance of maritime Antarctic rocks. Maritime Antarctic rocks are highly colonized by lichens which can be evidenced by just a glance of the landscape

birds might be present. Mosses, liverworts, and lichens are quite common in the ice-free zones, and even two flowering plants, *Deschampsia antarctica* and *Colobanthus quitensis*, can be found. Usually, rocks are profusely colonized by lichens (Fig. 10.4).

10.4 Rocks: An Oligotrophic Substrate for Life

Microbes can colonize different types of rocks, being quartz and sandstone the most common types of Antarctic rocks studied (Friedmann and Ocampo 1976; Pointing et al. 2009; Friedmann 1982; McKay and Friedmann 1985; Goordial et al. 2017; de la Torre et al. 2003; Siebert and Hirsch 1988; Chan et al. 2013). Lithobionts from granite boulder, gypsum, and volcanic andesite have also been analyzed (Table 10.1) but in less extent (Ascaso et al. 1990; Hughes and Lawley 2003).

While main lithobiontic niches in quartz rocks are either its cracks and cliffs (chasmoendolithic colonization), or the ventral surface of this translucent rock (hypolithic colonization), the more porous sandstones represent suitable niches for cryptoendolithic microbial communities (Makhalanyane et al. 2014; Pointing and Belnap 2012).

Microbes find in rocks a safe refuge against weather inclemencies such as temperature fluctuations, water scarcity, wind, and UV light incidence (Wierzchos et al. 2012; Friedmann 1982). Guillitte (1995) introduced the concept of bioreceptivity as the ability of a material to be colonized by living organisms, being influenced by different properties of the material. In the case of rocks, its bioreceptivity is highly influenced by its physical and chemical properties (Warscheid and Braams 2000). As will be further discussed, the ability of light to reach primary producers is of prime importance for the establishment of endolithic and hypolithic communities. Consequently, translucence and porosity of the rocks are key factors in the development of such communities. Porosity and fissures are also important features for water retention, which in cold dry desserts might be the difference between life and death.

The McMurdo Dry Valleys (MDV) is one of the most extreme cold habitats on Earth. It is considered to be a desert because of the low liquid water availability (Pointing and Belnap 2012). Precipitations are very scarce, with a mean annual record of 15 g cm^2, which mainly hits the ground as snow (Cowan and Tow 2004). Even though snow is a huge reservoir of water, high altitude, katabatic winds, and extreme low temperatures with annual means of −25 °C make snow melting a not so common event as most of it sublimes or is blown away by winds (Friedmann 1982; Friedmann et al. 1994). MDV also receive high radiation incidence due to high altitudes, snow reflection, and the decrease in total column ozone layer (Cary et al. 2010; Frederick and Snell 1988). For life, this high UV incidence is a big threat. In such extreme conditions, hypolithic and endolithic colonization is the strategy for survival.

Rocks are mainly an oligotrophic environment. Pointing et al. (2009) and Chan et al. (2013) determined the abiotic composition of Antarctic quartz and sandstone, reporting total organic carbon content to be negligible. This is consistent with the fact that some endoliths grow better in oligotrophic media and that the addition of rock extracts enhances their growth (Siebert and Hirsch 1988; Warscheid and Braams 2000). Despite being an oligotrophic environment, rocks are capable of sustaining microbial life. Community respiration was assessed in sandstone cryptoendoliths from University Valley, demonstrating relatively high respiration rates at temperatures as low as −20 °C indicating that the lithic community is metabolically active in this oligotrophic environment (Goordial et al. 2016). The metabolic potential of lithobiont communities in MDV has been assessed by (a) direct measurement of metabolic activities, (b) microarrays using GeoChip4, (c) descriptive metagenomics using Illumina technology, and (d) PCR amplification of specific functional genes (Chan et al. 2013; Friedmann et al. 1993; Le et al. 2016; Wei et al. 2016).

10.5 Who Are They?

The first studies (before the year 2000) carried out to characterize Antarctic lithobionts were mainly based on microscopic observations and isolation of culturable microorganisms (Table 10.1). Lichen-associated bacteria and cyanobacteria were the main group of bacteria found on the samples analyzed, being *Gloeocapsa*, *Chrococcidiopsis*, and *Plectonema* among the cyanobacteria identified and *Buellia*, *Lecidea*, *Acarospora*, and *Trebouxia* the lichens found (Friedmann and Ocampo 1976; Friedmann 1982; Broady 1981a, b). Once molecular biological methods based on nucleic acid analysis begun to be employed, the diversity of identified microorganisms strikingly increased as it will be further discussed. Nonetheless, a remarkable feature, regardless the methodology employed for the study, is the negligible presence of archaea in Antarctic rocks (Pointing et al. 2009; Goordial et al. 2017; Lee et al. 2012). The absence of archaeal sequences in endolith metagenomes has also been reported by Crits-Christoph et al. (2016) in a study performed in the Atacama Desert, and the authors suggest that this is a common feature among endolithic microbial communities in deserts, probably reflecting the extreme aridness of these niches.

According to either visual observation, microscopic analysis, or molecular approaches, it has been demonstrated that the composition of lithobiontic communities differs according to their location in the rock (e.g., hypolithic, endolithic, epilithic) (Makhalanyane et al. 2014; Pointing and Belnap 2012). Mosses, lichens, and cyanobacteria were found as epilithic colonizers, being lichens highly frequent in Maritime Antarctica (De Los Ríos et al. 2014; Broady 1981a, b). Nonetheless, very little is known about epilithic bacteria, and certainly, more work is required to have an accurate picture about the microbial community that inhabits this ecological niche.

Concerning hypolithic communities (considering a sensu lato definition) and according to a survey performed in the Dry Valleys, Cowan et al. (2010) differentiate this community as *Cyanobacteria*-dominated (Type I), fungal-dominated (Type II), and moss-dominated (Type III) communities. Interestingly, significant differences were found between hypolithic and soil bacterial communities despite their intimate contact (de los Ríos et al. 2014; Goordial et al. 2017; Wei et al. 2016; Van Goethem et al. 2016). Diverse studies identified *Cyanobacteria* (mainly

Oscillatoriales, Nostocales, and *Chroococcidiopsis*) as the most abundant hypolithic phylum accounting for up to 46% of the operational taxonomic units (OTUs) present in a survey performed in the Dry Valleys (Wei et al. 2016) and reviewed by Chan et al. (2013). Its abundance indicates that this group is the main primary producer of the system. *Acidobacteria* and *Actinobacteria* were also abundant although in a significant less extent. Conversely, Makhalanyane et al. (2014) found that Type II and Type III hypolithic communities in the Miers Valley are dominated by *Alphaproteobacteria*. The reasons of this difference await further investigation.

Endolithic colonization of quartz rocks may be usually visualized as colored bands running parallel to the rock surface, with layers of photosynthetic microorganisms alternating with fungi-dominated layers (De Los Ríos et al. 2014; Friedmann 1982). According to a pyrosequencing analysis of cryptoendolithic metagenomes from samples of Beacon Supergroup sandstone of the University Valley (Dry Valley, Ross Desert), about half of the community was eukaryotic, being *Ascomycota* the dominant phylum (45% abundance), although *Basidiomycota*, *Chlorophyta*, and *Streptophyta* were also detected. *Actinobacteria* dominated the bacterial community (19.8% of the microbial community) albeit *Proteobacteria*, *Bacteroidetes*, and *Firmicutes* were also present (Goordial et al. 2016). Interestingly, *Cyanobacteria* were minor component of this cryptoendolithic community. The photoautotroph community was dominated by *Trebouxia*, a lichen-associated alga that has been previously found in Antarctic rocks by de la Torre et al. (2003). Furthermore, a pyrosequencing analysis of two endolithic metagenomes of sandstones collected at the Victoria Valley identified *Actinobacteria* as the dominant bacterial phylum (Van Goethem et al. 2016), which is consistent with the data obtained by Goordial et al. (2016). *Gemmatimonadetes* and *Cyanobacteria* were also detected in both samples (Van Goethem et al. 2016).

10.6 Carbon Input into the Community

All lithobionts display one common characteristic: carbon input in the community mainly depends on photoautotrophs as primary producers. This dependence favors colonization of translucent rocks (Friedmann 1982). Light penetration into the lithic environment determines the depth to which phototrophic endolithic communities locate. Each millimeter beneath the surface light diminishes 70–90%. However, as little as 0.08% and 0.005% of the incident light proved to be sufficient for hypolithic and cryptoendolithic photosynthetic microorganisms to grow, respectively (Omelon 2016).

Depending on the lithobiont environment, phototrophic carbon fixation is performed by cyanobacteria, green algae, mosses, or a combination of them (Cowan et al. 2010). Direct measurement of carbon fixation was determined by Friedmann et al. (1993) in lichen-dominated cryptoendolithic samples from sandstones of

Southern Victoria Land. Interestingly, the authors showed that for photosynthesis to occur, temperature should be above −10 °C and humidity above 75%. Climate conditions in MDV limit this pathway to merely an average of 771 h per year (Friedmann et al. 1993). By estimating yearly gross productivity as 1215 mg C/m^2/y, net photosynthetic gain as 606 mg C/m^2/y, and net ecosystem productivity as 3 mg C/m^2/y, they conclude that only 0.025% of the gross productivity is incorporated into biomass. The authors attributed this fact to the metabolic demands that this harsh environment imposes which forces microorganisms to switch metabolism between activity and inactivity and to cope with freeze/thaw cycles and desiccation periods (Friedmann et al. 1993). The authors went further and estimated the contribution of lithic communities to the entire Ross Desert, representing microbial cells a net ecosystem productivity of 600–900 kg C/y, which might be dispersed by wind after exfoliation events. The difference between net photosynthetic gain and gross productivity (approximately 121,800–182,200 kg C/y) might be kept trapped in the rock, but it also might percolate and reach the soils, lakes, and rivers (Friedmann et al. 1993). This implies that primary producers in lithic ecosystem are of main importance for carbon input, not only to the rock itself, but to the entire Ross Dessert.

Chan et al. (2013) and Wei et al. (2016) used the GeoChip4-based functional gene array to assess functional traits of MDV hypolithic and endolithic communities. Chan et al. (2013) assessed the metabolic potential of endolithic communities of sandstone and hypolithic communities of quartz in the McKelvey Valley, while Wei et al. (2016) assessed the metabolic potential of hypolithic communities of quartz in the Miers Valley. Both lithic communities were mainly dominated by cyanobacteria. When hypolithic communities were compared, abundances of cyanobacteria were quite different (95% vs 46%) suggesting that factors other than substrate composition drive community dynamics. On the other hand, cyanobacteria abundance in endoliths was 64% and 56% for chasmoendoliths and cryptoendoliths, respectively, which might suggest that the type of rock (sandstone for endoliths vs quartz for hypoliths) has an impact in community dynamics. Their data suggest that carbon input in all the lithic communities was performed by both, photoautotrophic and chemoautotrophic pathways (Chan et al. 2013). Goordial et al. (2017) assessed the functional traits of cryptoendolithic sandstone communities of Beacon Supergroup using a descriptive metagenomic approach and direct enzymatic activity measurements. This cryptoendolithic community was lichen-dominated, while previously described functional analyses were done in cyanobacteria-dominated communities. The metagenome was enriched in genes involved in carbon fixation and photosystems, being the photoautotrophic community dominated almost entirely (99%) by the lichen-associated algae *Trebouxia*. Different algae were isolated, and their growth and photoautotrophic capacity were assessed, some of them proving to be metabolically active and capable of fixing carbon at up to −5 °C. Moreover, these authors found that the overall community was active even at −20 °C, as proved by respiration, and that at this temperature, photosynthesis could potentially occur since photosystem II is still active. Two isolates of endolithic *Stichococcus* displayed different growth patterns and photoautotrophic activities as a function of temperature suggesting that, at the community level, different species might contribute to carbon input over the range of temperatures that the community is faced to (Goordial et al. 2017).

10.7 Nitrogen Input into the Community

In the Dry Valleys, atmospheric nitrogen can be abiotically fixed and deposited by snowfall events, being a main source of nitrogen (Friedmann and Kibler 1980). Also, organic nitrogen would be available as ancient biomass legacy or due to air dispersion of biomass from streams and lakes (Moorhead et al. 1999). Recently, biological nitrogen fixation was demonstrated for hypolithic communities (Cowan et al. 2011), implying that diazotrophs play an important role in nitrogen input in lithic communities. The presence of the *nifH* gene, which codes for a structural protein of the nitrogenase, was assessed by PCR amplification, cloning, and sequencing. Nitrogen fixation potential was associated with *Cyanobacteria*, particularly Nostocales and *Proteobacteria*. While Nostocales are characterized by performing compartmentalized nitrogen fixation in heterocysts, *Proteobacteria* is a very diverse group, and nitrogen fixation potential cannot be adjudged to a specific genus. Nitrogenase activity, indirectly assessed by the acetylene reduction assay, confirmed that the community is capable of fixing atmospheric nitrogen. Nitrogen input was estimated to be 0.026–0.23 nmol N cm^{-2} h^{-1} which the authors translated to an estimate annual input of 0.38 kg in a 200 km^2 area of three McMurdo Dry Valleys (Cowan et al. 2011).

Nitrogen fixation potential was also evidenced by microarrays in sandstone endolithic communities and quartz hypolithic communities by Chan et al. (2013) and Wei et al. (2016). In McKelvey Valley, diazotrophy of lithic communities was adjudged to *Actinobacteria*, *Cyanobacteria*, and *Proteobacteria* (Chan et al. 2013). In Beacon Supergroup sandstone lithic communities, nitrogen fixation was adjudged to *Halobacteria* and *Methanopyri* archaea and to *Proteobacteria* (Wei et al. 2016). Interestingly, all the communities proved to have the pathways to potentially perform the complete nitrogen cycle.

On the contrary, cryptoendolithic communities in the Upper Dry Valleys seem to be unable to perform atmospheric nitrogen fixation. By a descriptive metagenomic approach, no *nifH* sequences were detected. Despite some nitrate and nitrite reductase genes were recovered, other genes involved in nitrogen cycling were absent (Goordial et al. 2017). It was previously shown that cryptoendolithic communities in the Upper Dry Valleys lack the ability to reduce acetylene and therefore to reduce N_2 (Friedmann and Kibler 1980). Altogether, these results suggest that these communities are not capable of performing a complete nitrogen cycle and that nitrogen input is probably due to abiotic nitrogen fixation, ancient heritage, or aerial dispersion of biomass from lakes and streams. Further investigation is required to find out which nitrogen sources are used by these lithobionts.

10.8 How to Cope with Stress?

Thrive with oligotrophic conditions is only one of the many challenges Antarctic lithobionts have to cope with. How do they withstand the extremely harsh environment they are faced to? The response seems to be not only finding protection in rocks but also by displaying an array of microbial adaptive mechanisms (De Los Ríos et al. 2014).

The first strategy is to "hibernate" and activate the metabolism only in the summer season when temperature rises, snow melts, and humidity increases to levels capable of sustaining life (Friedmann and Ocampo 1976; Friedmann 1982).

Once permissive climate conditions are reached, several microbial strategies are triggered. One of these strategies is the production of exopolysaccharides in order to increase water retention (Deming and Young 2017). Crust and biofilm matrices also improve water absorption and help to reduce diffusion of water vapor. Some organisms produce pigments which either absorb solar radiation in order to increase temperature, confer UV photoprotection, or act as photoprotective-quenching pigments (Dieser et al. 2010; Órdenes-Aenishanslins et al. 2016). Compounds present in some type of rocks, such as hematite and gypsum, may also protect microbia from radiation (Villar et al. 2005).

Expression of genes related to stress response is a widely used strategy (Goordial et al. 2017; Le et al. 2016). In that sense the capacity to effectively respond to osmotic stress has been well documented (Goordial et al. 2017; Chan et al. 2013; Le et al. 2016; Wei et al. 2016). An enhanced expression of genes involved in replication, recombination, and repair processes was also reported as a strategy used by hypolithic Antarctica communities to cope with environmental stresses (Le et al. 2016).

The presence of *pstB* homologous genes, involved in phosphate transport via the Pst system, has been found to be more abundant in cold desert hypoliths than in hot desert ones (Le et al. 2016). Genes involved in the response to phosphate limitation were also overrepresented in hypolithic communities when compared to soil communities (Wei et al. 2016). These results might suggest that phosphate metabolism is important for cold-adapted hypolithic bacteria. In that direction, our group found a high prevalence of polyphosphate (PolyP) granules in lithobionts isolated from basaltic andesite in King George Island (Carrasco Personal Communication). PolyP functions as storage of phosphorous and can substitute ATP as phosphoryl donor. Moreover, it has been hypothesized that PolyP is the precursor of ATP and the primitive energy donor in the origin of life (Albi and Serrano 2016). Some bacteria proved to accumulate PolyP when some nutrients are limiting (amino acids, nitrogen, sulfur) and using the accumulated PolyP as energy source. PolyP was also associated with metal ion chelation, leading to enhance heavy metal tolerance (Albi and Serrano 2016). The abundance of *pstB* homologs might indicate an active phosphorous internalization required for PolyP production.

According to Chan et al. (2013), expression of metal resistance systems is also a common trait among endoliths. Results obtained by our group (Carrasco Personal

Communication) are consistent with this observation since a high proportion of metal-resistant isolates were recovered from endolithic communities of diverse basaltic andesite rocks from King George Island. A plausible explanation is that the weathering effect that microorganisms have on the rock results in the leaching of high concentration of free metals, and therefore microbes able to avoid metal toxicity would be more competitive.

Interestingly, a large number of pathways for the production of antifungal and antibacterial compounds as well as genes involved in resistance to antibiotics were also found, suggesting competitive interactions at the community level (Goordial et al. 2017; Chan et al. 2013; Le et al. 2016).

10.9 Final Remarks

In the 1970s, when Imre Friedmann and Roseli Ocampo started this pioneer research area, the whole scientific community gains awareness of the amazing adaptability that microorganisms have to extreme conditions. Life inside rocks would have sound like a senseless idea at this moment. Nowadays lithic communities are being studied all over the world, contributing to our knowledge of such fascinating niche and the microbial community they harbor.

One of the many remarkable features of Antarctic lithobionts is the almost complete absence of Archaea in these communities. Considering that in many extreme environments Archaea proved to be present, their absence in these niches is an interesting fact. Also of interest is the high diversity of bacteria these niches harbor, being even more interesting when considering that bacterial communities of soil permafrost in which rock stands are significantly less diverse. As Wei et al. (2016) concluded, one can think of lithic communities as "islands of productivity within the generally depauperate landscape of desert soils."

To our knowledge there are no reports where a transcriptomic approach was used to assess lithobiont community functional traits. Transcriptomic analysis might require large samples in order to obtain sufficient RNA, at least with the technology we have nowadays. In the case of Antarctic rocks, the scenario is even more challenging considering all the drawbacks of Antarctic field expeditions and the impact this might have in RNA stability. Fortunately technology for assessing transcriptomes is evolving very fast, and hopefully, in a near future, efforts will be made in order to have an accurate snapshot of the yet poorly understood functional traits of these communities.

Acknowledgment Programa de Desarrollo de las Ciencias Básicas (PEDECIBA) and Agencia Nacional de Investigación e Innovación (POS_NAC_2016_1_ 129907).

References

Albi, T., & Serrano, A. (2016). Inorganic polyphosphate in the microbial world. Emerging roles for a multifaceted biopolymer. *World Journal of Microbiology and Biotechnology, 32*(2), 27.

Ascaso, C., & Wierzchos, J. (2002). New approaches to the study of Antarctic lithobiontic microorganisms and their inorganic traces, and their application in the detection of life in Martian rocks. *International Microbiology, 5*(4), 215–222.

Ascaso, C., Sancho, L. G., & Rodríguez-Pascual, C. (1990). The weathering action of saxicolous lichens in maritime Antarctica. *Polar Biology, 11*, 33–39.

Broady, P. A. (1981a). The ecology of chasmolithic algae at coastal locations of Antarctica. *Phycologia, 20*(3), 259–272.

Broady, P. A. (1981b). Ecological and taxonomic observations on subaerial epilithic algae from Princess Elizabeth Land and Mac.Robertson Land, Antarctica. *British Phycological Journal, 16*(3), 257–266.

Cary, S. C., et al. (2010). On the rocks: The microbiology of Antarctic Dry Valley soils. *Nature Reviews. Microbiology, 8*(2), 129–138.

Chan, Y., et al. (2013). Functional ecology of an Antarctic Dry Valley. *Proceedings of the National Academy of Sciences of the United States of America, 110*(22), 8990–8995.

Cowan, D. A., & Tow, L. A. (2004). Endangered antarctic environments. *Annual Review of Microbiology, 58*, 649–690.

Cowan, D. A., et al. (2010). Diverse hypolithic refuge communities in the McMurdo Dry Valleys. *Antarctic Science, 22*(6), 714–720.

Cowan, D. A., et al. (2011). Hypolithic communities: Important nitrogen sources in Antarctic desert soils. *Environmental Microbiology Reports, 3*(5), 581–586.

Crits-Christoph, A., et al. (2016). Phylogenetic and functional substrate specificity for endolithic microbial communities in hyper-arid environments. *Frontiers in Microbiology, 7*, 301.

De Los Ríos, A., Wierzchos, J., & Ascaso, C. (2014). Synthesis. The lithic microbial ecosystems of Antarctica's McMurdo Dry Valleys. *Antarctic Science, 26*(5), 459–477.

Deming, J. W., & Young, J. N. (2017). The role of exopolysaccharides in microbial adaptation to cold habitats. In R. Margesin (Ed.), *Psychrophiles: From biodiversity to biotechnology* (pp. 259–284). Cham: Springer International Publishing.

Diels, L. (1914). Die Algenvegetation der Südtiroler Dolomitriffe. Ein Beitrag zur Ökologie der Lithophyten. *Berichte der Deutschen Botanischen Gesellschaft, 32*, 502–526.

Dieser, M., Greenwood, M., & Foreman, C. M. (2010). Carotenoid pigmentation in Antarctic heterotrophic bacteria as a strategy to withstand environmental stresses. *Arctic, Antarctic, and Alpine Research, 42*(4), 396–405.

Frederick, J. E., & Snell, H. E. (1988). Ultraviolet radiation levels during the antarctic spring. *Science, 241*(4864), 438–440.

Friedmann, E. I. (1982). Endolithic microorganisms in the antarctic cold desert. *Science, 215*(4536), 1045–1053.

Friedmann, E. I., & Kibler, A. P. (1980). Nitrogen economy of endolithic microbial communities in hot and cold deserts. *Microbial Ecology, 6*(2), 95–108.

Friedmann, E. I., & Ocampo, R. (1976). Endolithic blue-green algae in the dry valleys: Primary producers in the antarctic desert ecosystem. *Science, 193*(4259), 1247–1249.

Friedmann, E. I., et al. (1993). Long-term productivity in the cryptoendolithic microbial community of the Ross Desert, Antarctica. *Microbial Ecology, 25*(1), 51–69.

Friedmann, E. I., Druk, A. Y., & McKay, C. P. (1994). Limits of life and microbial extinction in the antarctic desert. *Antarctic Journal of the United States, 29*(5), 176–179.

Goordial, J., et al. (2016). Nearing the cold-arid limits of microbial life in permafrost of an upper dry valley, Antarctica. *The ISME Journal, 10*(7), 1613–1624.

Goordial, J., et al. (2017). Comparative activity and functional ecology of permafrost soils and lithic niches in a hyper-arid polar desert. *Environmental Microbiology, 19*(2), 443–458.

Guillitte, O. (1995). Bioreceptivity: A new concept for building ecology studies. *Science of the Total Environment, 167*(1), 215–220.

Horowitz, N. H., Cameron, R. E., & Hubbard, J. S. (1972). Microbiology of the dry valleys of Antarctica. *Science, 176*(4032), 242–245.

Hughes, K. A., & Lawley, B. (2003). A novel Antarctic microbial endolithic community within gypsum crusts. *Environmental Microbiology, 5*(7), 555–565.

de la Torre, J. R., et al. (2003). Microbial diversity of cryptoendolithic communities from the McMurdo Dry Valleys, Antarctica. *Applied and Environmental Microbiology, 69*(7), 3858–3867.

Le, P. T., et al. (2016). Comparative metagenomic analysis reveals mechanisms for stress response in hypoliths from extreme hyperarid deserts. *Genome Biology and Evolution, 8*(9), 2737–2747.

Lee, C. K., et al. (2012). The inter-valley soil comparative survey: The ecology of Dry Valley edaphic microbial communities. *The ISME Journal, 6*(5), 1046–1057.

de los Ríos, A., Cary, C., & Cowan, D. (2014). The spatial structures of hypolithic communities in the Dry Valleys of East Antarctica. *Polar Biology, 37*(12), 1823–1833.

Makhalanyane, T. P., et al. (2013). Evidence for successional development in Antarctic hypolithic bacterial communities. *The ISME Journal, 7*(11), 2080–2090.

Makhalanyane, T. P., Pointing, S. B., & Cowan, D. A. (2014). Lithobionts: Cryptic and refuge niches. In D. A. Cowan (Ed.), *Antarctic terrestrial microbiology: Physical and biological properties of Antarctic soils* (pp. 163–179). Berlin/Heidelberg: Springer.

Marlow, J., Peckmann, J., & Orphan, V. (2015). Autoendoliths: A distinct type of rock-hosted microbial life. *Geobiology, 13*(4), 303–307.

McKay, C. P., & Friedmann, E. I. (1985). The cryptoendolithic microbial environment in the Antarctic cold desert: Temperature variations in nature. *Polar Biology, 4*, 19–25.

Moorhead, D. L., et al. (1999). Ecological legacies: Impacts on ecosystems of the McMurdo Dry Valleys. *Bioscience, 49*(12), 1009–1019.

Nienow, J. A., McKay, C. P., & Friedmann, E. I. (1988). The cryptoendolithic microbial environment in the Ross Desert of Antarctica: Light in the photosynthetically active region. *Microbial Ecology, 16*(3), 271–289.

Obbels, D., et al. (2016). Bacterial and eukaryotic biodiversity patterns in terrestrial and aquatic habitats in the Sor Rondane Mountains, Dronning Maud Land, East Antarctica. *FEMS Microbiology Ecology, 92*(6), fiw041.

Omelon, C. R. (2016). Endolithic microorganisms and their habitats. In H. C (Ed.), *Their world: A diversity of microbial environments. Advances in environmental microbiology*. Cham: Springer.

Órdenes-Aenishanslins, N., et al. (2016). Pigments from UV-resistant Antarctic bacteria as photosensitizers in dye sensitized solar cells. *Journal of Photochemistry and Photobiology B: Biology, 162*, 707–714.

Pointing, S. B., & Belnap, J. (2012). Microbial colonization and controls in dryland systems. *Nature Reviews Microbiology, 10*(8), 551–562.

Pointing, S. B., et al. (2009). Highly specialized microbial diversity in hyper-arid polar desert. *Proceedings of the National Academy of Sciences of the United States of America, 106*(47), 19964–19969.

Rogers, A. D. (2007). Evolution and biodiversity of Antarctic organisms: A molecular perspective. *Philosophical Transactions of the Royal Society of London. Series B, Biological Sciences, 362*(1488), 2191–2214.

Siebert, J., & Hirsch, P. (1988). Characterization of 15 selected coccal bacteria isolated from Antarctic rock and soil samples from the McMurdo-Dry Valleys (South-Victoria Land). *Polar Biology, 9*, 37–44.

Smith, M. C., et al. (2000). Sublithic bacteria associated with Antarctic quartz stones. *Antarctic Science, 12*(2), 177–184.

Van Goethem, M. W., et al. (2016). Characterization of bacterial communities in lithobionts and soil niches from Victoria Valley, Antarctica. *FEMS Microbiology Ecology, 92*(4), fiw051.

Villar, S. E., Edwards, H. G., & Seaward, M. R. (2005). Raman spectroscopy of hot desert, high altitude epilithic lichens. *Analyst, 130*(5), 730–737.

Vishniac, W. V., & Mainzer, S. E. (1973). Antarctica as a Martian model. *Life Sciences in Space Research, 11*, 25–31.

Warscheid, T., & Braams, J. (2000). Biodeterioration of stone: A review. *International Biodeterioration & Biodegradation, 46*(4), 343–368.

Wei, S. T., et al. (2016). Taxonomic and functional diversity of soil and hypolithic microbial communities in Miers Valley, McMurdo Dry Valleys, Antarctica. *Frontiers in Microbiology, 7*, 1642.

Wierzchos, J., de los Rios, A., & Ascaso, C. (2012). Microorganisms in desert rocks: The edge of life on earth. *International Microbiology, 15*(4), 173–183.

Wierzchos, J., et al. (2018). Endolithic microbial habitats as refuges for life in polyextreme environment of the Atacama Desert. *Current Opinion in Microbiology, 43*, 124–131.

Zucconi, L., et al. (2014). Mapping the lithic colonization at the boundaries of life in Northern Victoria Land, Antarctica. *Polar Biology, 39*, 91–102.

Chapter 11
Marine Fungi Associated with Antarctic Macroalgae

Mayara B. Ogaki, Maria T. de Paula, Daniele Ruas, Franciane M. Pellizzari, César X. García-Laviña, and Luiz H. Rosa

Abstract Fungi are well known for their important roles in terrestrial ecosystems, but filamentous and yeast forms are also active components of microbial communities from marine ecosystems. Marine fungi are particularly abundant and relevant in coastal systems where they can be found in association with large organic substrata, like seaweeds. Antarctica is a rather unexplored region of the planet that is being influenced by strong and rapid climate change. In the past decade, several efforts have been made to get a thorough inventory of marine fungi from different environments, with a particular emphasis on those associated with the large communities of seaweeds that abound in littoral and infralittoral ecosystems. The algicolous fungal communities obtained were characterized by a few dominant species and a large number of singletons, as well as a balance among endemic, indigenous, and cold-adapted cosmopolitan species. The long-term monitoring of this balance and the dynamics of richness, dominance, and distributional patterns of these algicolous fungal communities is proposed to understand and model the influence of climate change on the maritime Antarctic biota. In addition, several fungal isolates from marine Antarctic environments have shown great potential as producers of bioactive natural products and enzymes and may represent attractive sources of biotechnological products.

M. B. Ogaki · M. T. de Paula · D. Ruas · L. H. Rosa (✉)
Departamento de Microbiologia, Universidade Federal de Minas Gerais, Belo Horizonte, MG, Brazil
e-mail: lhrosa@icb.ufmg.br

F. M. Pellizzari
Department of Biological Sciences, Paraná State University, Paranaguá, Brazil

C. X. García-Laviña
Biochemistry and Molecular Biology, Faculty of Sciences, Universidad de la República, Montevideo, Uruguay

S. Castro-Sowinski (ed.), *The Ecological Role of Micro-organisms in the Antarctic Environment*, Springer Polar Sciences,
https://doi.org/10.1007/978-3-030-02786-5_11

Keywords Marine fungi · Algicolous fungal communities · Bioactive natural products · Ecological role · Dynamics of richness

11.1 Introduction

Marine ecosystems represent an unexplored environment regarding microbial life, where about one million microorganisms (Bacteria, Archaea, viruses, and fungi) are found per milliliter of seawater (Glöckner et al. 2012). In the oceans around the world, microbes are key players in the biogeochemical cycling of carbon, nitrogen, phosphorus, silica, iron, and other trace elements (Nelson et al. 1996; Moore et al. 2002; Morel and Price 2003; Voss et al. 2013). They are at the base of the trophic net as primary producers, primary consumers, and decomposers and also are associated with macroorganisms as seaweeds and animals in both beneficial and harmful ways (Richmond 2004; Ramanam et al. 2016).

The role that fungi play in marine ecosystems is not as well-known and long documented as their role in terrestrial habitats. In fact, the field of marine mycology did not flourish until the publication of the book *Fungi in Oceans and Estuaries* of Johnson and Sparrow (1961) and the work of Kohlmeyer and Kohlmeyer compiled in their book *Marine Mycology* (1979). In this last publication, marine fungi were categorized into two major ecological groups: obligate and facultative marine fungi. Obligate marine fungi are those growing and sporulating exclusively in marine or estuarine habitats, while facultative marine fungi are those from freshwater and terrestrial milieus able to grow and possibly to sporulate in the marine environment. The dimension of marine fungal diversity has been estimated at about 10,000 species (Raghukumar 2017). However, only 1112 marine species have recently been listed by Jones et al. (2015), belonging mostly to *Ascomycota* (943 species), followed by *Basidiomycota* (96 species), asexual morphs (43 species), *Chytridiomycota* (26 species), *Zygomycota* (3 species), and *Blastocladiomycota* (1 species). It is clear that marine fungi do not represent a coherent phylogenetic group. Therefore, it seems likely that fungal adaptation to marine environments was the result of multiple terrestrial-marine transitions (Richards et al. 2012).

The distribution of fungi in marine environments is not uniform and is influenced primarily by the availability of dissolved oxygen and the presence of organic matter (Kohlmeyer et al. 2004). For instance, the open ocean is considered to be a fungal desert (Kohlmeyer and Kohlmeyer 1979). In this oligotrophic environment, scarce nutrients and enzymes secreted by fungi are likely to be lost by rapid diffusion in the water column. This is why osmotrophs are not favored in liquid environments as other heterotrophic microorganisms with a phagotrophic grazing behavior are (Richards et al. 2012). Nonetheless, the non-abundant and non-diverse fungal assemblages of these vast environments of the planet are considered to be dominated by ascomycetous and basidiomycetous yeasts (Bass et al. 2007; Kohlmeyer and Kohlmeyer 1979), even though sequence-based surveys are pointing out the presence of an overlooked diversity of deep-branching fungi related to chytrids (Richards et al. 2012).

In contrast to the open ocean, coastal environments sustain diverse fungal communities of filamentous and yeast forms belonging mostly to Dikarya. Most of them are found to be associated with a wide variety of substrata, including silt sediments, sand, corals, calcareous structures produced by mollusks and barnacles, living animals, decaying leaves of mangroves, wood, intertidal grasses, and seaweeds (Hyde et al. 1998; Kohlmeyer et al. 2004). These substrates from intertidal zones can also carry dormant propagules of terrestrial or freshwater fungi that are passively washed into the marine milieu. A surface sterilization of the substrate is considered sufficient to ascertain the true obligate or facultative marine nature of the fungi recovered. Otherwise, the term "marine-derived fungi" is used (Jones et al. 2015; Raghukumar 2017).

Macroalgae are particularly abundant in well-lit coastal environments and are considered the second largest source of marine-derived fungi, after corals, in these environments (Bugni and Ireland 2004). Among them, green algae grow abundantly in the intertidal zone, while red and brown algae prefer subtidal conditions. About a hundred fungal species have been described in association with seaweeds as parasites, saprobes, or endophytes (Raghukumar 2017; Suryanarayanan 2012). These are collectively known as "algicolous fungi," a relatively unexplored group of fungi with great metabolic potential as producers of bioactive compounds (Furbino et al. 2014; Godinho et al. 2013).

In the past decade, important efforts have been made to unveil the diversity of marine fungi from different marine Antarctic ecosystems, with a particular emphasis on those associated with Antarctic seaweeds (Furbino et al. 2014, 2017; Godinho et al. 2013; Gonçalves et al. 2013, 2017; Loque et al. 2010). These advances will be discussed below.

11.2 Marine Antarctic Fungi

Antarctica is considered to be one of the harshest and most pristine ecosystems in the world where low temperature, low levels of organic nutrients, low water availability, strong winds, and strong UV radiation are found (Furbino et al. 2014). Still, under these conditions, microbial communities thrive (Santiago et al. 2017), and within these communities, many fungal species have been recovered from different terrestrial and marine substrates (Ruisi et al. 2007).

There are only few reports regarding marine substrates different from algae, including wood baits, invertebrate animals, and the deep-sea and marine sediments (Table 11.1). In samples of immersed wooden baits in Penguin Bay, Grasso et al. (1997) described nontypical marine fungal species, belonging to *Ascomycota* and their anamorphs. Henriquez et al. (2014) isolated from marine sponges 100 ascomycetes and, among them, members of the genera *Pseudogymnoascus* that produced bioactive compounds of antimicrobial and antitumoral activities. Using culture-independent techniques, Bass et al. (2007) evaluated water samples from the Drake Passage at different depths (250–500 m and 200–3000 m) and concluded that fungi are relatively rare in these deep-sea habitats, recovering mostly yeasts. Gonçalves et al. (2013) sampled core sediments from 100, 500, 700, and 1100 m in the

Table 11.1 Marine Antarctic fungi from different habitats

Region	Substrate	Fungal taxa	Reference
Penguin Bay (Adelie Cove), South Georgia	Wooden baits	*Trichocladium achrasporum*	Grasso et al. (1997)
		Trichocladium lignicola	
		Phoma sp.	
		Trichocladium constrictum	
Drake Passage	Water column at different depth	Uncultured fungi	Bass et al. (2007)
Bransfield Strait of Admiralty Bay, King George Island	Sediment	*Penicillium solitum*	Gonçalves et al. (2013)
King George Island	Sponges	*Epicoccum* sp.	Henriquez et al. (2014)
		Phoma sp.	
		Trichocladium sp.	
		Aureobasidium pullulans	
		Cladosporium sp.	
		Pseudogymnoascus sp.	
		Penicillium sp.	
		Thelebolus sp.	
		Uncultured *Pseudeurotium* sp.	
Maxwell Bay, King George Island	Oligochaete *Grania* sp.	*Cystobasidium slooffiae*	Herrera et al. (2017)
		Rhodotorula glutinis	
		Rhodotorula mucilaginosa	
Bransfield and Gerlache Straits	Seawater	*Acremonium* sp.	Gonçalves et al. (2017)
		Aspergillus pseudoglaucus	
		Cladosporium sphaerospermum	
		Cystobasidium slooffiae	
		Exophiala xenobiotica	
		Glaciozyma antarctica	
		Graphium rubrum	
		Lecanicillium attenuatum	
		Metschnikowia australis	
		Penicillium chrysogenum	
		Penicillium citreosulfuratum	
		Purpureocillium lilacinum	
		Simplicillium aogashimaense	

Admiralty Bay of King George Island and recovered only isolates of *Penicillium solitum* with amylasic and esterasic activities. Herrera et al. (2017) found different species of yeasts associated with the gut of *Grania* sp., an oligochaete that lives in marine interstitial habitats feeding on the debris of seaweeds. The authors suggested a possible mutualistic relationship between them. Recently, seawater samples at different depths of the Gerlache and Bransfield Straits in the Northern Antarctic Peninsula were collected by Gonçalves et al. (2017), and different yeasts and filamentous fungi were obtained, at high densities in some places. The authors brought the attention on underwater currents as a dispersal mechanism used by marine fungi across the Antarctic Ocean.

Antarctica is being influenced by strong and rapid climate change, mainly in the Peninsula region where temperatures have risen by 3 °C in 50 years and there is also evidence that indicates a warming of the circumpolar current and a loss of ice cover, particularly in the Antarctic islands (Bridge and Spooner 2012). This raises the question on how profoundly the effects of global change could affect the marine biota of Antarctica. In this regard, following changes in the patterns of richness, dominance, and distribution of marine fungal communities could help to understand the influence of these changes. A particular emphasis on the balance between endemic, indigenous, and cosmopolitan fungal species has been given in association with Antarctic macroalgae by Furbino et al. (2014), which may represent models to analyse possibles climate changes in the Antarctic Ocean. Endemic species are characterized as true psychrophilic fungi that actively grow and reproduce only in Antarctica (Ruisi et al. 2007), while indigenous species are taxa recurrently recovered from a particular substrate in Antarctica but also appear in non-Antarctic sites (Arenz et al. 2014). Finally, cosmopolitan species are those species widely distributed that are able to grow at least under Antarctic summer conditions, showing a mesophilic or psychrophilic behavior (Furbino et al. 2014). A decrease in endemic species associated with an increase of cosmopolitan taxa within the fungal communities associated with endemic macroalgae is considered to reflect the influence of climate change in the maritime Antarctic Peninsula.

11.3 Algicolous Antarctic Fungi

Marine seaweeds are the key primary producers in littoral and infralittoral ecosystems of the Antarctic. Indeed, it has been suggested that algal beds cover about 30% of the bottom surface in the maritime Antarctic, with an estimate of 74,000 tons of wet biomass around Admiralty Bay alone (Nedzarek and Rakusa-Suszczewski 2004). These communities are characterized by a high degree of endemism and the presence of cold-adapted species, mostly of red and brown algae (Wiencke and Clayton 2002; Wiencke and Amsler 2012; Wiencke et al. 2014). In a recent study, Pellizzari et al. (2017) identified 104 macroalgal taxa (28 Phaeophyceae, 24 Chlorophyta, and 52 Rhodophyta) in the South Shetland archipelago, demonstrating that their diversity is probably increasing because of more efficient sampling techniques or because of changes in biogeographical distribution. All these Antarctic seaweeds shelter a community of symbiont, saprobe, and parasitic fungi. The structure of the fungal assemblage depends not only on the algal host species, but also on other factors like geographic isolation or the capability to tolerate or detoxify the array of antifungal metabolites produced by the seaweed (Suryanarayanan 2012). In addition, macroalgae from Antarctica show a wide plasticity regarding morphofunctional groups (filamentous, balloon-like, fleshy, terete, foliaceous, and calcareous algae) that could influence the development of particular fungal lineages.

Efforts have been made to get a thorough inventory of marine fungi associated with seaweeds from the Antarctic. The list of fungi recovered is summarized in Table 11.2 in relation to their algal hosts that are shown in Fig. 11.1. Loque et al. (2010) obtained 75 fungal isolates associated with three marine algae, with the fungus *Pseudogymanoascus pannorum* and the yeast *Metschnikowia australis* as the prevalent species. Godinho et al. (2013) and Furbino et al. (2014) surveyed the diversity of fungi associated with different sets of macroalgae. Both works obtained inventories of around 50 taxa and arrived to the same conclusion that fungal assemblages from a particular host were comprised by a few dominant species and a high number of singletons. Duarte et al. (2016) obtained 25 different taxa of yeasts from 9 algal species. Finally, Furbino et al. (2017) recovered 44 fungal isolates from 7 different seaweed species, with some of the isolates displaying agarolytic and carrageenolytic activities. This led the authors to hypothesize that the Antarctic macroalgae shelter saprobes fungi capable of producing enzymes with the potential to degrade algal biomass.

Fig. 11.1 Macroalgal species collected in Antarctica and selected as hosts to recover algicolous fungi: (**a**) *Palmaria decipiens*, (**b**) *Gigartina skottsbergii*, (**c**) *Georgiella confluens*, (**d**) *Curdiea racovitzae*, (**e**) *Iridaea cordata*, (**f**) *Pyropia endiviifolia*, (**g**) *Ulva flexuosa*, (**h**) *Acrosiphonia arcta*, (**i**) *Monostroma hariotii*, (**j**) *Adenocystis utricularis*, (**k**) *Cystosphaera jacquinotii*, (**l**) *Phaeurus antarcticus*, (**m**) *Ascoseira mirabilis*, (**n**) *Desmarestia* sp., and (**o**) *Himantothallus grandifolius*

Table 11.2 List of fungal taxa described in association with macroalgae from Antarctica

Island	Macroalgae group	Macroalgae host	Fungal taxa	Reference
King George Island	Phaeophyceae	*Adenocystis utricularis*	*Metschnikowia australis*	Loque et al. (2010)
			Pseudogymnoascus pannorum	
			Antarctomyces psychrotrophicus	
			Oidiodendron sp.	
			Penicillium sp.	
			Phaeosphaeria herpotrichoides	
			Rhodotorula mucilaginosa	
		Ascoseira mirabilis	*Metschnikowia australis*	Furbino et al. (2017)
			Antarctomyces pellizariae	
			Leucosporidiella fragaria	
			Beauveria bassiana	
			Cladosporium sp.	
			Leucosporidiella muscorum	
			Penicillium chrysogenum	
		Desmarestia menziesii	*Metschnikowia australis*	Godinho et al. (2013)
			Penicillium sp.	
		Cystosphaera jacquinotii	*Metschnikowia australis*	Duarte et al. (2016)
			Mrakia sp.	
		Himantothallus grandifolius	*Metschnikowia australis*	Duarte et al. (2016)
			Cryptococcus carnescens	
			Dioszegia xingshanensis	
			Cryptococcus victoriae	
			Cryptococcus sp.	
			Holtermanniella festucosa	
			Mrakia sp.	
			Holtermanniella nyarrowii	
			Leucosporidiella fragaria	
			Leucosporidiella muscorum	
			Rhodotorula glacialis	

(continued)

Table 11.2 (continued)

Island	Macroalgae group	Macroalgae host	Fungal taxa	Reference
		Desmarestia anceps	*Pseudogymnoascus pannorum*	Loque et al. (2010)
			Aureobasidium pullulans	
			Metschnikowia australis	Duarte et al. (2016)
			Ustilaginaceae	
	Rhodophyta	*Palmaria decipiens*	*Metschnikowia australis*	Loque et al. (2010) and Furbino et al. (2017)
			Rhodotorula mucilaginosa	Loque et al. (2010)
			Cryptococcus carnescens	
			Penicillium sp.	Furbino et al. (2017)
		Pyropia endiviifolia	*Metschnikowia australis*	Furbino et al. (2014)
			Cladosporium sp.	
			Pseudogymnoascus sp.	
			Penicillium sp.	
			Dipodascus australiensis	
		Gigartina skottsbergii	*Metschnikowia australis*	Duarte et al. (2016)
			Glaciozyma litorale	
			Glaciozyma martinii	
			Mrakia sp.	
			Rhodotorula glacialis	
	Chlorophyta	*Monostroma hariotii*	*Metschnikowia australis*	Furbino et al. (2014)
			Guehomyces pullulans	
			Cryptococcus albidosimilis	
			Rhodotorula laryngis	
			Cryptococcus victoriae	
			Pseudogymnoascus sp.	
			Rhodotorula mucilaginosa	
			Cystofilobasidium infirmominiatum	
			Meyerozyma guilliermondii	
			Cryptococcus adeliensis	
			Rhodotorula minuta	

(continued)

Table 11.2 (continued)

Island	Macroalgae group	Macroalgae host	Fungal taxa	Reference
		Acrosiphonia arcta	*Pseudogymnoascus* sp.	Godinho et al. (2013)
			Metschnikowia australis	
			Penicillium sp.	
			Candida sake	
			Cladosporium sp.	
			Cladosporium tenuissimum	
			Debaryomyces hansenii	
			Mortierella sp.	
			Phoma sp.	
			Thelebolus globosus	
Deception	Chlorophyta	*Monostroma hariotii*	*Aspergillus tabacinus*	Furbino et al. (2014)
			Pseudogymnoascus sp.	
			Penicillium sp.	Godinho et al. (2013)
			Pseudogymnoascus destructans	
			Meyerozyma guilliermondii	
			Cryptococcus cf. *laurentii*	
			Cordycipitaceae sp.	
			Helotiales sp.	
			Hyaloscyphaceae sp.	
			Mrakia sp.	Duarte et al. (2016)
	Phaeophyceae	*Adenocystis utricularis*	*Debaryomyces hansenii*	Godinho et al. (2013)
			Meyerozyma caribbica	
			Penicillium sp.	
			Aspergillus conicus	
			Pseudogymnoascus sp.	
			Penicillium citrinum	
	Rhodophyta	*Iridaea cordata*	*Metschnikowia australis*	Furbino et al. (2017)
			Pseudogymnoascus sp.	
			Doratomyces sp.	
			Penicillium sp.	
		Pyropia endiviifolia	*Penicillium* sp.	Furbino et al. (2014)
			Meyerozyma guilliermondii	
			Pseudogymnoascus sp.	
			Verticillium sp.	
			Aspergillus sp.	
			Lecanicillium sp.	
		Palmaria decipiens	*Cryptococcus magnus*	Duarte et al. (2016)
			Dioszegia athyri	
			Rhodotorula marina	
			Ustilaginaceae	
			Tilletiopsis washingtonensis	

(continued)

Table 11.2 (continued)

Island	Macroalgae group	Macroalgae host	Fungal taxa	Reference
Elephant	Chlorophyta	*Monostroma hariotii*	*Penicillium steckii*	Furbino et al. (2014)
			Penicillium sp.	
			Aspergillus sp.	
			Cladosporium sp.	
			Penicillium citrinum	
			Penicillium crustosum	
	Chlorophyta	*Ulva intestinalis*	*Penicillium* sp.	Godinho et al. (2013)
			Penicillium discolor	
			Antarctomyces psychrotrophicus	
			Cryptococcus victoriae	
			Engyodontium sp.	
			Geomyces luteus	
			Helotiales sp.	
			Mycoarthris cf. *corallines*	
			Thelebolus globosus	
	Rhodophyta	*Palmaria decipiens*	*Penicillium* sp.	Godinho et al. (2013)
			Geomyces sp.	
			Acremonium sp.	
			Fusarium sp.	
			Yamadazyma mexicana	
			Aspergillus sp.	
			Chaetomium sp.	
			Penicillium spinulosum	
		Pyropia endiviifolia	*Cadophora malorum*	Furbino et al. (2014)
			Penicillium sp.	
			Pseudogymnoascus sp.	
			Thelebolus globosus	
			Aspergillus sp.	
			Antarctomyces psychrotrophicus	
			Cladosporium lignicola	
			Aspergillus protuberus	
			Mortierella antarctica	
			Oidiodendron truncatum	
	Phaeophyceae	*Phaeurus antarcticus*	*Penicillium* sp.	Godinho et al. (2013)
			Pseudogymnoascus sp.	
			Aspergillus terreus	
			Eurotium herbariorum	
			Eurotium repens	
			Penicillium steckii	

(continued)

Table 11.2 (continued)

Island	Macroalgae group	Macroalgae host	Fungal taxa	Reference
Livingston	Rhodophyta	*Curdiea racovitzae*	*Metschnikowia australis*	Furbino et al. (2017)
		Gigartina skottsbergii	*Metschnikowia australis*	Furbino et al. (2017)
			Penicillium sp.	
Robert	Phaeophyceae	*Ascoseira mirabilis*	*Metschnikowia australis*	Duarte et al. (2016)
		Adenocystis utricularis	*Metschnikowia australis*	Duarte et al. (2016)
			Glaciozyma litorale	
			Sporidiobolus pararoseus	
			Pseudozyma sp.	
			Ustilaginaceae sp.	
			Pseudozyma tsukubaensis	
		Desmarestia menziesii	*Candida sake*	Duarte et al. (2016)
			Metschnikowia australis	
			Glaciozyma litorale	
			Mrakia sp.	
	Rhodophyta	*Gigartina skottsbergii*	*Metschnikowia australis*	Duarte et al. (2016)
			Mrakia sp.	
			Leucosporidiella muscorum	
		Iridaea cordata	*Metschnikowia australis*	Duarte et al. (2016)
			Glaciozyma litorale	
Nelson	Phaeophyceae	*Desmarestia menziesii*	*Metschnikowia australis*	Duarte et al. (2016)
			Mrakia sp.	
Half Moon	Rhodophyta	*Curdiea racovitzae*	*Metschnikowia australis*	Duarte et al. (2016)
			Sporidiobolus pararoseus	
		Gigartina skottsbergii	*Metschnikowia australis*	Duarte et al. (2016)
			Mrakia sp.	
		Iridaea cordata	*Metschnikowia australis*	Duarte et al. (2016)
Robert	Rhodophyta	*Georgiella confluens*	*Metschnikowia australis*	Furbino et al. (2017)
			Cladosporium sp.	
			Coprinellus radians	
			Penicillium sp.	
			Rhodotorula mucilaginosa	

Some fungal species were recovered from several algal species and appear to be endemic to the Antarctic. This is the case of the ascomycetous yeast *M. australis* that was obtained in all the abovementioned surveys associated to very different kinds of seaweeds. This particular species was also recovered from Antarctic seawater (Fell and Hunter 1968), from the stomach of the Antarctic krill *Euphausia superba* (Donachie and Zdanowski, 1998), in Antarctic freshwater and marine sediments (Vaz et al. 2011), and it was the only fungus obtained from the intravesicular liquid of *Adenocystis utricularis* (Loque et al., 2010).

Antarctomyces includes only two known species, which are considered endemic to Antarctica: *A. psychrotrophicus* and *A. pellizariae*. The former one has been isolated from the macroalgal species *Ascoseira mirabilis* (Furbino et al., 2017), *Ulva intestinalis* (Godinho et al., 2013), and *Pyropia endiviifolia* (Furbino et al., 2014) but also from other Antarctic environments such as soils (Stchigel et al. 2001), freshwater lakes (Gonçalves et al. 2012), and lichen thalli (Santiago et al. 2015) and as a symbiotic endophyte of the Antarctic grass *Deschampsia antarctica* (Rosa et al. 2009). *A. pellizariae* represents a new blue-pigmented species recently reported in Antarctic snow (de Menezes et al. 2017).

Pseudogymnoascus pannorum and its anamorphic stage *G. pannorum* are another species considered to be endemic to Antarctica (Arenz et al. 2014) that were isolated from *Adenocystis utricularis* and *Desmarestia anceps* (Loque et al., 2010). However, other species of this genus are considered indigenous because of their ubiquitous distribution in cold regions. In general, *Pseudogymnoascus* species are considered truly psychrophilic and halotolerant, being able to utilize different carbon sources, with particular cellulolytic and keratinolytic activities (Mercatini et al. 1989). Several isolates of *Pseudogymnoascus* not assigned to a particular species were isolated from other algal species in other marine Antarctic surveys (Furbino et al. 2014, 2017; Godinho et al. 2013). Other indigenous fungi associated with Antarctic seaweeds are *Cadophora malorum*, *Cryptococcus victoriae*, *Cryptococcus adeliensis*, and *Mortierella antarctica* (Furbino et al., 2014).

The cosmopolitan genus *Penicillium* has been recovered from several Antarctic seaweeds (Furbino et al. 2014, 2017; Godinho et al. 2013; Loque et al. 2010) but also in other Antarctic environments such as soils (Azmi and Seppelt 1998), lakes (Ellis-Evans 1996), wood (Arenz et al. 2006), marine sediments (Gonçalves et al. 2013), in permafrost (Zucconi et al. 2012). According to Bugni and Ireland (2004), *Penicillium* represents one of the most common genera isolated from macroalgae. However, the endophytic nature of the association was questioned because members of this genus are usually washed out when surface-sterilization protocols are used on the host (Zuccaro et al. 2003). Conversely, the fact that some *Penicillium* isolates associated with seaweeds displayed agarolytic and carrageenolytic activities suggests that their presence is not merely a contamination of propagules from other environments but rather indicates their possible role as latent saprobes on algal tissues (Furbino et al. 2017). Other cold-adapted cosmopolitan species associated with Antarctic seaweeds are *Cryptococcus albidosimilis*, *Guehomyces pullulans*, *Meyerozyma guilliermondii*, *Phoma herbarum*, *Rhodotorula laryngis*, *Rhodotorula mucilaginosa*, and *Rhodotorula minuta* (Furbino et al., 2014).

11.4 Biotechnological Potential of Antarctic Algicolous Fungi

Algicolous fungi are considered particularly capable of producing secondary metabolites with novel bioactivities of pharmaceutical and agricultural interest (Bugni and Ireland 2004; Suryanarayanan 2012). In fact, marine fungi associated with diverse green, red, and brown seaweeds have been reported to produce strong antioxidants as well as antialgal, antifungal, and antiinsect metabolites. These metabolites may help in deterring colonization of algal thalli by other microbes, in warding off herbivores, and in protecting the algal host from stresses (Suryanarayanan et al. 2010, 2012).

Particular interest has been given to Antarctic fungi because their ability to survive in extremely harsh conditions was considered suggestive of the presence of unusual biochemical pathways that could lead to new bioactive compounds (Santiago et al. 2012). Despite this, only few studies were performed dealing with the metabolic capabilities of marine Antarctic fungi in general and Antarctic algicolous fungi in particular (Furbino et al. 2014; Godinho et al. 2013; Gonçalves et al. 2015). Among those studies, Godinho et al. (2013) reported the bioactive compounds produced by two distinct *Penicillium* sp. isolates recovered from the macroalgae *Monostroma hariotii* and *Palmaria decipiens* (Fig. 11.2). Extracts displayed high and selective antifungal activities against the plant pathogen *Cladosporium sphaerospermum* and trypanocidal activities against *Trypanosoma cruzi*, the etiological agent of Chagas disease, with NMR spectral data suggesting the presence of highly functionalized aromatic compounds. The work of Furbino et al. (2014) also showed that several algicolous isolates of *Pseudogymnoascus* spp., *Dipodascus australiensis*, *Guehomyces pullulans*, and *Metschnikowia australis* were able to

Fig. 11.2 Algicolous *Penicillium* isolates and their algal hosts. (**a**) *Monostroma hariotii*, (**b**) *Palmaria decipiens*, (**c**) *Penicillium* sp. colonies, and (**d**, **e**) conidiophores of *Penicillium* sp. (optical and scanning electron microscopy, respectively)

produce bioactive natural products with selective antifungal activities against *Candida albicans*, *Candida krusei*, and *C. sphaerospermum*. Additionally, the authors demonstrated that *Penicillium steckii* isolated from *M. hariotii* was able to produce antiviral compounds that inhibited the yellow fever virus (Furbino et al. 2014).

Algicolous fungi from Antarctica could also be of industrial interest because of their hydrolytic enzymes. Cold-adapted enzymes have considerable potential application in the food, fine chemical, and bioethanol industries because they have high specific activities at low and moderate temperatures and are inactivated by moderate temperature increases (Gerday et al. 2000). In this sense, Furbino et al. (2017) reported that *Beauveria bassiana*, *Penicillium chrysogenum*, *Penicillium* sp., *Pseudogymnoascus* sp., *Cladosporium* sp., and *Doratomyces* sp. displayed carrageenolytic and agarolytic activities. These results suggest that the Antarctic macroalgae shelter fungal saprobes are able to produce enzymes with potential to degrade the algal biomass and release essential minerals in the ocean surrounding. Finally, Gonçalves et al. (2013) recovered isolates of *Penicillium solitum* from Antarctic marine sediments with amylasic and esterasic activities. All these studies highlight the need to discover and preserve this valuable fungal germplasm given that some cold-requiring species may not be able to persist in a warming environment.

11.5 Conclusion and Perspectives

Antarctica, especially the maritime region, has been considered to be influenced by strong and rapid climate change. Fungi and other microbes living in marine environments are likely to be perturbed by these changes. Given this situation, several efforts have been made to inventory and preserve *ex situ* the fungal components of different communities, particularly those associated with endemic seaweeds. The long-term monitoring of the balance and dynamics of richness, dominance, and distributional patterns among endemic, indigenous, and cosmopolitan fungal taxa might be used to understand and model the influence of climate change on the maritime Antarctic biota. In this regard, species like *M. australis, A. psychrotrophicus*, and species of *Pseudogymnoascus* that were isolated from different seaweeds are considered endemic to Antarctica. Instead, other algicolous fungi as members of the genera *Penicillium*, *Cryptococcus*, and *Rhodotorula* are considered cold-adapted cosmopolitan species. A decrease in the above mentioned endemic species associated with an increase of cosmopolitan taxa within the fungal communities would reflect the influence of climate change in the Peninsula region of Antarctica. This is why it is of fundamental interest to continue monitoring these fungal communities.

In addition, several fungal isolates from marine Antarctic environments have shown great potential as producers of bioactive natural products and enzymes and may represent attractive sources of biotechnological products. However, the diversity, ecological role, and biotechnological application of Antarctic marine fungi are still quite unexplored. Further research is necessary to unveil the fungal diversity associated with different substrates and environments from the Antarctic. The obtained fungal assemblages should be preserved in collections to serve as models in ecological, evolutionary, and biotechnological approaches.

References

Arenz, B. E., Held, B. W., Jurgens, J. A., Farrell, R. L., & Blanchette, R. A. (2006). Fungal diversity in soils and historic wood from the Ross Sea Region of Antarctica. *Soil Biology and Biochemistry, 38*(10), 3057–3064.

Arenz, B. E., Blanchette, R. A., Farrell, R. L. (2014). Fungal diversity in Antarctic soils. In *Antarctic terrestrial microbiology* (pp. 35–53). Germany: Springer.

Azmi, O. R., & Seppelt, R. D. (1998). The broad-scale distribution of microfungi in the Windmill Islands region, continental Antarctica. *Polar Biology, 19*(2), 92–100.

Bass, D., Howe, A., Brown, N., Barton, H., Demidova, H., Michele, H., Li, L., Sanders, H., Watkinson, S., Willcock, S., & Richards, T. A. (2007). Yeast forms dominate fungal diversity in the deep oceans. *Proceedings of the Royal Society B, 274*, 3069–3307.

Bridge, P. D., & Spooner, B. M. (2012). Non-lichenized Antarctic fungi: Transient visitors or members of a cryptic ecosystem? *Fungal Ecology, 5*(4), 381–394.

Bugni, T. S., & Ireland, C. M. (2004). Marine-derived fungi: A chemically and biologically diverse group of microorganisms. *Natural Product Reports, 21*(1), 143–163.

de Menezes, G. C., Godinho, V. M., Porto, B. A., Gonçalves, V. N., & Rosa, L. H. (2017). *Antarctomyces pellizariae* sp. nov., a new, endemic, blue, snow resident psychrophilic ascomycete fungus from Antarctica. *Extremophiles, 21*, 259–269.

Donachie, S. P., & Zdanowski, M. K. (1998). Potential digestive function of bacteria in krill Euphausia superba stomach. *Aquatic Microbial Ecology, 14*, 129–136.

Duarte, A. W. F., Passarini, M. R. Z., Delforno, T. P., Pellizzari, F. M., Cipro, C. V. Z., Montone, R. C., Petry, M. V., Putzke, J., Rosa, L. H., & Sette, L. D. (2016). Yeasts from macroalgae and lichens that inhabit the South Shetland Islands, Antarctica. *Environmental Microbiology Reports, 8*, 874–888.

Ellis-Evans, J. C. (1996). Microbial diversity and function in Antarctic freshwater ecosystems. *Biodiversity and Conservation, 5*, 1395–1431.

Fell, J. W., & Hunter, I. L. (1968). Isolation of heterothallic yeast strains of Metschnikowia Kamienski and their mating reactions with *Chlamydozyma wickerham* spp. *Antonie Van Leeuwenhoek, 34*, 365–376.

Furbino, L. E., Godinho, V. M., Santiago, I. F., Pellizari, F. M., Alves, T. M., Zani, C. L., Junior, P. A. S., Romanha, A. J., Carvalho, A. G. O., Gil, L. H. V. G., Rosa, A. C., Minnis, A. M., & Rosa, L. H. (2014). Diversity patterns, ecology and biological activities of fungal communities associated with the endemic macroalgae across the Antarctic Peninsula. *Microbial Ecology, 67*, 775–787.

Furbino, L., Pellizzari, F. M., Neto, P. C., Rosa, C. A., & Rosa, L. H. (2017). Isolation of fungi associated with macroalgae from maritime Antarctica and their production of agarolytic and carrageenolytic activities. *Polar Biology*. https://doi.org/10.1007/s00300-017-2213-1.

Gerday, C., Aittaleb, M., Bentahir, M., Chessa, J. P., Claverie, P., Collins, T., & Hoyoux, A. (2000). Cold-adapted enzymes: From fundamentals to biotechnology. *Trends in Biotechnology, 18*, 103–107.

Glöckner, F. O., Stal, L. J., Sandaa, R. A., Gasol, J. M., O'Gara, F., Hernandez, F., Labrenz, M., Stoica, E., Varela, M. M., Bordalo, A., & Pitta, P. (2012). In J. B. Calewaert & N. McDonough (Eds.), *Marine microbial diversity and its role in ecosystem functioning and environmental change, Marine Board Position Paper 17*. Ostend: Marine Board-ESF.

Godinho, V. M., Furbino, L., Santiago, I. F., Pelizzari, F. M., Yokoya, N. S., Pupo, D., Dicla, A., Alves, T. M., Junior, P. A., Romanha, A. J., Zani, C. L., Cantrell, C. L., Rosa, C. A., & Rosa, L. H. (2013). Diversity and bioprospecting of fungal communities associated with endemic and cold-adapted macroalgae in Antarctica. *ISME, 7*, 77–145.

Gonçalves, V. N., Vaz, A. B., Rosa, C. A., & Rosa, L. H. (2012). Diversity and distribution of fungal communities in lakes of Antarctica. *FEMS Microbiology Ecology, 82*(2), 459–471.

Gonçalves, V. N., Campos, L. S., Melo, I. S., Pellizari, V. H., Rosa, C. A., & Rosa, L. H. (2013). *Penicillium solitum*: A mesophilic, psychrotolerant fungus present in marine sediments from Antarctica. *Polar Biology, 36*, 1823–1831.

Gonçalves, V. N., Carvalho, C. R., Johann, S., Mendes, G., Alves, T. M., Zani, C. L., Junior, P. A. S., Murta, S. M. F., Romanha, A. J., Cantrell, C. L., Rosa, C. A., & Rosa, L. H. (2015). Antibacterial, antifungal and antiprotozoal activities of fungal communities present in different substrates from Antarctica. *Polar Biology, 38*, 1143–1152.

Gonçalves, V. N., Vitoreli, G. A., Menezes, G. C. A., Mendes, C. R. B., Secchi, E. R., Rosa, C. A., & Rosa, L. H. (2017). Taxonomy, phylogeny and ecology of cultivable fungi present in seawater gradients across the Northern Antarctica Peninsula. *Extremophiles, 21*, 1005–1015.

Grasso, S., Bruni, V., & Maio, G. (1997). Marine fungi in Terra Nova Bay (Ross Sea, Antarctica). *The New Microbiologica, 20*, 371–376.

Henríquez, M., Vergara, K., Norambuena, J., Beiza, A., Maza, F., Ubilla, P., Araya, I., Chávez, R., San-Martín, A., Darias, J., Darias, M. J., & Vaca, I. (2014). Diversity of cultivable fungi associated with Antarctic marine sponges and screening for their antimicrobial, antitumoral and antioxidant potential. *World Journal of Microbiology and Biotechnology, 30*, 65–76.

Herrera, L. M., García-Laviña, C. X., Marizcurrena, J. J., Volonterio, O., de León, R. P., & Castro-Sowinski, S. (2017). Hydrolytic enzyme-producing microbes in the Antarctic oligochaete *Grania* sp. (Annelida). *Polar Biology, 40*, 947–953.

Hyde, K. D., Jones, E. B. G., Leano, E., Pointing, S. B., Poonyth, A. D., & Vrijmoed, L. L. P. (1998). Role of fungi in marine ecosystems. *Biodiversity and Conservation, 7*, 1147–1161.

Johnson, T. W., & Sparrow, F. K. (1961). Fungi in oceans and estuaries. Fungi in oceans and estuaries. *Science, 137*, 662–663.

Jones, G. E. B., Suetrong, S., Sakayaroj, J., Bahkali, A. H., Abdel-Wahab, M. A., Boekhout, T., & Pang, K. (2015). Classification of marine *Ascomycota*, *Basidiomycota*, *Blastocladiomycota* and *Chytridiomycota*. *Fungal Diversity, 73*, 1–72.

Kohlmeyer, J., & Kohlmeyer, E. (1979). *Marine mycology: The higher fungi*. New York: Academy Press.

Kohlmeyer, J., Volkmann-Kohlmeyer, B., & Newell, S. Y. (2004). Marine and estuarine mycelial Eumycota and Oomycota. In G. M. Mueller, G. G. Bills, & M. S. Foster (Eds.), *Biodiversity of fungi: Inventory and monitoring methods*. New York: Elsevier Academic Press.

Loque, C. P., Medeiros, A. O., Pellizzari, F. M., Oliveira, E. C., Rosa, C. A., & Rosa, L. H. (2010). Fungal community associated with marine macroalgae from Antarctica. *Polar Biology, 33*, 641–648.

Mercantini, R., Marsella, R., & Cervellati, M. C. (1989). Keratinophilic fungi isolated from Antarctic soil. *Mycopathologia, 106*, 47–52.

Moore, J. K., Doney, S. C., Glover, D. M., & Fung, I. Y. (2002). Iron cycling and nutrient-limitation patterns in surface waters of the World Ocean. *Deep Sea Research, Part II, 49*, 463–507.

Morel, F. M. M., & Price, N. M. (2003). The biogeochemical cycles of trace metals in the oceans. *Science, 300*, 944.

Nedzarek, A., & Rakusa-Suszczewski, S. (2004). Decomposition of macroalgae and the release of nutrient Admiralty Bay, King George, Antarctica. *Polar Biosci, 17*, 26–35.

Nelson, D. M., DeMaster, D. J., Dunbar, R. B., & Smith, W. O. J. (1996). Cycling of organic carbon and biogenic silica in the Southern Ocean: Estimates of water-column and sedimentary fluxes on the Ross Sea continental shelf. *Journal of Geophysical Research, 101*, 18519–18532.

Pellizzari, F., Silva, M. C., Silva, E. M., Medeiros, A., Oliveira, M. C., Yokoya, N. S., Rosa, L. H., & Colepicolo, P. (2017). Diversity and spatial distribution of seaweeds in the South Shetland

Islands, Antarctica: An updated database for environmental monitoring under climate change scenarios. *Polar Biology, 40*, 1671.

Raghukumar, S. (2017). *Fungi in coastal and oceanic marine ecosystems. Marine Fungi* (p. 378). Germany: Springer.

Ramanan, R., Kim, B. H., Cho, D. H., Oh, H. M., & Kim, H. S. (2016). Algae-bacteria interactions: Evolution, ecology and emerging applications author links open overlay. *Biotechnology Advances, 34*, 14–39.

Richards, T. A., Jones, M. D., Leonard, G., & Bass, D. (2012). Marine fungi: Their ecology and molecular diversity. *Annual Review of Marine Science, 4*, 495–522.

Richmond, A. (2004). *Handbook of microalgal culture: Biotechnology and applied phycology* (p. 566). Oxford: Blackwell Science Ltd.

Rosa, L. H., Vaz, A. B., Caligiorne, R. B., Campolina, S., & Rosa, C. A. (2009). Endophytic fungi associated with the Antarctic grass *Deschampsia antarctica* Desv (Poaceae). *Polar Biology, 32*, 161–167.

Ruisi, S., Barreca, D., Selbmann, L., Zucconi, L., & Onofri, S. (2007). Fungi in Antarctica. *Reviews in Environmental Science and Biotechnology, 6*, 127–141.

Santiago, I. F., Alves, T. M., Rabello, A., Junior, P. A. S., Romanha, A. J., Zani, C. L., Rosa, C. A., & Rosa, L. H. (2012). Leishmanicidal and antitumoral activities of endophytic fungi associated with the Antarctic angiosperms *Deschampsia antarctica* Desv. and *Colobanthus quitensis* (Kunth) Bartl. *Extremophiles, 16*, 95–103.

Santiago, I. F., Soares, M. A., Rosa, C. A., & Rosa, L. H. (2015). Lichensphere: A protected natural microhabitat of the non-lichenised fungal communities living in extreme environments of Antarctica. *Extremophiles, 19*, 1087–1097.

Santiago, I. F., Rosa, C. A., & Rosa, L. H. (2017). Endophytic symbiont yeasts associated with the Antarctic angiosperms *Deschampsia antarctica* and *Colobanthus quitensis*. *Polar Biology, 40*, 177–183.

Stchigel, A. M., Josep, C. A. N. O., Mac Cormack, W., & Guarro, J. (2001). *Antarctomyces psychrotrophicus* gen. et sp. nov., a new ascomycete from Antarctica. *Mycological Research, 105*, 377–382.

Suryanarayanan, T. S. (2012). Fungal endosymbionts of seaweeds. In *Biology of marine fungi* (pp. 53–69). Germany: Springer.

Suryanarayanan, T. S., Venkatachalam, A., Thirunavukkarasu, N., Ravishankar, J. P., Doble, M., & Geetha, V. (2010). Internal mycobiota of marine macroalgae from the Tamilnadu coast: Distribution, diversity and biotechnological potential. *Botanica Marina, 53*, 457–468.

Vaz, A. B., Rosa, L. H., Vieira, M. L., Garcia, V. D., Brandão, L. R., Teixeira, L. C., & Rosa, C. A. (2011). The diversity, extracellular enzymatic activities and photoprotective compounds of yeasts isolated in Antarctica. *Brazilian Journal of Microbiology, 42*, 937–947.

Voss, M., Bange, H. W., Dippner, J. W., Middelburg, J. J., Montoya, J. P., & Ward, B. (2013). The marine nitrogen cycle: Recent discoveries, uncertainties and the potential relevance of climate change. *Phil Trans R Soc B, 368*, 0121.

Wiencke, C., & Amsler, C. D. (2012). *Seaweeds and their communities in polar regions. Seaweed biology: Novel insights into ecophysiology, ecology and utilization* (p. 493). Germany: Springer.

Wiencke, C., & Clayton, M. N. (2002). Antarctic seaweeds. In J. W. Wagele (Ed.), *Synopses of the Antarctic benthos* (p. 239). Germany: Lichtensein.

Wiencke C, Amsler CD, Clayton MN (2014) Macroalgae. De Broyer C, Koubbi P, Griffiths HJ, Raymond B, Udekemd'Acoz CD Biogeographic Atlas of the Southern Ocean. Scientific Committee on Antarctic Research, Cambridge, UK, 66–73.

Zuccaro, A., Schulz, B., & Mitchell, J. I. (2003). Molecular detection of ascomycetes associated with *Fucus serratus*. *Mycological Research, 107*, 1451–1466.

Zucconi, L., Selbmann, L., Buzzini, P., Turchetti, B., Guglielmin, M., Frisvad, J. C., & Onofri, S. (2012). Searching for eukaryotic life preserved in Antarctic permafrost. *Polar Biology, 35*, 749–757.

Chapter 12
Microbial Role in the Ecology of Antarctic Plants

Júnia Schultz and Alexandre Soares Rosado

Abstract Antarctica is a true mosaic of extremes. It is a continent of superlatives, with low temperatures, freezing and thawing cycles, high salinity and intense solar radiation, among other environmental extremes. These unique conditions exert evolutionary pressure, which selects the biological community to develop in place, such as the microbial community. Ice-free soils represent a very small proportion of the total land area of Antarctica, and in these areas, a scarce Antarctic vegetation grows. The vegetation of these ice-free habitats is characterized by low coverage and low productivity, being mainly composed of lower plants, with only two vascular plant species, *Deschampsia antarctica* and *Colobanthus quitensis*. Climate change in Antarctica may present new threats to terrestrial ecosystems particularly by increasing the distribution of the native plants but also increasing the successful establishment of non-native species. It is known that the vegetation cover has an important role in the microbial diversity of Antarctic soils due that these microorganisms produce molecules that cooperate with the establishment and development of plants in harsh conditions and vice versa. Our chapter selects and discusses some of the few studies that describe microbe-plant interactions in Antarctica and how these interactions can modulate the distribution, diversity and abundance of native vascular plants and microbial diversity in Antarctica.

Keywords Rhizospheric microbes · *Deschampsia antarctica* · *Colobanthus quitensis* · Microbial diversity · Development of plants

J. Schultz · A. S. Rosado (✉)
Laboratory of Molecular Microbial Ecology, Institute of Microbiology Paulo de Góes, Federal University of Rio de Janeiro, Rio de Janeiro, Brazil
e-mail: asrosado@micro.ufrj.br

S. Castro-Sowinski (ed.), *The Ecological Role of Micro-organisms in the Antarctic Environment*, Springer Polar Sciences,
https://doi.org/10.1007/978-3-030-02786-5_12

12.1 Introduction

Antarctica is known for its environmental extremes and is considered one of Earth's most inhospitable and severe continents because of its climatic conditions (Cowan and Tow 2004; Margesin and Miteva 2011; Cowan 2014). It is also known as a continent of superlatives due to its negative temperatures, low availability of water and nutrients, high salinity, intense radiation, freezing and thawing cycles and long periods without light (Dieser et al. 2010; Cowan 2014).

Antarctica is characterized as a pristine continent possessing a mosaic of ecosystems with different habitats and peculiar characteristics, such as dry valleys, subglacial lakes, maritime coasts, glaciers, marine fumaroles and polar geothermal environments (Herbold et al. 2014; Goordial et al. 2016). These ecosystems, especially the Antarctic Peninsula, have been strongly influenced by global warming, which has reportedly increased the temperature of Antarctica in recent decades (Convey and Smith 2006). In the Antarctic maritime region, observations have shown that the increasing temperature is linked to an increase in glacier melting, exposing more land areas to environmental conditions (Convey and Smith 2006).

One of the consequences of the thawing and increase in the exposed land is the significant increase in the distribution and abundance of the vascular plant species colonizing Antarctica (Aislabie et al. 2009); these currently comprise four species, *Deschampsia antarctica* Desv. (Poaceae) and *Colobanthus quitensis* (Kunth) Bartl (Caryophyllaceae), which are endemic species, and *Poa annua* L. (Poaceae) and *Poa pratensis* L. (Poaceae), which were introduced to the continent accidentally (Pertierra et al. 2013; Teixeira et al. 2013). All these species are commonly found growing together and are well adapted to the Antarctic maritime region (Avery et al. 2003; Teixeira et al. 2010, 2013).

The structure and diversity of the microbial community associated with the rhizosphere of these Antarctic plants differ from those of the microbial community in the adjacent soil (Teixeira et al. 2013). The microbial community associated with these plants offers bioproducts that assist in the development and health of the plants in the hostile environment of Antarctica. These microorganisms provide molecules that favour plant growth; provide nutrients, such as iron, nitrogen and phosphorus; and inhibit the attack by other microorganisms (Mendes et al. 2013).

These microorganisms are sensitive to environmental disturbances, and environmental changes can cause disturbances in the microbial diversity of Antarctic soils, altering the relationship between plants and associated ecosystems (Yergeau and Kowalchuk 2008; Teixeira et al. 2010). In this context, we will show the geographical distribution of the plant species found in the Antarctic continent, reporting their descriptions, their physiology and the general strategies they have developed for colonization. We will also highlight the microbial diversity associated with the rhizosphere of these plant species, listing the microbial groups and their main characteristics and reporting the differences in the plant-associated microbes and the microbes of the surrounding soil. Finally, we will discuss the ecology and interactions between the plants and the microbial community inhabiting the rhizosphere and the soil. Results were obtained through independent cultivation methods, next-generation sequencing and bioinformatics tools, which have provided accurate information on ecological dynamics.

12.2 Antarctic Vascular Plants

The Antarctic continent is predominantly covered by permanent ice and snow, with only 1% free areas available for plant colonization. Most ice- and snow-free areas are found on the Antarctic Peninsula and associated islands and on the continent's coastal regions, which do anthropogenic influences and global warming (Convey et al. 2009) heavily affect.

Antarctic vegetation consists mainly of lower plant groups, such as mosses, lichens, ferns, hepatics, fungi and algae (Table 12.1), which are adapted to survive under severe environmental conditions, tolerating dehydration, a high incidence of UV light and low temperatures (Bargagli 2008; Teixeira et al. 2010). There are no trees or shrubs, but Antarctica has species of vascular plants, two of which are native to the continent and have already been mentioned, *Deschampsia antarctica* and *Colobanthus quitensis* (Table 12.1).

The species *Deschampsia antarctica* L. (Poaceae) is also known by the popular name Antarctic hair grass (Fig. 12.1a, b). According to Chwedorzewska (2008), this grass has dark green and well-developed leaves and sturdy and upright stalks, but under adverse environmental conditions, the plant has a much smaller size and loose and olive-coloured locks. Birds, such as seagulls and skuas, carry out seed dispersal, but grass growth by clonal growth has also been observed (Convey 1996; Parnikoza et al. 2007).

D. antarctica is a diploid (2n = 26) species, which has been well adapted to the extreme conditions of Antarctica. It performs a variety of metabolic processes at very low temperatures and high levels of UV radiation and desiccation, allowing its development, including growth and flowering in harsh conditions. In addition, *D. antarctica* can photosynthesize at the freezing point (Bravo and Griffith 2005; Chew et al. 2012). Furthermore, according to Bravo et al. (2001), during the summer, the species accumulates high levels of sucrose and proline, which are osmo-protectants that improve tolerance to environmental stresses.

Another native species is *Colobanthus quitensis* (Kunth) Bartl (Caryophyllaceae), popularly known as Antarctic pearlwort (Fig. 12.1c, d). It grows perennially in

Table 12.1 Richness of plant species found in the biogeographical zones of Antarctica

Antarctic zone	Flowering plants	Ferns and allies	Mosses	Liverworts	Lichens
Continental Antarctica	0	0	25	1	150
Maritime Antarctica	2	0	100	25	250
Sub-Antarctica	60	16	250	85	250

Adapted from Convey (2013)

Fig. 12.1 The vascular plants occurring in Antarctica. Representatives of *Deschampsia antarctica* (**a**, **b**), *Colobanthus quitensis* (**c**, **d**), *Poa annua* (**e**) and *Poa pratensis* (**f**). (Photos courtesy of Alexandre Soares Rosado (**a**–**d**), Daiane Valente (**e**) and Luis R. Pertierra (**f**))

compact pillows and has a small size, light green colouring and yellow flowers (Mantovani and Vieira 2000). There is almost no evidence of vegetative reproduction in *C. quitensis*, and all the studied populations seemingly originated from seeds, which can survive for a long time and be dispersed by birds (Parnikoza et al. 2015). Antarctic populations of *C. quitensis* have been described as morphologically and physiologically plants adapted to succeed in this cold environment, maintaining high photosynthetic rates at low temperatures (Xiong et al. 1999).

The grasses *Poa annua*, known as annual meadow grass, and *Poa pratensis*, known as blue grass (Poaceae) (Fig. 12.1e, f), are the most widespread and abundant non-native vascular plants in the Antarctic and subantarctic regions (Shaw et al. 2010; Molina-Montenegro et al. 2012; Pertierra et al. 2017). Both are from European origin and probably were inadvertently introduced in Antarctica. They are currently present in the major groups of the subantarctic islands, with more recent occurrences reported on King George Island, Cierva Point and other locations in the Antarctic Peninsula (Shaw et al. 2010; Molina-Montenegro et al. 2012, 2014).

These two non-native species of angiosperms were accidentally introduced to the Antarctic continent, and probably due to human activities the abundance of these invasive species increased, consequently increasing the competitiveness for nutrients and niches between them and the native species (Convey 2011; Molina-Montenegro et al. 2012). Thus, the arrival of these invasive species to Antarctica may influence the biogeochemical cycles, causing an ecological imbalance in the region (Convey and Smith 2006).

The distribution of the four vascular plant species in Antarctica and their abundance are associated with three main factors: (1) climatic factors, such as the frequency and duration of freezing and thawing cycles, temperature, ultraviolet irradiation, exposure to wind and availability of water; (2) edaphic factors, such as the soil characteristics, origin and topography; and (3) biotic factors, such as animals and other plants that interfere with their distribution (Nedzarek and Chwedorzewska 2004; Chwedorzewska 2008; Teixeira et al. 2013). The four described plant species are distributed along the ice-free areas and can be found mainly in the maritime region in the northern and western regions of the Antarctic continent, where climatic conditions are less harsh (Fig. 12.2) (Lewis Smith 2003; Chwedorzewska 2009), especially on King George Island, which is located in the Antarctic Peninsula (Olech and Chwedorzewska 2011). In addition, non-native plants are mainly distributed in locations with strong human influence with the biome, such as research stations and monitoring centres (Chew et al. 2012; Molina-Montenegro et al. 2012).

As result of the climate change caused by the global warming, populations of vascular species are spreading to places where they have not previously been observed. The increase in water flow due to melting snow and rising temperatures makes these previously unsuitable areas already suitable for the development of these vascular plants (Convey and Smith 2006; Chew et al. 2012; Abraham et al. 2013).

12.3 Microorganisms Associated with Antarctic Plants

Even the harsh environmental limitations for life in Antarctica, a high microbial diversity is frequently reported by scientific researchers, showing that, unlike plants, microorganisms easily colonize the Antarctic continent (Barrientos et al. 2008; Ganzert et al. 2011; Cury et al. 2015; Niederberger et al. 2015).

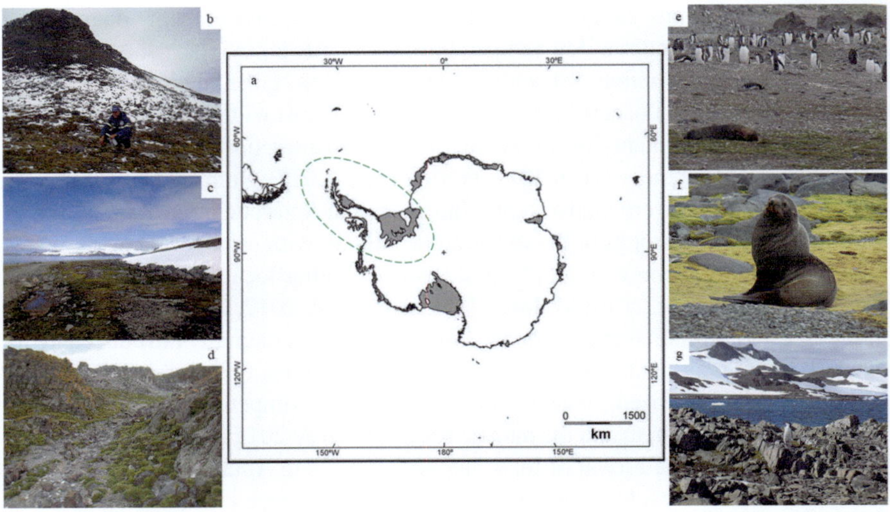

Fig. 12.2 General features of Antarctica (**a**). Map of the Antarctic continent. The grey shading indicates ice-free areas, and the green dashed circle indicates the region with the occurrence of vascular plants in the Antarctic Peninsula. The images in (**b–d**) show ice-free and covered by native vascular plants areas. The images in (**e–g**) show the presence of animals, which influences the growth and dispersion of plants (Photograph credits: Alexandre Soares Rosado – Photos (**b–f**); Amanda Gonçalves Bendia – Photo (**c**); Yuri Pinheiro – Photo (**g**)). (Source: Adapted from Simões et al. 2004)

The microorganisms that survive such conditions, which are detrimental to most life on Earth, are called extremophilic microorganisms. They are known as extremophiles due to their ability to adapt at adverse conditions by the modification of their macromolecular structures and biochemical processes at a molecular level. These modifications allow them to cope with the extreme physicochemical parameters of the environment (Antranikian et al. 2005; Zhang and Kim 2010).

Many reports describe the occurrence of complex microbial communities in different extreme Antarctic environments, including bacteria and archaea at geothermal sites (Amenábar et al. 2013; Herbold et al. 2014), arid soils and rocks in the dry valleys (Cary et al. 2010; Makhalanyane et al. 2013). In addition, many studies faced the identification of Antarctic psychrophilic microorganisms (Yergeau et al. 2007a, b; Bottos et al. 2014; Jesus et al. 2015); however, a few studies considering the microbial diversity found in the rhizosphere of native Antarctic plants have been conducted.

Barrientos et al. (2008) reported the characterization of bacteria from the rhizosphere of *D. antarctica*. Using culture-dependent methods, 16S rRNA sequencing for microbial identification, determination of resistance to antibiotics, heavy metal tolerance and the searching for microorganisms that produce enzymes with industrial interest, the authors faced a first approximation of the microbiome associated with this plant. They observed that most of the isolates ($n = 70$) belonged to the genera *Pseudomonas*, *Flavobacterium* and *Arthrobacter*, and these isolates produced different enzymes, such as alkaline and acid phosphatases, esterases and lipases. In addition, 66 out of 70 bacterial isolates were resistant to antibiotics such as ampicillin, actinomycin-D, gentamicin, tetracycline and streptomycin, suggesting that antibiotic resistance is widely spread among plant-associated microbes.

Teixeira et al. (2010, 2013) addressed the microbial diversity of rhizosphere soils of *D. antarctica* and *C. quitensis* in Admiralty Bay King George Island, Maritime Antarctica, by PCR/DGGE of the 16S rRNA gene and pyrosequencing. According to their results, the dominant microbial groups in the rhizospheric soil belonged to the phyla *Firmicutes*, *Actinobacteria* and *Proteobacteria*, in which *Firmicutes* was the most dominant phylum. The authors reported that approximately 70% of the sequences belonged to the class *Clostridia*, suggesting that anaerobic bacteria are abundant in the rhizosphere of Antarctic vascular plants, whereas the class *Bacilli* corresponded to 15% of all the sequences of *Firmicutes* (Table 12.2).

Interestingly, the bacterial communities found in the rhizospheric soil of both native Antarctic vascular plant species were very different from those reported in adjacent non-rhizospheric soils and soils from other regions of Antarctica that do not have plant history (Teixeira et al. 2013) (Fig. 12.3). Yergeau et al. (2007a) found that Antarctic soils support fewer abundance of microorganisms than those with vegetation cover. Thus, the information supports that the structure, abundance, diversity and activity of the soil microbial community in Antarctica are influenced by the presence of vegetation (Yergeau et al. 2007a).

Antarctic plants also offer a unique niche for the establishment of fungal communities (Arenz et al. 2014). The report from Upson et al. (2008) showed the presence of dark septate endophytes (DSE) in the roots of *C. quitensis* and *D. antarctica* collected in the Leonie Islands, Antarctic Peninsula. Upson et al. (2009) showed the occurrence of *Rhizoscyphus ericae*, *Gyoerffyella* sp., *Leptodontidium orchidicola*, *Tapesia* sp. and *Mollisia* sp. In addition, Rosa et al. (2009) identified endophytic species of *Alternaria* and *Phaeosphaeria*, as well as other unidentified ascomycetes, from the leaves of *D. antarctica.*

These rhizospheric microorganisms, bacteria and fungi may produce different compounds that confer beneficial properties to plants. Next sections describe how the microbe-plant interaction occurs with emphasis in this interaction in the extreme environment of Antarctica and, in addition, describe the consequences in the ecology of plant.

12.4 The Ecological Role of Microbe in Plant Growth and Health

Plant growth-promoting microorganisms (PGPM), as nitrogen-fixing bacteria and mycorrhizal fungi, phytohormone-producing microbes and biocontrollers (microorganisms that produce plant protection products against pathogens) have been widely studied due to their beneficial effects on plant growth and health

Table 12.2 Reports of microorganisms already found in the rhizopheric soil and associated with native vascular plants in Antarctica

Antarctic vascular plant	Microorganism/microbial group	Relative abundance (%) (estimative)	Reference
Deschampsia antarctica	*Actinobacteria*	65	Teixeira et al. (2013)
	Firmicutes	12	
	Gammaproteobacteria	8	
	Betaproteobacteria	5	
	Alphaproteobacteria	5	
	Bacteroidetes	2	
	Acidobacteria	2	
	Epsilonproteobacteria	1	
Colobanthus quitensis	*Actinobacteria*	62	Teixeira et al. (2013)
	Firmicutes	10	
	Gammaproteobacteria	9	
	Betaproteobacteria	8	
	Alphaproteobacteria	6	
	Bacteroidetes	2	
	Acidobacteria	2	
	Epsilonproteobacteria	1	
Deschampsia antarctica	*Pseudomonas* sp. Da-bac TI-8	–	Berríos et al. (2013)
Deschampsia antarctica	*Alternaria triticina*	–	Rosa et al. (2009)
	Alternaria alternata		
	Ascomycete sp.		
	Phaeosphaeria dennisiana		
	Entrophospora sp.		
Deschampsia antarctica and *Colobanthus quitensis*	*Dothidiomycetidae*	–	Upson et al. (2009)
	Helotiaceae		
	Rhizoscyphus ericae		
	Helotiales		
	Gyoerffyella sp.		
	Leptodontidium orchidicola		
	Tapesia sp.		
	Mollisia sp.		
Deschampsia antarctica and *Colobanthus quitensis*	Dark septate endophytes (DSE)	–	Upson et al. (2008)
Deschampsia antarctica	*Pseudomonas* sp.	–	Barrientos-Díaz et al. (2008)
	P. tolaasii	–	
	P. trivialis	–	
	P. panacis	–	
	Flavobacterium sp.	–	
	Arthrobacter sp.	–	

(continued)

Table 12.2 (continued)

Antarctic vascular plant	Microorganism/microbial group	Relative abundance (%) (estimative)	Reference
Deschampsia antarctica and *Colobanthus quitensis*	*Firmicutes*	40	Teixeira et al. (2010)
	Clostridia	70	
	Bacilli	15	
	Proteobacteria	30	
	Gammaproteobacteria	30	
	Alphaproteobacteria	25	
	Epsilonproteobacteria	20	
	Betaproteobacteria	15	
	Deltaproteobacteria	5	
	Actinobacteria	25	
	Bifidobacteriales	80	
	Chloroflexi	5	
	Gemmatimonadetes		
	Verrucomicrobia		
	Bacteroidetes		
	Planctomycetes		

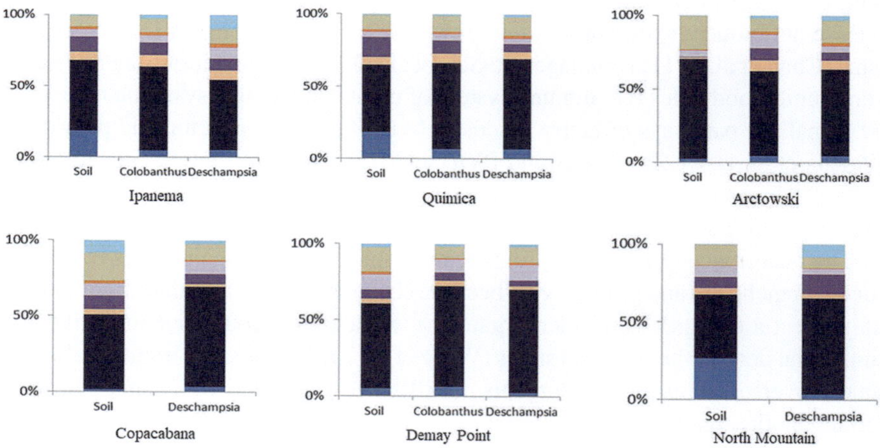

Sample sites	Na	Ca	Mg	K	H+Al	Al	S	pHwater	Corg	P	K	N-NH$_4$	N-NO$_3$
	Cmol$_c$ dm^{-3}							1:2.5	g kg^{-1}	mg l^{-1}		µg g^{-1}	
Arctowski	0.49	12	7	0.02	20	3.55	19.51	4.3	0.97	1005	9	590.11	13.98
Ipanema	0.74	11	6	0.02	11	0	17.76	5.7	5.11	786	9	2.31	0.051
Química	0.31	9.4	5.6	0.32	1.3	0	15.63	6.6	2.28	620	128	7.21	0.891
Demay Point	0.32	5.3	8.3	1.3	22.6	14.9	15	4.9	5.3	464.2	429	263.91	10.51
North Mountain	0.25	28.5	4.5	0.12	0	0.9	33.4	8.2	1.9	349	47	n.d.[#]	n.d.
Copacabana	0.47	4.2	1.3	0.64	7.4	0.9	6.6	4.8	1.2	649.1	250	361.56	14.07

n.d.: not detected

Bacteroidetes
Epsilon-
Beta-
Actinobacteria
Firmicutes
Gamma-
Alpha-
Acidobacteria

Fig. 12.3 Structure and relative abundance of the microbial community found in the rhizosphere of *D. antarctica* and *C. quitensis* compared with vegetation-free soils from Admiralty Bay, Maritime Antarctica. (Source: Adapted from Teixeira et al. 2013)

(Teplitski et al. 2011). These microbes can directly or indirectly influence the development and productivity (biomass) of plants (Schnitzer et al. 2011).

Among positive effects performed by PGPM, a few microorganisms produce chemicals involved in the acquisition of nutrients (e.g., the acquisition of iron by bacterial siderophores or phosphorus by the secretion of organic acids from mycorrhizal fungi) and/or the development of roots (e.g., by the production of phytohormones). For example, the production of phytohormones increases the area of roots, thus the scanning area for the uptake of nutrients (Johnson and Graham 2013). In addition, the supply of nutrients to plants by nitrogen-fixing microbes results in a significant increase in the plant nutritional status (Mendes et al. 2013). The microbial community also plays a key role in the release of nutrients from soil minerals and the nutritional enhancement of nutrient-poor soils, thus favouring plant growth; notably, the microorganisms themselves benefit from this effect (Mapelli et al. 2012).

An iron acquisition strategy by plants involves the use of microbial iron-chelating molecules, known as siderophores (Lemanceau et al. 2009). These are small and high-affinity iron-chelating molecules that transport iron across microbial and plant cell membranes (Hider and Kong 2010). In addition, microorganisms present in the rhizosphere can produce volatile organic compounds that modulate plant growth and mediate plant-microorganism interactions (Effmert et al. 2012).

Rhizospheric microorganisms can also provide protection against soil pathogens (biocontrol), competing for trace elements, nutrients and space, and/or interfering with the virulence mechanisms developed by pathogens. In addition, they may also release antibiotics, which inhibit the growth or activity of competing microorganisms (Chan et al. 2011; Brakhage and Schroeckh 2011). Interestingly, the microbial community modulates the immune system of plants, inducing a systemic resistance by signalling pathways effective against a broad range of opportunistic pathogens (Pineda et al. 2010; Zamioudis and Pieterse 2012). Quorum sensing-related molecules produced by rhizobacteria can also trigger a series of plant responses, activating defence-related genes (Hartmann and Schikora, 2012).

The occurrence of a diverse group of rhizospheric microorganisms is important for the plant health and productivity, because changes in biotic and abiotic conditions may affect a few groups of microorganisms, while other groups remain active and functional despite the imposed stress (Wagg et al. 2011). Thus, environments with a much diverse group of microorganism will be less affected by changes in their biotic/abiotic conditions, including climate changes (Awasthi et al. 2014). In this way, the microbial diversity acts as an insurance for maintaining the plant stability and the ecosystem functioning (Wagg et al. 2011; Eisenhauer et al. 2012).

Such AM fungi, with many functions, could benefit the host from AM fungi-rich communities, theoretically providing additional services to host plants that may increase their productivity (Koide 2000).

In exchange for all the benefits that microorganisms offer to plants, they benefit from the bioproducts exuded by the plant root, such as the secretion of different carbon sources produced by photosynthesis. Plant exudates may create a habitat with higher concentrations of nutrients and energy for metabolic activities, but also these compounds chemically modify the soil (Achouak and Haichar 2013). Root exudate composition varies according to plant species, environmental nutritional status, plant health, stresses and the microbial community (Lankau 2011), affecting the species-specific association between microorganisms and plants (Bais et al. 2006; Broeckling et al. 2008).

Most of the microbial species found in the rhizosphere are organotrophic, that is, they need to assimilate organic compounds to obtain the energy for their functioning (Mendes et al. 2013). The availability and accessibility of organic compounds are limited in most soils, and carbon availability is the most common limiting factor in soil for bacterial growth (Rousk and Baath 2007). In particular, Antarctic soils are considered nutrient-poor ones, with low content of organic matter, except in areas with the presence of birds and vegetation cover (Teixeira et al. 2013).

12.5 The Ecological Consequences of the Microbe-Plant Interaction in Antarctica

Antarctic plants have developed a set of mechanisms to grow and survive in this hostile environment (Rothschild and Mancinelli 2001; Lewis Smith 2003). These mechanisms include the adjustment in cellular and physiological molecular responses such as the change in the membrane lipid composition and the production of antioxidants, osmocompatible solutes and stress proteins, among others (Gill and Tuteja 2010).

In addition to these biochemical and physiological changes, external factors also aid in the adaptation and dissemination of plants in Antarctica (Gielwanowska et al. 2011; Chew et al. 2012). For example, microorganisms produce molecules that actively cooperate in the establishment and development of plants (Barea et al. 2013). According to Bais et al. (2006) and Ho et al. (2017), the beneficial interaction between plants and microbes mainly occurs in the rhizosphere, and a certain groups of Antarctic microorganisms are very important for maintaining plant health. Therefore, a systemic and integrated view is necessary for a complete understanding of the microbial community structure and plant dynamics.

PGPM have also adapted to live and perform all their functions in extreme conditions, and they also contribute to the increase in Antarctic plant biomass (Teixeira et al. 2010; Jorquera et al. 2012). Microorganisms may also assist in the bioremediation of contaminated Antarctic environments. Pollutants may negatively influence the development and health of plants, and a reduced group of microbes have the ability to degrade toxic compounds, allowing plant growth and development (Peixoto et al. 2011; Fukuhara et al. 2013). The Antarctic environment near the Scientific Bases are usually contaminated with petroleum hydrocarbons, and microorganisms from these locations have developed the ability to survive and degrade these toxic compounds as shown by Jesus et al. (2015) and Cury et al. (2015) (Table 12.3).

Table 12.3 Characteristics of the Antarctic rhizosphere microorganisms associated with the native vascular plants of Antarctica

Antarctic vascular plant	Antarctic rhizosphere microorganisms associated	Characteristics of the microorganism/microbial group	Reference
Deschampsia antarctica	*Pseudomonas* sp.	Growth in oil	Aislabie et al. (2006)
		Bioremediation abilities	
		Phytohormone producer	Berríos et al. (2013)
		Inorganic phosphate solubilizer	
		Enzyme producer	
		Organic acid producer	
		Antimicrobial properties	Matthijs et al. (2014)
		Iron-binding siderophore producer	Barry and Challis (2009)
		Increased resistance to thermal stress	Mishra et al. (2011)
Deschampsia antarctica	*Pseudomonas* sp.	Enzyme producer	Barrientos-Díaz et al. (2008)
	P. tolaasii	Resistance to antibiotics and heavy metals	
	P. trivialis	Tolerance to different temperatures	
	P. panacis	Growth in oil	Pajuelo et al. (2011)
	Flavobacterium sp.	Bioremediation abilities	
	Arthrobacter sp.	Antimicrobial properties	Tomova et al. (2015)
			Presta et al. (2016)
Deschampsia antarctica	*Alternaria triticina*	Saprophytic and pathogenic capacities	Rosa et al. (2009)
	Alternaria alternata		
	Phaeosphaeria dennisiana	Leaf fungal pathogen	Rosa et al. (2009)
	Entrophospora sp.	Arbuscular mycorrhizal fungi	Sieverding and Oehl (2006)

Regarding the interactions between Antarctic plants and rhizospheric microbes, there are a few reports describing the microbial communities associated with *Deschampsia antarctica* and *Colobanthus quitensis*. These plants modulate the bacterial diversity and abundance in Antarctica (Teixeira et al. 2013), but also the microbial communities affect the development of plants in this extreme environment.

The microbial communities found in the Antarctic non-covered and vegetative-covered soils are significantly different (Yergeau et al. 2007a), and also the microbial communities found in the rhizosphere of vascular plants are different from the ones found in the adjacent soil (Teixeira et al. 2013). Thus, the presence of plant cover is an important factor in structuring the microbial community and vice versa. Antarctic vascular plants develop in a tundra-like soil, and their tufts create microhabitats, with a favourable microclimate for storing water and providing shelter against severe environmental conditions, thus enabling microbial colonization (Parnikoza et al. 2011; Yergeau 2014).

Roots exude different organic compounds commonly used by soil microorganisms as nutrients and energy source (Haichar et al. 2008). There are higher levels of nitrate, total dissolved nitrogen, dissolved organic carbon and free amino acids in soils under the influence of *D. antarctica* and *C. quitensis* rather than in soils dominated by lichen or moss or soils without vegetation, which promotes the establishment and selection of microbial groups (Kowalchuk et al. 2010; Yergeau 2014).

The rhizospheric microorganisms of Antarctic plants play an important role in the biogeochemical cycle of nutrients directly associated with the development of the plant, especially phosphorus, which is an essential element for the growth and development of both plants and microorganisms (Berríos et al. 2013). This nutrient is not always available for assimilation by the plant, and the bioavailability of phosphorus is already low in soils derived from volcanic ash, as the Antarctic soils. This kind of soil is damaging to plants, and the right microbial community can help to cope with this harmful stress (Gyaneshwar et al. 2002).

A group of PGPM has the ability to solubilize phosphorus from soil and to deliver this nutrient to the plant in an easy absorbing chemical form (Cakmakci et al. 2006). Based on Stevenson (2005), the solubilization of phosphorus compounds in the Antarctic soils is mainly due to the microbial production of organic acids that lower the soil pH. For example, the phosphorus-solubilizing bacterium *Pseudomonas* sp. Da-bac TI-8, which was isolated from the rhizosphere of *D. antarctica*, produces and secretes gluconic acid and phytohormones, behaving as a psychrotolerant PGPM (Berríos et al. 2013).

Probably, Teixeira et al. (2013) conducted the most complete characterization of the interaction between microbes and Antarctic plants. They used real-time quantitative PCR (qPCR) with different primers, 16S rRNA gene PCR amplification and microarray analysis in combination with an analysis of the physicochemical parameters of the soil samples to unveil the microbial communities. The authors found differences between the structure and composition of the microbial community at the phylum and class levels in the rhizosphere of *D. antarctica* and *C. quitensis* compared with soils without vegetation from six sampling sites (Arctowski, Quimica, Ipanema, North Mountain, Demay Point, Copacabana) around Admiralty Bay, King George Island, in the Antarctica maritime region.

The results showed that there was a statistically significant difference between the bacterial communities of the soil and the plant rhizosphere in four of the six analysed sites. In the samples from soil without vegetation, *Firmicutes* and *Acidobacteria* were the most abundant phyla, while the rhizosphere samples were dominated by the phylum *Actinobacteria*. It can be noticed that, although soil characteristics influence the structure of the microbial community, the microbial communities studied by Teixeira et al. (2013) were predominantly shaped by the presence of plants. Yergeau et al. (2007a) reported the same strong influence of plants at other sites in the maritime Antarctic.

Teixeira et al. (2013) did not detect clear differences between the microbial communities in the rhizosphere of *C. quitensis* and *D. antarctica* in most of the sampling sites, except in the region of Arctowski Station on King George Island. In samples collected in this Station, they observed a clear segregation between the microbial communities associated with the rhizosphere of the two Antarctic plants, indicating a rhizospheric effect (Fig. 12.3). A similar result was reported by Teixeira et al. (2010), who used the technique of pyrosequencing for the first time in Antarctic samples.

12.6 Concluding Remarks

Antarctica has a wide variety of lower plants but only two species of endemic vascular plants, *Deschampsia antarctica* and *Colobanthus quitensis*. Virtually all the vegetation is found in the maritime region, mainly in the Antarctic Peninsula, where there are extensive free-ice areas.

Compared with other environments, the relationship between Antarctic plants and the microbial communities associated with their roots, or present in the soils, has not been fully studied. However, the information supports that the microbiome under the influence of those vascular plants affects and shapes the structure, composition, diversity and development of Antarctic plants.

Plant-microbe interactions are essential for plant performance and ecology in the Antarctic continent. However, there is no doubt that more studies are needed to expand our understanding regarding the delicate and important interaction between Antarctic microorganisms and plants and especially to understand the effects that the human presence and global climate changes can cause in this fascinating, dangerous and at the same time threatened extreme environment.

References

Abraham, J. P., Baringer, M., Bindoff, N. L., et al. (2013). A review of global ocean temperature observations: Implications for ocean heat content estimates and climate change. *Reviews of Geophysics, 51*, 450483.

Achouak, W., & Haichar, E. (2013). Shaping of microbial community structure and function in the rhizosphere by four diverse plant species. In F. J. de Bruijn (Ed.), *Molecular microbial ecology of the rhizosphere* (Vol. 1, pp. 161–167). Hoboken: Wiley Blackwell.

Aislabie, J., Saul, D. J., & Foght, J. M. (2006). Bioremediation of hydrocarbon-contaminated polar soils. *Extremophiles, 10*(3), 171–179.

Aislabie, J., Jordan, S., Ayton, J., Klassen, J. L., Barker, G. M., et al. (2009). Bacterial diversity associated with ornithogenic soil of the Ross Sea region, Antarctica. *Canadian Journal of Microbiology, 55*, 21–36.

Amenábar, M. J., Flores, P. A., Pugin, B., Boehmwald, F. A., & Blamey, J. M. (2013). Archaeal diversity from hydrothermal systems of Deception Island, Antarctica. *Polar Biology, 36*, 373–380.

Antranikian, G., Vorgias, C. E., & Bertoldo, C. (2005). Extreme environments as a resource for microorganisms and novel biocatalysts. *Advances in Biochemical Engineering/Biotechnology, 96*, 219–262.

Arenz, S., Meydan, S., Starosta, A. L., Berninghausen, O., Beckmann, R., Vázquez-Laslop, N., & Wilson, D. N. (2014). Drug-sensing by the ribosome induces translational arrest via active site perturbation. *Molecular Cell, 56*, 446–452.

Avery, L. M., Smith, R. I. L., & West, H. M. (2003). Response of rhizosphere microbial communities associated with Antarctic hairgrass (*Deschampsia antarctica*) to UV radiation. *Polar Biology, 26*, 525–529.

Awasthi, A., Singh, M., Soni, S. K., Singh, R., & Kalra, A. (2014). Biodiversity acts as insurance of productivity of bacterial communities under abiotic perturbations. *The ISME Journal, 8*, 2445–2452.

Bais, H. P., Weir, T. L., Perry, L. G., Gilroy, S., & Vivanco, J. M. (2006). The role of root exudates in rhizosphere interactions with plants and other organisms. *Annual Review of Plant Biology, 57*, 233–266.

Barea, J. M., Pozo, M. J., Azcón, R., & Azcón-Aguilar, C. (2013). Microbial interactions in the rhizosphere. In F. J. de Bruijn (Ed.), *Molecular microbial ecology of the rhizosphere* (pp. 29–44). Hoboken: Wiley-Blackwell.

Bargagli, R. (2008). Environmental contamination in Antarctic ecosystems. *Science of the Total Environment, 400*, 212–226.

Barrientos, L., Gidekel, M., & Gutiérrez-Moraga, A. (2008). Characterization of rhizospheric bacteria isolated from *Deschampsia antarctica* Desv. *World Journal of Microbiology and Biotechnology, 24*, 2289–2296.

Barry, S. M., & Challis, G. L. (2009). Recent advances in siderophore biosynthesis. *Current Opinion in Chemical Biology, 13*, 205–215.

Berríos, G., Cabrera, G., Gidekel, M., & Gutiérrez-Moraga, A. (2013). Characterization of a novel Antarctic plant growth-promoting bacterial strain and its interaction with Antarctic hair grass (*Deschampsia antarctica* Desv). *Polar Biology, 36*, 349–362.

Bottos, E. M., Woo, A. C., Zawar-Reza, P., Pointing, S. B., & Cary, S. C. (2014). Airborne bacterial populations above desert soils of the McMurdo Dry Valleys, Antarctica. *Microbial Ecology, 67*, 120–128.

Brakhage, A. A., & Schroeckh, V. (2011). Fungal secondary metabolites - strategies to activate silent gene clusters. *Fungal Genetic Biology, 48*, 15–22.

Bravo, L. A., & Griffith, M. (2005). Characterization of antifreeze activity in Antarctic plants. *Journal of Experimental Botany, 56*, 1189–1196.

Bravo, L. A., Ulloa, N., Zuniga, G. E., Casanova, A., Corcuera, L. J., & Alberdi, M. (2001). Cold resistance in Antarctic angiosperms. *Physiologia Plantarum, 111*, 55–65.

Broeckling, C. D., Broz, A. K., Bergelson, J., Manter, D. K., & Vivanco, J. M. (2008). Root exudates regulate soil fungal community composition and diversity. *Applied and Environmental Microbiology, 74*, 738.

Cakmakci, R., Donme, F., & Aydin Sahin, A. F. (2006). Growth promotion of plants by plant growth promoting rhizobacteria under greenhouse and two field different soil conditions. *Soil Biology and Biochemistry, 38*, 1482–1487.

Cary, S. C., McDonald, I. R., Barrett, J. E., & Cowan, D. A. (2010). On the rocks: The microbiology of Antarctic Dry Valley soils. *Nature Reviews Microbiology, 8*, 129138.

Chan, K. G., Atkinson, S., Mathee, K., et al. (2011). Characterization of N-acylhomoserine lactone-degrading bacteria associated with the *Zingiber officinale* (ginger) rhizosphere: Co-existence of quorum quenching and quorum sensing in *Acinetobacter* and *Burkholderia*. *BMC Microbiology, 11*, 51.

Chew, O., Lelean, S., John, U. P., & Spangenberg, G. C. (2012). Cold acclimation you induce rapid and dynamic changes file in freeze tolerance in the cryophile mechanisms *Deschampsia antarctica* E. Desv. *Plant Cell & Environment, 35*, 829–837.

Chwedorzewska, K. J. (2008). *Poa annua* L. in Antarctic – Searching for the source of introduction. *Polar Biology, 31*, 263–268.

Chwedorzewska, K. J. (2009). Terrestrial Antarctic ecosystems in the changing world: an overview. *Polish Polar Research, 30*, 263–276.

Convey, P. (1996). The influence of environmental characteristics on life history attributes of Antarctic terrestrial biota. *Biological Reviews, 71*, 191–225.

Convey, P. (2011). Antarctic terrestrial biodiversity in a changing world. *Polar Biology, 34*, 1629–1641.

Convey, P. (2013). Antarctic ecosystems. In S. A. Levin (Ed.), *Encyclopedia of biodiversity* (Vol. 1, 2nd ed., pp. 179–188). Waltham: Academic.

Convey, P., & Smith, R. I. L. (2006). Responses of terrestrial Antarctic ecosystems to climate change. *Plant Ecology, 182*, 1–10.

Convey, P., Bindschadler, R., Di Prisco, G., Fahrbach, E., Gutt, J., Hodgson, D. A., Mayewski, P. A., Summerhayes, C. P., Turner, J., & ACCE Consortium. (2009). Antarctic climate change and the environment. *Antarctic Science, 21*, 541–563.

Cowan, D. A. (Ed.). (2014). *Antarctic terrestrial microbiology*. Berlin: Springer.

Cowan, D. A., & Tow, L. (2004). Endangered Antarctic environments. *Annual Review of Microbiology, 58*, 649–690.

Cury, J. C., Jurelevicius, D. A., Villela, H. D. M., Jesus, H. E., Peixoto, R. S., Schaeder, C. E. G. R., Bícego, M. C., Seldin, L., & Rosado, A. S. (2015). Microbial diversity and hydrocarbon depletion in low and high diesel-polluted soil samples from Keller Peninsula, South Shetland Islands. *Antarctic Science, 11*, 1–11.

Dieser, M., Greenwood, M., & Foreman, C. M. (2010). Carotenoid pigmentation in Antarctic heterotrophic bacteria as a strategy to withstand environmental stresses. *Arctic, Antarctic, and Alpine Research, 42*, 396–405.

Effmert, U., Kalderas, J., Warnke, R., & Piechulla, B. (2012). Volatile mediated interactions between bacteria and fungi in the soil. *Journal of Chemical Ecology, 38*, 665–703.

Eisenhauer, N., Scheu, S., & Jousset, A. (2012). Bacterial diversity stabilizes community productivity. *PLoS One, 7*, 1–5.

Fukuhara, Y., Horii, S., Matsuno, T., Matsumiya, Y., Mukai, M., & Kubo, M. (2013). Distribution of hydrocarbon-degrading bacteria in the soil environment and their contribution to bioremediation. *Applied Biochemistry and Biotechnology, 170*, 329–339.

Ganzert, L., Lipski, A., Hubberten, H. W., & Wagner, D. (2011). The impact of different soil parameters on the community structure of dominant bacteria from nine different soils located on Livingston Island, South Shetland Archipelago, Antarctica. *FEMS Microbiology Ecology, 76*, 476–491.

Giełwanowska, I., Pastorczyk, M., & Kellmann-Sopyła, W. (2011). Influence of environmental changes on polar physiology and development of vascular plants. *Papers on Global Change IGBP, 18*, 53–62.

Gill, S. S., & Tuteja, N. (2010). Reactive oxygen species and antioxidant machinery in abiotic stress tolerance in crop plants. *Plant Physiology and Biochemistry, 48*, 909–930.

Goordial, J., Raymond-Bouchard, I., Riley, R., Ronholm, J., Shapiro, N., Woyke, T., & Whyte, L. (2016). Improved high-quality draft genome sequence of the Eurypsychrophile *Rhodotorula* sp. JG1b, isolated from permafrost in the hyperarid upper-elevation McMurdo Dry Valleys, Antarctica. *Genome Announcements, 4*, e00069–e00016.

Gyaneshwar, P., Kumar, G. N., Parekh, L. J., & Poole, P. S. (2002). Role of soil microorganisms in improving P nutrition of plants. *Plant and Soil, 245*, 83–93.

Haichar, F. Z., Marol, C., Berge, O., Rangel-Castro, J. I., Prosser, J. I., et al. (2008). Plant host habitat and root exudates shape soil bacterial community structure. *The ISME Journal, 2*, 1221–1230.

Hartmann, A., & Schikora, A. (2012). Quorum sensing of bacteria and trans-kingdom interactions of N-acyl homoserine lactones with eukaryotes. *Journal of Chemical Ecology, 38*, 704–713.
Herbold, C. W., McDonald, I. R., & Cary, C. (2014). Microbial ecology of geothermal habitats in Antarctica. In D. A. Cowan (Ed.), *Antarctic terrestrial microbiology: Physical and biological properties of Antarctic soils* (pp. 181–215). New York: Springer.
Hider, R. C., & Kong, X. (2010). Chemistry and biology of siderophores. *Natural Product Reports, 27*, 637–657.
Ho, A., Di Lonardo, D. P., & Bodelier, P. L. E. (2017). Revisiting life strategy concepts in environmental microbial ecology. *FEMS Microbiology Ecology, 93*, 1–14.
Jesus, H. E., Peixoto, R. S., Cury, J. C., Van Elsas, J. D., & Rosado, A. S. (2015). Evaluation of soil bioremediation techniques in an aged diesel spill at the Antarctic Peninsula. *Applied Microbiology and Biotechnology, 99*, 10815–10827.
Johnson, N. C., & Graham, J. H. (2013). The continuum concept remains a useful framework for studying mycorrhizal functioning. *Plant and Soil, 363*, 411–419.
Jorquera, M. A., Shaharoona, B., Nadeem, S. M., de la Luz Mora, M., & Crowley, D. E. (2012). Plant growth-promoting rhizobacteria associated with ancient clones of creosote bush (Larrea tridentata). *Microbial Ecology, 64*, 1008–1017.
Koide, R. T. (2000). Functional complementarity in the arbuscular mycorrhizal symbiosis. *New Phytologist, 147*(2), 233–235.
Kowalchuk, G. A., Yergeau, E., Leveau, J. H. J., Sessitch, A., & Bailey, M. (2010). Plant-associated microbial communities. In W.-T. Liu & J. K. Jansson (Eds.), *Environmental molecular microbiology* (pp. 133–147). Norwich: Caister Academic Press.
Lankau, R. A. (2011). Intraspecific variation in allelochemistry determines an invasive species' impact on soil microbial communities. *Oecologia, 165*, 453–463.
Lemanceau, P., Expert, D., Gaymard, F., Bakker, P., & Briat, J. F. (2009). Role of iron in plant–microbe interactions. *Advances in Botanical Research, 51*, 491–549.
Lewis Smith, R. I. (2003). The enigma of *Colobanthus quitensis* and *Deschampsia antarctica* in Antarctica. In A. H. I. Huickes, W. W. C. Gieskes, R. L. M. Schorno, S. M. van der Vies, & W. I. Volff (Eds.), *Antarctic biology in a global context* (pp. 234–239). Leiden: Backham Publishers.
Makhalanyane, T. P., Valverde, A., Birkeland, N. K., Cary, S. C., Marla Tuffin, I., & Cowan, D. A. (2013). Evidence for successional development in Antarctic hypolithic bacterial communities. *The ISME Journal, 7*, 2080–2090.
Mantovani, A., & Vieira, R. C. (2000). Leaf micromorphology of Antarctic pearlwort *Colobanthus quitensis* (Kunth) Bartl. *Polar Biology, 23*, 531–538.
Mapelli, F., Marasco, R., Balloi, A., Rolli, E., Cappitelli, F., Daffonchio, D., & Borin, S. (2012). Mineral-microbe interactions: Biotechnological potential of bioweathering. *Journal of Biotechnology, 157*, 473–481.
Margesin, R., & Miteva, V. (2011). Diversity and ecology of psychrophilic microorganisms. *Research in Microbiology, 162*, 346–361.
Matthijs, S., Wauven, C. V., Cornu, B., Lumeng, Y., Cornelis, P., Thomas, C. M., & Ongena, M. (2014). Antimicrobial properties of *Pseudomonas* strains producing the antibiotic mupirocin. *Research in Microbiology, 165*, 695–704.
Mendes, R., Garbeva, P., & Raaijmakers, J. M. (2013). The rhizosphere microbiome: Significance of plant beneficial, plant pathogenic, and human pathogenic microorganisms. *FEMS Microbiology Reviews, 37*, 634–663.
Mishra, P. K., Bisht, S. C., Ruwari, P., Selvakumar, G., Bisht, J. K., Bhatt, J. C., et al. (2011). Alleviation of cold stress effects in wheat (*Triticum aestivum L.*) seedlings by application of psychrotolerant pseudomonads from N.W. Himalayas. *Archives of Microbiology, 193*, 497–513.
Molina-Montenegro, M. A., Carrasco-Urra, F., Rodrigo, C., Convey, P., Valladares, F., & Gianoli, E. (2012). Occurrence of the non-native annual bluegrass on the Antarctic mainland and its negative effects on native plants. *Conservation Biology, 26*, 717–723.

Molina-Montenegro, M. A., Carrasco-Urra, F., Acuña-Rodríguez, I., Oses, R., Torres-Díaz, C., & Chedorzewska, K. (2014). Assessing the importance of human activities for the establishment of the invasive *Poa annua* in Antarctica. *Polar Research, 33*, e21425.

Nędzarek, A., & Chwedorzewska, K. J. (2004). Nutrients content in water flash chosen sites of Deschampsia Antarctica (King George Island, Antarctica). *Folia Universitatis Agriculture Stetinensis, 234*, 299–304.

Niederberger, T. D., Sohm, J. A., Gunderson, T. E., Parker, A. E., Tirindelli, J., & Capone, D. G. (2015). Microbial community composition of transiently wetted Antarctic Dry Valley soils. *Frontiers in Microbiology, 6*, 1–12.

Olech, M., & Chwedorzewska, K. J. (2011). The first appearance and establishment of alien vascular plant in natural habitats on the fore field of retreating glacier in Antarctica. *Antarctic Science, 23*, 153–154.

Pajuelo, E., Rodríguez-Llorente, I. D., Lafuente, A., & Caviedes, M. Á. (2011). Legume–*Rhizobium* symbioses as a tool for bioremediation of heavy metal polluted soils. In M. Khan, A. Zaidi, R. Goel, & J. Musarrat (Eds.), *Biomanagement of metal-contaminated soils. Environmental pollution* (pp. 95–123). Dordrecht: Springer.

Parnikoza, I., Maidanuk, D. N., & Kozeretska, I. A. (2007). Are *Deschampsia antarctica* Desv. and *Colobanthus quitensis* (Kunth) Bartl. migratory relicts? *Cytology and Genetics, 41*, 36–40.

Parnikoza, I., Kozeretska, I., & Kunakh, V. (2011). Vascular plants of the maritime Antarctic: Origin and adaptation. *American Journal of Plant Sciences, 2*, 381–395.

Parnikoza, I. Y. U., Abakumov, E. V., Dykyy, I. V., Pilipenko, D. V., Shvydun, P. P., Kozeretska, I. A., & Kunakh, V. A. (2015). Influence of birds on the spatial distribution of *Deschampsia antarctica* E. Desv. on Galindez Island (Argentine Islands, maritime Antarctic). *Vestnik Sankt-Peterburgskogo Universiteta, 1*, 78–97.

Peixoto, R., Chaer, G. M., Carmo, F. L., Araujo, F. V., Paes, J. E., Volpon, A., Santiago, G. A., & Rosado, A. S. (2011). Bacterial communities reflect the spatial variation in pollutant levels in Brazilian mangrove sediment. *Antonie Van Leeuwenhoek, 99*, 341–354.

Pertierra, L. R., Lara, F., Benayas, J., & Hughes, K. A. (2013). Poa pratensis L., current status of the longest-established nonnative vascular plant in the Antarctic. *Polar Biology, 36*, 1473–1481.

Pertierra, L. R., Hughes, K. A., Tejedo, P., Enríquez, N., Luciañez, M. J., & Benayas, J. (2017). Eradication of the non-native *Poa pratensis* colony at Cierva Point, Antarctica: A case study of international cooperation and practical management in an area under multi-party governance. *Environmental Science & Policy, 69*, 50–56.

Pineda, A., Zheng, S.-J., van Loon, J. J. A., Pieterse, C. M. J., & Dicke, M. (2010). Helping plants to deal with insects: The role of beneficial soil-borne microbes. *Trends in Plant Science, 15*(9), 507–514.

Presta, L., Inzucchi, I., Bosi, E., Fondi, M., Perrin, E., Miceli, E., Turino, M. L., Giudice, A. L., Pascale, D., & Fani, R. (2016). Draft genome sequence of *Flavobacterium* sp. strain TAB 87, able to inhibit the growth of cystic fibrosis bacterial pathogens belonging to the *Burkholderia cepacia* complex. *Genome Announcements, 4*, e00410–e00416.

Rosa, L., Vaz, A., Caligiorne, R. B., Campolina, S., & Rosa, C. A. (2009). Endophytic fungi associated with the Antarctic grass *Deschampsia antarctica* Desv. (Poaceae). *Polar Biology, 32*, 161–167.

Rothschild, L. J., & Mancinelli, R. L. (2001). Life in extreme environments. *Nature, 409*, 1092–1101.

Rousk, J., & Bååth, E. (2007). Fungal biomass production and turnover in soil estimated using the acetate-in-ergosterol technique. *Soil Biology and Biochemistry, 39*, 2173–2177.

Schnitzer, S. A., Klironomos, J. N., HilleRisLambers, J., et al. (2011). Soil microbes drive the classic plant diversity-productivity pattern. *Ecology, 92*, 296–303.

Shaw, J., Spear, D., Greve, M., & Chown, S. L. (2010). Taxonomic homogenization and differentiation across Southern Ocean islands differ among insects and vascular plants. *Journal of Biogeography, 37*, 217–228.

Sieverding, E., & Oehl, F. (2006). Revision of *Entrophospora* and description of *Kuklospora* and *Intraspora*, two new genera in the arbuscular mycorrhizal Glomeromycetes. *Journal of Applied Botany and Food Quality, 80*, 69–81.

Simões, J. C., Arigony-Neto, J., & Bremer, U. F. (2004). O uso de mapas antárticos em publicações. *Pesq Antart Bras, 4*, 191–197.

Stevenson, F. J. (2005). *Cycles of soil: Carbon, nitrogen, phosphorus, sulfur, micronutrients*. New York: Wiley.

Teixeira, L. C., Peixoto, R. S., Cury, J. C., Sul, W. J., Pellizari, V. H., et al. (2010). Bacterial diversity in rhizosphere soil from Antarctic vascular plants of Admiralty Bay, maritime Antarctica. *The ISME Journal, 4*, 989–1001.

Teixeira, L. C. R. S., Yergeau, E., Balieiro, F. C., Piccolo, M. C., Peixoto, R. S., Rosado, A. S., & Greer, C. W. (2013). Plant and bird presence strongly influences the microbial communities in soils of Admirality Bay. Maritime Antarctica. *PLoS One, 8*, e66109.

Teplitski, M., Warriner, K., Bartz, J., & Schneider, K. R. (2011). Untangling metabolic and communication networks: Interactions of enterics with phytobacteria and their implications in produce safety. *Trends in Microbiology, 19*, 121–127.

Tomova, I., Stoilova-Disheva, M., Lazarkevich, I., & Vasileva-Tonkova, E. (2015). Antimicrobial activity and resistance to heavy metals and antibiotics of heterotrophic bacteria isolated from sediment and soil samples collected from two Antarctic islands. *Frontiers in Life Science, 8*, 348–357.

Upson, R., Newsham, K. K., & Read, D. J. (2008). Root-fungal associations of *Colobanthus quitensis* and *Deschampsia antarctica* in the maritime and subantarctic. *Arctic, Antarctic, and Alpine Research, 40*, 592–599.

Upson, R., Newsham, K. K., Bridge, P. D., Pearce, D. A., & Read, D. J. (2009). Taxonomic affinities of dark septate root endophytes of *Colobanthus quitensis* and *Deschampsia antarctica*, the two native Antarctic vascular plant species. *Fungal Ecology, 2*, 184–196.

Wagg, C., Jansa, J., Schmid, B., & van der Heijden, M. G. A. (2011). Belowground biodiversity effects of plant symbionts support above ground productivity. *Ecology Letters, 14*, 1001–1009.

Xiong, F. S., Ruhland, C. T., & Day, T. A. (1999). Photosynthetic temperature response of the Antarctic vascular plants Colobanthus quitensis and Deschampsia Antarctica. *Physiologia Plantarum, 106*, 276–286.

Yergeau, E. (2014). Fell-field soil microbiology. In D. Cowan (Ed.), *Antarctic terrestrial microbiology*. Berlin: Springer.

Yergeau, E., & Kowalchuk, G. A. (2008). Responses of Antarctic soil microbial communities and associated functions to temperature and freeze-thaw cycle frequency. *Environmental Microbiology, 10*, 2223–2235.

Yergeau, E., Bokhorst, S., Huiskes, A. H., Boschker, H. T., Aerts, R., et al. (2007a). Size and structure of bacterial, fungal and nematode communities along an Antarctic environmental gradient. *FEMS Microbiology Ecology, 59*, 436–451.

Yergeau, E., Newsham, K. K., Pearce, D. A., & Kowalchuk, G. A. (2007b). Patterns of bacterial diversity across a range of Antarctic terrestrial habitats. *Environmental Microbiology, 9*, 2670–2682.

Zamioudis, C., & Pieterse, C. M. J. (2012). Modulation of host immunity by beneficial microbes. *Molecular Plant-Microbe Interactions, 25*, 139–150.

Zhang, C., & Kim, S. K. (2010). Research and application of marine microbial enzymes: Status and prospects. *Marine Drugs, 8*, 1920–1934.

Chapter 13
Microbial Symbionts of Antarctic Marine Benthic Invertebrates

Angelina Lo Giudice, Maurizio Azzaro, and Stefano Schiaparelli

Abstract The microbial colonization of living surfaces may be affected by several environmental and biological factors and may play an important role in the development and evolution of the holobiont. Antarctica, as an extreme and isolated environment, offers a unique opportunity to study the peculiar and often strict interactions that are established between a benthic host and its symbionts. Despite this, to date the association between microbes and Antarctic benthic invertebrates has been only seldom investigated, resulting in fragmented and poor information. This chapter will be devoted to showcase our current knowledge on prokaryotic (Bacteria and Archaea) and eukaryotic (yeasts and diatoms) microbial symbionts of Antarctic benthic invertebrate hosts, including mainly Porifera and, at to a lesser extent, Cnidaria, Echinodermata and Annelida.

Keywords Microbial symbionts · Benthic invertebrates · Microbial diversity · Host-specificity · Microbiome

13.1 Introduction

In polar marine ecosystems, temporal fluctuations in temperature and sea-ice coverage primarily affect both the quantity and timing of resource supply to consumers. Inputs of nutrients from sediment and primary producers dramatically change with the season, but, despite the extremely limiting environmental conditions, this

---

A. Lo Giudice (✉) · M. Azzaro
Institute for the Biological Resources and Marine Biotechnologies, National Research Council, U.O.S. Messina, Messina, Italy
e-mail: angelina.logiudice@cnr.it

S. Schiaparelli
Department of Earth, Environment and Life Sciences, University of Genoa, Genoa, Italy

Italian National Antarctic Museum (MNA, Section of Genoa), University of Genoa, Genoa, Italy

S. Castro-Sowinski (ed.), *The Ecological Role of Micro-organisms in the Antarctic Environment*, Springer Polar Sciences,
https://doi.org/10.1007/978-3-030-02786-5_13

phenomenon promotes stable and biodiverse food webs (Calizza et al. 2018). The biology of polar benthic species becomes mainly limited by food, and organisms concentrate foraging and reproductive activities in summer months, when inputs from pelagic and sympagic algae, whose availability is closely related to sea-ice dynamics, increase. Hence, polar shelf environments and their benthic communities are subject to many disturbances, particularly in the nearshore shallows. Carbon supply and energy sources may also vary in frequency and intensity seasonally, as well as geographically and according to bathymetry. Thus, nearshore ecosystem becomes thus shaped by winds, wave action, temperature and localized anoxia, freshening and pollution, among other influences. However, ice-mediated disturbances are without a doubt the most influencing ones. While the West Antarctic Peninsula is rapidly warming, resulting in sea-ice cover decrease, the sea-ice cover of East Antarctica unexpectedly tends to increase, in both cases influencing coastal benthic food webs. As reviewed by Barnes and Conlan (2007), ice occurs in four main forms (e.g. ice foot, ice scour, anchor ice and fast ice) that, singularly or in combination, influence benthic communities in a very different temporal and spatial manner, also in relation to climate change (Clark et al. 2017). Ice develops around all shores during winter, and then it plays an abrasion action of the sediments (thus disturbing the inhabiting communities) in spring and summer, forming anchor ice on the seafloor. Even if Antarctic benthic biota may experience periodic disruption by ice scour, benthic communities appear to be relatively stable, being structured mainly by predatory and competitive interactions (Dayton et al. 1969). However, in recent years, a variety of biotic interactions has been evidenced, changing the historical view of Antarctic benthic communities dominated by predation and competition (Schiaparelli 2014).

The benthic environment of the Antarctic continental shelf remains the coldest and most stable marine habitat on Earth, where many endemic macroinvertebrates have evolved strictly cold-adapted genotypes since the mid-Cenozoic (Abele et al. 2017). The Antarctic marine benthic fauna is thus considered to be the most isolated on the planet, having a predominantly circumpolar distribution, probably related to the Antarctic Circumpolar Current (Clarke and Johnston 2003), and most invertebrate species are highly endemic of the Southern Ocean. Antarctic benthic communities are typically composed of a large mixture of sessile suspension feeders and mobile macroinvertebrates (Fig. 13.1). Sessile fauna include mainly Porifera, which may form the so-called sponge fields, but also Anthozoa, with subclasses Hexacorallia (Actiniaria) and Octocorallia (Alcyonacea, Pennatulacea), followed by Hydrozoa, Bryozoa, Brachiopoda, Polychaeta and Ascidiacea (Gili et al. 2006; Clark et al. 2004, 2015). The occurrence of all the above taxa may be patchy and is generally characterized by a significant spatial heterogeneity, generated by variations in ice cover, sediment dynamics, and hydrodynamic and trophic factors acting at all spatial scales (Clarke and Crame 1989). Mobile invertebrates, co-occurring with sessile fauna or dominating in some areas, are mainly represented by Echinodermata and Peracarida and to a lesser extent by Pycnogonida (Fig. 13.2), Ostracoda, Caridea, Bivalvia (Fig. 13.3) and Nemertea (Fig. 13.4). Among them, benthic filter-feeding organisms are of particular interest as they are able to filter

Fig. 13.1 A typical assemblage of macroinvertebrates found in Terra Nova Bay (Adelie Cove) dominated by suspension feeders

large volumes of waters to collect edible particles, including microbes, suspended in the water column.

Antarctic marine benthic invertebrates represent a particularly attractive model for the study of microbial symbiosis as they inhabit a geographically isolated ecosystem and must cope with very harsh environmental conditions (i.e. near-freezing temperatures and cyclical sea-ice formation). According to the recently proposed "hologenome concept" (Rosenber and Zilber-Rosenberg 2016), the holobiont and the hologenome (that includes also the genetic information of the major microbial symbionts) acts as a unique biological entity, playing a fundamental role in the adaptation and evolution of the holobiont itself. In fact, the holobiont could maintain its unique features (in terms of morphology, development, behaviour, physiology and resistance to diseases), thanks to the transmission between generations of the host genome and the associated microbiome. Changes in either genome can result in variations that can be selected for or against (Rosenber and Zilber-Rosenberg 2018).

The association between marine invertebrates and microorganisms, both prokaryotic (i.e. Bacteria and Archaea) and eukaryotic (i.e. yeasts and diatoms), in the Antarctic environment, has been rarely investigated, and, to date, it is limited to a few works on Porifera and sporadic reports on some species within the Cnidaria, Echinodermata and Oligochaeta. This chapter will be aimed at reviewing our current knowledge on microbes associated with Antarctic benthic invertebrates as a yet underexplored source of microbial diversity.

Fig. 13.2 An unidentified pycnogonid photographed at 25 m in Thetys Bay (Terra Nova Bay)

Fig. 13.3 The Antarctic scallop *Adamussium colbecki* (E. A. Smith, 1902) forms large beds in Terra Nova Bay (~25 m of depth)

Fig. 13.4 *Parborlasia corrugatus* (McIntosh, 1876) is a very large nemertean, which may attain more than 1 m in length. This specimen was photographed at 25 m in Thetys Bay (Terra Nova Bay)

13.2 Prokaryotes Associated with Antarctic Benthic Invertebrates

As described in the following sections, available data on the association between Antarctic benthic hosts (listed in Table 13.1) and prokaryotic symbionts derive from the analysis of a number of sponge species, the soft coral *Alcyonium antarcticum*, the sea urchin *Sterechinus neumayeri* and the oligochaete *Grania* sp.

*13.2.1 Antarctic Sponges (*Phylum *Porifera)*

Sponges are among the most abundant macroinvertebrates within the Antarctic benthic communities, playing a key role in community dynamics. They serve as food sources for several predators (e.g. echinoderms, nemerteans and gastropods; Cerrano et al. 2000a; Schiaparelli et al. 2003), as well as substrates for epibionts and endobionts (e.g. isopods, amphipods, gastropods, bivalves and polychaetes), colonizing outer surfaces and interstices of ostia and oscula (McClintock et al. 2005). Among marine microorganisms, Bacteria, Archaea and benthic diatoms (for these latter, see Sect 13.3) have been frequently detected on outer surfaces (e.g. Hamilton et al. 1997; Amsler et al. 2000; Cerrano et al. 2000b; Papaleo et al. 2012; Mangano et al. 2014; Rodríguez-Marconi et al. 2015). Sponges from different oceans generally have a

Table 13.1 Antarctic benthic invertebrate hosts analysed for their associated bacterial community

Host phylum	Host species	Sampling site(s)	Reference(s)
Annelida	*Grania* sp.	Maxwell Bay[a]	Herrera et al. (2017)
Cnidaria	*Alcyonium antarcticum* Wright & Studer, 1889	McMurdo Sound[b]	Webster and Bourne (2007)
Echinodermata	*Sterechinus neumayeri* (Meissner, 1900)	Maxwell Bay[a]	González-Aravena et al. (2016)
Porifera	*Clathria* sp.	Fildes Bay[a]	Rodríguez-Marconi et al. (2015)
	Haliclona sp.	Fildes Bay[a]	Rodríguez-Marconi et al. (2015)
	Haliclonissa verrucosa Burton, 1932	Terra Nova Bay[b]	Papaleo et al. (2012)
	Hemigellius pilosus (Kirkpatrick, 1907)	Terra Nova Bay[b]	Mangano et al. (2014)
	Homaxinella balfourensis (Ridley & Dendy, 1886)	McMurdo Sound[b]; Weddell Sea	Webster et al. (2004) and Xin et al. (2011)
	Hymeniacidon torquata Topsent, 1916	Fildes Bay[a]	Rodríguez-Marconi et al. (2015)
	Kirkpatrickia variolosa (Kirkpatrick, 1907)	McMurdo Sound[b]; Fildes Bay[a]	Webster et al. (2004) and Rodríguez-Marconi et al. (2015)
	Latrunculia (*Latrunculia*) *apicalis* Ridley & Dendy, 1886	McMurdo Sound[b]	Rodríguez-Marconi et al. (2015)
	Leucetta antarctica Dendy, 1918	Fildes Bay[a]	Rodríguez-Marconi et al. (2015)
	Lissodendoryx (*Ectyodoryx*) *nobilis* (Ridley & Dendy, 1886)	Terra Nova Bay[b]	Papaleo et al. (2012) and Mangano et al. (2009, 2018)
	Megaciella annectens (Ridley & Dendy, 1886)	Fildes Bay[a]	Rodríguez-Marconi et al. (2015)
	Mycale (*Oxymycale*) *acerata* Kirkpatrick, 1907	McMurdo Sound[b]	Rodríguez-Marconi et al. (2015)
	Myxilla (*Myxilla*) *mollis* Ridley & Dendy, 1886	Weddell Sea	Xin et al. (2011)
	Myxilla sp.	Fildes Bay[a]	Rodríguez-Marconi et al. (2015)
	Myxodoryx hanitschi (Kirkpatrick, 1907)	Terra Nova Bay[b]	Mangano et al. (2018)
	Phorbas glaberrimus (Topsent, 1917)	Terra Nova Bay[b]	Mangano et al. (2018)
	Radiella antarctica Plotkin & Janussen, 2008	Weddell Sea	Xin et al. (2011)
	Sphaerotylus antarcticus Kirkpatrick, 1907	McMurdo Sound[b]	Rodríguez-Marconi et al. (2015)
	Anoxycalyx (*Scolymastra*) *joubini* (Topsent, 1916)	Terra Nova Bay[b]	Mangano et al. (2009) and Papaleo et al. (2012)
	Rossella nuda Topsent, 1901	Weddell Sea	Xin et al. (2011)
	Rossella racovitzae Topsent, 1901	Weddell Sea	Xin et al. (2011)

[a]King George Island, South Shetlands
[b]Ross Sea

microbial phylogenetic signature distinctly different from that of marine plankton and sediments and often species-specific (Hentschel et al. 2002). The host can be directly involved in the selection of symbiotic bacteria by the production of bioactive metabolites (Soldatou and Baker 2017). Hence, it is plausible to assume that the host maintains stable the associated microbial community to gain benefits from such interaction. At this regard, a number of studies have highlighted that sponge-associated microbial communities might be sponge-specific and different from that occurring in the bulk environment, also in polar regions (Webster et al. 2004; Rodríguez-Marconi et al. 2015). Furthermore, Mangano et al. (2009, 2018) suggested that the interpopulation interactions occurring among bacterial populations inhabiting the same/different host species, as well as the production of N-acyl homoserine lactones involved in the quorum sensing (intrapopulation interactions), might play a key role in profiling the bacterial community associated with Antarctic sponges.

Even though sponges dominate vast areas of the Antarctic shelves, information on sponge-associated prokaryotic communities remains quite scarce and fragmentary. In a pioneer study, Webster et al. (2004) analysed the prokaryotic community (i.e. Bacteria and Archaea) associated with five Antarctic sponge species (i.e. *Kirkpatrickia variolosa*, *Latrunculia apicalis*, *Homaxinella balfourensis*, *Mycale acerata* and *Sphaerotylus antarcticus*) (Fig. 13.5) at McMurdo Sound, Ross Sea, Scott Base (NZ), McMurdo Station intake jetty (US), and Cape Armitage, which lies approximately halfway between the two bases. The phylogenetic affiliation of sponge-derived Bacteria and Archaea was assessed by 16S rRNA sequencing of cloned DNA fragments, and DGGE was used to determine the stability of the prokaryotic associations within each sponge species and across spatial scales. The bacterial communities primarily clustered within the *Gamma*- (e.g. genera *Vibrio* and *Alteromonas*) and *Alphaproteobacteria* (mainly *Roseobacter* spp.) and the CF group of *Bacteroidetes* (mainly *Polaribacter* spp.). Many of the sponge-derived bacterial sequences were closely related to sequences previously retrieved from Antarctic seawater and sea ice rather than from microorganisms associated with sponge from temperate and tropical areas. This finding suggested that many of the Antarctic sponge-derived bacteria were either not specialized symbionts or that the diversity of symbionts was restricted to species that can cope with −2 °C and these are related to other bacteria that inhabit this extreme environment (Webster et al. 2004). Bacterial DGGE (Fig. 13.6) analysis from sponge and seawater samples, at each site, revealed that the composition of the bacterial communities was common to sponges belonging to the same species, regardless of the collection site. Archaeal sequences were retrieved in three of the five Antarctic sponges analysed (i.e. *M. acerata*, *L. apicalis* and *K. variolosa*). They clustered closely together within the Crenarchaeota and were phylogenetically similar to an uncultured archaeon from the tropical sponge *Rhopaloeides odorabile*. The detection of previously undescribed archaeal strains in Antarctic sponges by Webster et al. (2004) first extended their distribution in marine systems to cold water environments.

More recently, Rodríguez-Marconi et al. (2015) provided new insights into the characterization of whole microbial communities (by considering Bacteria, Archaea and Eukarya) in Antarctic sponges by analysing eight different sponge species (e.g.

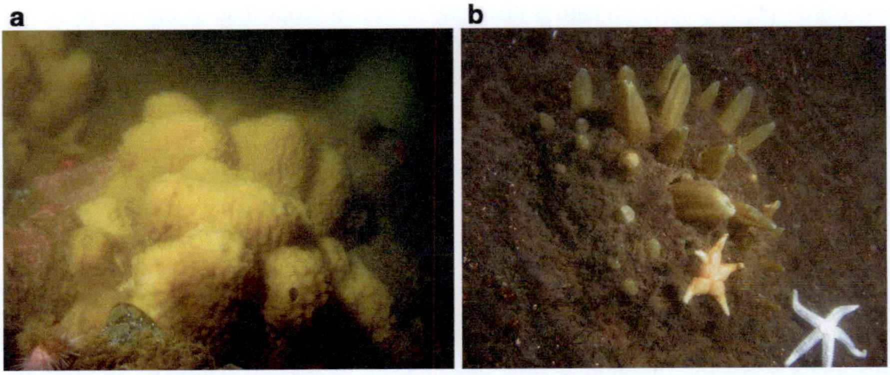

Fig. 13.5 The Porifera *Mycale* (*Oxymycale*) *acerata* Kirkpatrick, 1907 (**a**) and *Sphaerotylus antarcticus* Kirkpatrick, 1907 (**b**) are among the commonest species found on hard bottom in Thetys Bay

Fig. 13.6 Thetys Bay, in Terra Nova Bay (Ross Sea), is characterized by extreme seasonality in sea-ice coverage and productivity in the Ross Sea. The benthic community is rich and diverse in this nearshore area

Myxilla sp., *Clathria* sp., *Kirkpatrickia variolosa*, *Haliclona* sp., *Megaciella annectens*, *Hymeniacidon torquata* and *Leucetta antarctica*) collected at Fildes Bay (King George Island, South Shetlands), using high-throughput sequencing technologies of ribosomal genes. Overall, the associated bacterial community was dominated by *Proteobacteria*, followed by *Bacteroidetes*, *Verrucomicrobia* and *Planctomycetes*, confirming previous observation by Webster et al. (2004). However, Rodríguez-Marconi et al. (2015) detected a few taxa present at minor proportions, increasing to 25 the number of phyla detected in Antarctic sponges. Interestingly, sponges and seawater community did not share most phylotypes, suggesting that sponges could be highly host specific and that they could represent a diverse reservoir of microorganisms in the Antarctic marine ecosystem. Microbial communities associated with Antarctic sponges displayed also a higher diversity rather than their counterparts from the surrounding seawater. Finally, by comparing at phylum level the bacterial 16S rRNA gene inventories from polar, temperate and tropical ecosystems, Rodríguez-Marconi et al. (2015) observed a number of differences, suggesting a particular signature for Antarctic sponges. Interestingly, as previously reported by Webster et al. (2004), Rodríguez-Marconi et al. (2015) retrieved also bacterial ammonia oxidizers *Nitrosomonadales* and nitrifiers *Methylophilales*, as well as anaerobic ammonia-oxidizing Archaea (e.g. Nitrosopumilales among Thaumarchaeota), reinforcing the idea that prokaryotes associated with Antarctic sponges could make an important contribution to the marine nitrogen cycling. In particular, the sponges *L. antarctica* and *M. annectens* presented relatively high abundances of sequences affiliated to the Nitrosopumilales, supporting that Thaumarchaeota is the major within the Archaea domain in sponges worldwide.

Further studies on prokaryotic symbionts of Antarctic marine sponges focused on cultivable bacteria. Papaleo et al. (2012) and Mangano et al. (2009, 2014) analysed the bacterial communities associated with sponges from the Thetys Bay (Terra Nova Bay, Ross Sea) (Fig. 13.7). The phylogenetic analysis, performed by the 16S rRNA gene sequencing, affiliated 140 bacterial strains from *Anoxycalyx joubini*, *Lissodendoryx nobilis* and *Haliclonissa verrucosa* to 15 genera, mainly within the

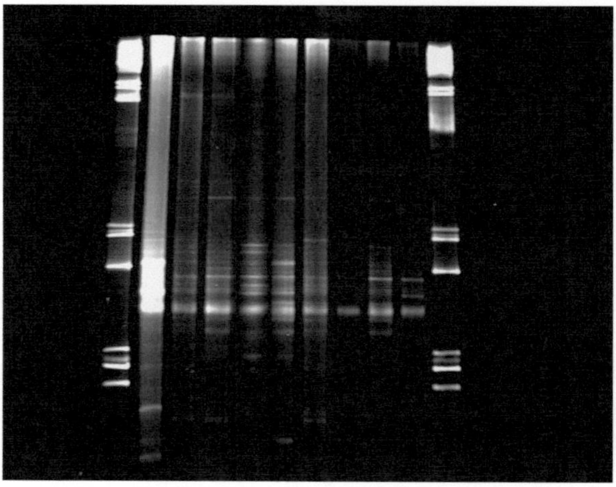

Fig. 13.7 Example of denaturing gradient gel electrophoresis (DGGE) patterns of bacterial communities

Gammaproteobacteria and *Actinobacteria*, followed by *Alphaproteobacteria*, CF group of *Bacteroidetes* and *Firmicutes* (Papaleo et al. 2012). Interestingly, the three sponges shared only members from three genera (*Psychrobacter*, *Pseudoalteromonas* and *Arthrobacter*). The genera *Marinobacter*, *Colwellia*, *Rhodococcus*, *Gillisia*, *Staphylococcus* and *Oceanobacillus* were isolated only from *H. verrucosa*, whereas *Sphingopyxis*, *Octadecabacter* and *Pseudomonas* were obtained from *A. joubini*. Among the *Alphaproteobacteria*, *Sulfitobacter* and *Octadecabacter* were shared by *H. verrucosa* and *A. joubini*.

Working with the sponge *Hemigellius pilosus*, Mangano et al. (2014) and Papaleo et al. (2012), based on 16S rRNA gene sequencing, identified members mainly from *Gammaproteobacteria* and *Actinobacteria*, followed by few members affiliated to *Alphaproteobacteria* and CFB group of *Bacteroidetes*. *Gammaproteobacteria* clustered in two main groups: (i) two subclusters formed by *Pseudoalteromonas* and *Colwellia* affiliates and (ii) isolates that were strongly related to the genera *Psychrobacter* and *Shewanella*. The next abundant group, *Actinobacteria*, clustered in two groups: (i) isolates strongly related to the genera *Arthrobacter* and *Citricoccus* and (ii) isolates branched in two groups formed by *Leifsonia* and *Microbacterium*. *Bacteroidetes* was represented by a single strain in the genus *Bizionia*. Only one isolate was affiliated to *Alphaproteobacteria*, showing a high degree of sequence identity with a bacterium previously isolated from the Antarctic sponge *A. joubini* (Antarctic bacterium TB59; Mangano et al. 2009).

Xin et al. (2011) focused their research on the isolation and characterization of Gram-positive bacteria from deep-sea (up to 4790 m) Antarctic sponge species (i.e. *Homaxinella balfourensis*, *Rossella nuda*, *Rossella racovitzae*, *Myxilla mollis*, *Radiella antarctica*) collected in the Weddell Sea (Antarctica). Interestingly, among *Actinobacteria* (24 strains), the genera *Dietzia* and *Brevibacterium* were first reported in association with marine sponges worldwide. Furthermore, the genus *Streptomyces* predominated and was shared among sponge species. Overall, Gram-positive bacteria seemed to be rather unspecialized with respect to their sponge host.

13.2.2 *The Soft-Coral Alcyonium Antarcticum* (Phylum *Cnidaria)*

Octocorallia (Alcyonaria) are one of the major components of benthic Antarctic communities in terms of both abundance and diversity. In particular, *Alcyonium antarcticum* (Fig. 13.8) is widely distributed in Antarctic and sub-Antarctic areas in depths varying between 25 and 642 m (Casas et al. 1997). Octocoral colonies are composed of polyps connected by a fleshy tissue. The lack of calcium carbonate solid skeletons is compensated by the presence of internal minute spiky sclerites, rendering shape and support (Núñez-Pons et al. 2013). Specialist consumers such as pycnogonids attack them.

Fig. 13.8 A close-up view of *Alcyonium antarcticum* Wright & Studer, 1889 (Thetys Bay, 25 m depth)

The research in coral microbiology has been frequent in tropical and temperate climates. However, the information regarding the association between Antarctic corals and prokaryotes is scarce (Webster and Bourne 2007). First investigations were performed by DGGE, culture-based analysis, 16S rRNA gene clone-library analysis and FISH, analysing the microbial communities associated with *A. antarcticum* from different sites at McMurdo Sound. Most detected microorganisms belonged to *Gammaproteobacteria*, although other phylogenetic groups including *Alpha*- and *Betaproteobacteria*, *Bacteroidetes*, *Firmicutes*, *Actinomycetales*, *Planctomycetes* and *Chlorobi* and bacteria from the functional group of sulphate-reducing bacteria were also present. Multivariate (nMDS) analysis of DGGE banding patterns and principal component analysis of quantitative FISH data revealed no differences in the community composition between the sampling sites, providing evidence to support specific coral-microbial interactions. Following a culture-dependent approach, ten isolates were obtained and identified as *Gammaproteobacteria* (seven isolates in the genera *Pseudomonas*, *Psychrobacter*, *Shewanella*), *Bacteroidetes* (two isolates in the genera *Psychroserpens* and *Algoriphagus*), and *Actinobacteria* (a single isolate in the genus *Agrococcus/Coryn ebacterium*), with high 16S rRNA gene sequence identity to previously reported psychrophiles or sequences detected from the Antarctic environment. Interestingly, a stable microbial community was shown to exist across replicate coral samples within a site and between sites.

13.2.3 The Sea Urchin Sterechinus neumayeri *(Phylum Echinodermata)*

The Antarctic sea urchin *Sterechinus neumayeri* (Fig. 13.9) is commonly distributed around the Antarctic continent, from the shallow subtidal zone down to 500 m depth, and plays a key role in the ecosystem structure. Due to its abundance and its circum-Antarctic distribution, *S. neumayeri* represents a suitable model for studies in reproductive biology, embryology, ecology, physiology and toxicology (Diaz et al. 2018). Only recently, González-Aravena et al. (2016) reported on the isolation of 42 bacterial strains from the Antarctic sea urchin *Sterechinus neumayeri*. Consistently with studies of several marine samples from Antarctica, they belonged to *Gammaproteobacteria* (genera *Pseudoalteromonas*, *Psychrobacter*, *Shewanella* and *Pseudomonas*), *Bacteroidetes* (class *Flavobacteria*) and *Actinobacteria*. The authors suggested their role as reservoir of resistance genes for antibiotic and heavy metal resistance.

13.2.4 The Marine Worm Grania *sp. (Phylum Annelida)*

Annelids represent an important component of the Southern Ocean benthic communities in both deep and nearshore areas. These organisms show a wide bathymetric distribution and a large-scale biogeography. Within the class Oligochaeta, Herrera et al. (2017) analysed the cultivable fraction of the bacterial community

Fig. 13.9 The regular sea urchin *Sterechinus neumayeri* (Meissner, 1900) may use algae and debris as camouflage in order to avoid detection by predators (Thetys Bay, 15 m depth)

associated with the gut of *Grania* sp. from Artigas Beach at Maxwell Bay (King George Island, South Shetland Islands, Maritime Antarctica). *Flavobacterium*, *Pseudomonas*, *Psychrobacter* and *Salinibacterium* were the main bacterial genera retrieved, in addition to an *Enterobacteriaceae* isolate. Results from substrate (e.g. cellulose, starch, proteins and triacylglycerols) hydrolysis tests suggested that associated bacteria may be strictly involved in the degradation of algae, which constitute the oligochaete main food source, thus contributing to the nutrient recycling in the Antarctic ecosystem.

13.3 Eukaryotes Associated with Antarctic Benthic Invertebrates

To the best of our knowledge, the association between Antarctic benthic invertebrates and microeukaryotes (both autotrophic and heterotrophic) has been investigated mainly for sponges. Few reports exist on the importance of benthic diatoms (and other small planktonic organisms, such as protozoans) in the nutrition, for example, of hydroids (Gili et al. 1996), brittle sea stars (Kellogg et al. 1982) and ascidians (Tatián et al. 2004).

Eukaryote communities within the sponge tissues appear to be, in general, less rich and diverse than bacterial/archaeal communities but more diverse than their planktonic counterparts (Rodríguez-Marconi et al. 2015). In this chapter, main findings on diatoms and yeasts associated with Antarctic benthic invertebrates will be reviewed, even if additional microeukaryotes (such as filamentous fungi, dinoflagellates and protozoans) have been seldom reported (Webster et al. 2004; Rodríguez-Marconi et al. 2015).

13.3.1 Diatoms

The presence of diatom species in Antarctic sponges has been mainly related to the filtration of phytoplankton cells from the water column. However, species-specific relationships between diatoms and Antarctic sponges were sometimes reported, as it was observed for *Melosira* sp. (class Bacillariophyceae) in *Anoxycalyx joubini* (Cerrano et al. 2000b), *Porannulus contentus* (class Bacillariophyceae) in *Mycale acerata* (Hamilton et al. 1997; Cerrano et al. 2004b) and *Hyalodiscus* sp. (subclass Coscinodiscophyceae), living epibionthically on the spicule fur of *Sphaerotylus antarcticus* (Totti et al. 2005).

Antarctic sponges host large populations of planktonic and benthic diatoms (Hamilton et al. 1997; Amsler et al. 2000; Bavestrello et al. 2000; Cerrano et al. 2000b), which play different ecophysiological roles and establish a complex pattern of relationships with their host, ranging from mutualism to parasitism in the harsh

Antarctic environment (Cerrano et al. 2004a). Diatoms are primarily a food source for sponges (Ahn et al. 2016). Gaino et al. (1994), by using a scanning electron microscope (SEM), observed a consistent presence of benthic diatoms underneath the exopinacoderm of the Antarctic sponges *Phorbas glaberrimus* and *Tedania charcoti* (Terra Nova Bay, Ross Sea), in addition to diatoms entering the sponge via the traditional inhalant water current system. The authors suggested that diatoms might be strictly involved in physiological and adaptive modifications of the sponges. Their accumulation in the outermost surface of the sponge could be responsible for strengthening the cortex, while their storage in the mesohyl matrix could represent an important alimentary source to be utilized by the sponge during the oligotrophic months. In fact, diatoms produce extracellular polysaccharides that could be used by sponges as a supplemental energy substrate (Bavestrello et al. 2000). Data obtained by Cerrano et al. (2004a) clearly indicated that diatom concentration in tissues of six species of Antarctic sponges (i.e. *Dendrilla antarctica*, *Homaxinella flagelliformis*, *Kirkpatrickia variolosa*, *Suberites montiniger*, *Haliclona dancoi*, *Haliclona penicillata*) was a seasonal phenomenon, with the massive presence of diatoms (mainly the planktonic species *Fragilariopsis curta*) related to the algal bloom in the water column in summer, thus confirming their role as a food source. The lack of frustules at the beginning of summer was a proof that diatom frustules were probably expelled through the excurrent canals or dissolved inside the sponge cells during the cold season. This was also the case of *Mycale acerata* that incorporated diatoms (mainly *Porannulus contentus*, *Fragilariopsis curta*, *Thalassiosira* cf. *gracilis*, *T. perpusilla* and *Plagiotropis* sp.) from the water column and utilized them as a food source, accumulating frustules inside the choanosome (Cerrano et al. 2004b).

On the other hand, the establishment of mono- or pauci-specific assemblages (characterized by integral diatom shells and by the presence of cytoplasm inside the shells) inside a single sponge has been reported (Bavestrello et al. 2000; Cerrano et al. 2000b), suggesting that diatoms are able to complete their entire biological cycle inside the sponge body. For example, Cerrano et al. (2000b) reported on healthy diatoms belonging to the genus *Melosira* embedded in the tissues of the hexactinellid sponge *Anoxycalyx joubini*. The SEM analysis showed that those zones where the dermal layer of the sponge was completely destroyed were replaced by a dense assemblage of this diatom species, supporting the hypothesis that the embedded living diatoms may have a negative impact on sponge tissues, leading to degenerative processes. Thus, diatoms could act as real parasites (eventually resulting in tissue degeneration) which probably utilize the products of the metabolism of their host as an energy source. Such unusual symbiotic behaviour shifting to parasitism could be a successful strategy that allows the mixotrophic diatoms to survive in Antarctic winter darkness when light is insufficient (Bavestrello et al. 2000).

In a recent study, Rodríguez-Marconi et al. (2015) adopted advanced biomolecular techniques to describe the microbial communities within Antarctic sponges. The presence of a highly complex associated microbial eukaryote community was observed. Bacillariophyta, Syndiniales and Dinophyceae were major contributors to Antarctic sponge-associated microbial eukaryotes, with evident differences in their composition between distinct sponge species. In particular, diatom-assigned sequences were similar to known Antarctic taxa, such as *Thalassiosira antarctica* (also retrieved by Webster et al. 2004) and *Porosira glacialis* (phylum Bacillariophyta). Interestingly, the unique Bacillariophyta phylotype dominating the planktonic community was different from the most abundant one in sponges, suggesting once again a sponge-specific association.

13.3.2 Yeasts

A number of yeast species have been found to successfully colonize the Antarctic continent. Among them, some are indigenous to the Antarctic and obligate psychrophiles, whereas others reach Antarctica by the wind and ocean currents as well as by migratory birds, humans and other animals who occasionally visit this habitat (Duarte et al. 2013). Yeasts from cold environments may have a potential role as nutrient recyclers, hydrolysing natural compounds by the secretion of hydrolytic enzymes (Buzzini et al. 2012; Duarte et al. 2013). In general, the association between yeasts and marine benthic invertebrates has been overlooked worldwide, especially in Antarctic seas. A list of yeasts isolated from Antarctic invertebrates is shown in Table 13.2. Vaca et al. (2013) reported the first description of cultivable psychrotolerant yeasts associated with 11 Antarctic marine sponge species from Fildes Bay (King George Island, South Shetland Islands). In total, 20 yeast isolates were obtained, mainly from the sponge genera *Tedania* and *Hymeniacidon*. All ascomycetous yeasts (13 isolates) were identified as *Metschnikowia australis*, which seems to be ubiquitous in the Antarctic marine environments (it was also retrieved in sediment and seawater and in association with macroalgae) (Vaca et al. 2013). Among *Basidiomycota* (seven isolates), three main species were observed, i.e. *Cystofilobasidium infirmominiatum* (four isolates), *Rhodotorula pinicola* (one isolate) and *Leucosporidiella creatinivora* (one isolate), with these two latter that were first reported for the Antarctic environment. More interestingly, none of the basidiomycetous yeasts reported by Vaca et al. (2013) were previously associated with marine sponges. An additional strain belonged to a new species within the *Leucosporidiella* genus and was then further taxonomically characterized (Laich et al. 2014).

Concomitantly to Vaca et al. (2013), Duarte et al. (2013) explored the diversity of yeasts recovered from several kinds of marine matrices, both abiotic and biotic (including sea stars, algae, sea urchin, sea squirt, sponges, sea isopod and sea snail), from Admiralty Bay (King George Island, South Shetland Islands). With respect to benthic organisms, associated yeasts were mainly *Basidiomycota* in the genera *Rhodotorula* (i.e. *R. laryngis* and *R. mucilaginosa*), and *Cryptococcus* (i.e. *C. albidosimilis*, *C. adeliensis* and *C. laurentii*), followed by affiliates to the species *Bullera pseudoalba*, *Cystofilobasidium infirmominiatum* and *Guehomyces pullulans*. Additionally, few members of *Ascomycota* were also retrieved being identified as *Debaryomyces hansenii*, *Meyerozyma guilliermondii* and *Wickerhamomyces anomalus*. The predominance of *Rhodotorula* spp. and *Cryptococcus* spp. in cold

Table 13.2 Cultivable yeasts associated with Antarctic benthic invertebrate hosts from Admiralty Bay and Fildes Bay in the King George Island (South Shetlands)

Yeast division	Host phylum	Host species	Sampling site	Yeast identity	Reference
Ascomycota	Arthropoda	Unidentified isopod	Admiralty Bay	*Meyerozyma guilliermondii*	Duarte et al. (2013)
	Echinodermata	Unidentified sea star	Admiralty Bay	*Meyerozyma guilliermondii*	Duarte et al. (2013)
	Mollusca	*Nacella concinna* (Strebel, 1908)	Admiralty Bay	*Wickerhamomyces anomalus*	Duarte et al. (2013)
		Unidentified sea snail	Admiralty Bay	*Meyerozyma guilliermondii*	Duarte et al. (2013)
	Porifera	*Dendrilla* sp.	Fildes Bay	*Metschnikowia australis*	Vaca et al. (2013)
		Hymeniacidon spp.	Fildes Bay	*Metschnikowia australis*	Vaca et al. (2013)
		Hymeniacidon spp.	Fildes Bay	*Metschnikowia australis*	Vaca et al. (2013)
		Tedania spp.	Fildes Bay	*Metschnikowia australis*	Vaca et al. (2013)
		Unidentified sponge	Admiralty Bay	*Debaryomyces hansenii*	Duarte et al. (2013)
Basidiomycota	Echinodermata	Unidentified sea star	Admiralty Bay	*Cryptococcus adeliensis*, *C. albidosimilis*, *Cystofilobasidium infirmominiatum*, *Guehomyces pullulans*	Duarte et al. (2013)
	Mollusca	*Nacella concinna*	Admiralty Bay	*Cryptococcus laurentii*, *Rhodotorula mucilaginosa*	Duarte et al. (2013)
		Unidentified sea urchin	Admiralty Bay	*Cryptococcus laurentii*, *Rhodotorula larynges*, *R. mucilaginosa*	Duarte et al. (2013)
	Porifera	*Hymeniacidon* spp.	Fildes Bay	*Cystofilobasidium infirmominiatum*, *Leucosporidiella* sp., *Rhodotorula pinicola*	Vaca et al. (2013)
		Tedania spp.	Fildes Bay	*Cystofilobasidium infirmominiatum*, *Leucosporidiella creatinivora*	Vaca et al. (2013)
		Unidentified sponge	Admiralty Bay	*Bullera pseudoalba*, *Cryptococcus laurentii*, *Rhodotorula mucilaginosa*	Duarte et al. (2013)

environments could derive from their ability to produce polysaccharides and utilize available nutrients in oligotrophic systems (needing the minimal nutritional requirements) and, as for prokaryotes, the increase in the proportion of unsaturated fatty acids compared to the saturated fatty acids in the plasmatic membrane (Margesin and Miteva 2011).

13.4 Concluding Remarks

The study of the association between Antarctic benthic invertebrates and their microbial symbionts is still at its infancy, and it is evident that Antarctic holobionts have not been given the attention they deserve as a source of still undescribed microbial biodiversity. If we relate the very low number of host phyla that have been considered to date for microbial association to the wide benthic diversity in Antarctic nearshore areas, current data on microbial symbiosis appear certainly insufficient. Main results have been obtained by microscopic observations (i.e. for diatoms) and cultivation approaches (i.e. for bacteria and yeasts), while biomolecular techniques to investigate more in depth the whole microbial communities (including Archaea) have been only rarely applied. As it was observed for temperate and tropical climates, also in the case of Antarctic Porifera, as the most frequently considered host phylum, the associated communities generally differ from those in the water column, thus displaying host specificity and suggesting a positive interaction between the holobionts and their microbial symbionts.

In compliance with the Protocol on Environmental Protection to the Antarctic Treaty, future investigations should be addressed to enlarge the plethora of invertebrate hosts to be analysed for associated microbes. This will allow to individuate differences/similarities between the microbial communities associated with hosts belonging to different phyla but living in the same area and to understand if the composition of the microbial communities is mainly driven by the bulk environment or strictly linked to the hosts-microbes concomitant evolution. At this regard, it should be fundamental also investigating on the mechanisms beyond the interactions and the eventual selection of symbionts by the host.

Acknowledgement This chapter was supported by grants from the Italian Programma Nazionale di Ricerche in Antartide, Project PNRA16_00020, "Antarctic Porifera: Hot-spots for prokaryotic diversity and biotechnological potentialities" (PEA Code 2016/AZ1.08). All images but Fig. 13.1 have been made by Stefano Schiaparelli during different dives in Tethys Bay (Terra Nova Bay) (© PNRA). The PNRA project GEOSMART (2013/AZ2.06, PI Paolo Montagna) is acknowledged for the ROV image of Fig. 13.1, taken by Simonepietro Canese in Adelie Cove (Terra Nova Bay).

References

Abele, D., Vazquez, S., Buma, A. G. J., Hernandez, E., Quiroga, C., Held, C., Frickenhaus, S., Harms, L., Lopez, J. L., Helmke, E., & Mac Cormack, W. P. (2017). Pelagic and benthic communities of the Antarctic ecosystem of Potter Cove: Genomics and ecological implications. *Marine Genomics, 33*, 1–11.

Ahn, I. Y., Moon, H. W., Jeon, M., & Kang, S.-H. (2016). First record of massive blooming of benthic diatoms and their association with megabenthic filter feeders on the shallow seafloor of an Antarctic fjord: Does glacier melting fuel the bloom? *Ocean Science Journal, 51*, 273–279.

Amsler, C. D., Moeller, C. B., McClintock, J. B., Iken, K. B., & Baker, B. J. (2000). Chemical defenses against diatom fouling in Antarctic marine sponges. *Biofouling, 16*, 29–45.

Barnes, D. K. A., & Conlan, K. E. (2007). Disturbance, colonization and development of Antarctic benthic communities. *Philosophical Transactions of the Royal Society B, 362*, 11–38.

Bavestrello, G., Cerrano, C., Cattaneo-Vietti, R., Gaino, E., Penna, A., & Sarà, M. (2000). Parasitic diatoms inside Antarctic sponges. *The Biological Bulletin, 198*, 29–33.

Buzzini, P., Branda, E., Goretti, M., & Turchetti, B. (2012). Psychrophilic yeasts from worldwide glacial habitats: Diversity, adaptation strategies and biotechnological potential. *FEMS Microbiology Ecology, 82*, 217–241.

Calizza, E., Careddu, G., Sporta Caputi, S., Rossi, L., & Costantini, M. L. (2018). Time- and depth-wise trophic niche shifts in Antarctic benthos. *PLoS One, 13*(3), e0194796.

Casas, C., Ramil, F., & van Ofwegen, L. P. (1997). Octocorallia (Cnidaria: Anthozoa) from the Scotia Arc, South Atlantic Ocean. I. The genus *Alcyonium* Linnaeus, 1758. *Zoologische Mededelingen (Leiden), 71*, 299–311.

Cerrano, C., Bavestrello, G., Calcinai, B., Cattaneo-Vietti, R., & Sarà, A. (2000a). Asteroids eating sponges from Tethys Bay, East Antarctica. *Antarctic Science, 12*, 425–426.

Cerrano, C., Arillo, A., Bavestrello, G., Calcinai, A., Cattaneo-Vietti, R., Cancinai, B., Cattaneo-Vietti, R., Penna, A., Sarà, M., & Totti, C. (2000b). Diatom invasion in the Antarctic hexactinellid sponge *Scolymastra joubini*. *Polar Biology, 23*, 441–444.

Cerrano, C., Calcinai, B., Cucchiari, E., Di Camillo, C., Nigro, M., Regoli, F., Sarà, A., Schiaparelli, S., Totti, C., & Bavestrello, G. (2004a). Are diatoms a food source for Antarctic sponges? *Chemistry and Ecology, 20*, 57–64.

Cerrano, C., Calcinai, B., Cucchiari, E., Di Camillo, C., Totti, C., & Bavestrello, G. (2004b). The diversity of relationships between Antarctic sponges and diatoms: The case of *Mycale acerata* Kirkpatrick, 1907 (Porifera, Demospongiae). *Polar Biology, 27*, 231–237.

Clark, G. F., Raymond, B., Riddle, M. J., Stark, J. S., & Johnston, E. L. (2015). Vulnerability of Antarctic shallow invertebrate-dominated ecosystems. *Austral Ecology, 40*, 482–491.

Clark, G. F., Stark, J. S., Palmer, A. S., Riddle, M. J., & Johnston, E. L. (2017). The roles of sea-ice, light and sedimentation in structuring shallow Antarctic benthic communities. *PLoS One, 12*, e0168391.

Clarke, A., & Crame, J. A. (1989). The origin of the Southern Ocean marine fauna. In J. A. Crame (Ed.), *Origins and evolution of the Antarctic biota* (Vol. 47, pp. 253–268). London: Special Publications of Geological Society.

Clarke, A., & Johnston, N. M. (2003). Antarctic marine benthic diversity. *Oceanography and Marine Biology Annual Review, 41*, 47–114.

Clarke, A., Aronson, R., Crame, J., Gili, J., & Blake, D. (2004). Evolution and diversity of the benthic fauna of the Southern Ocean continental shelf. *Antarctic Science, 16*, 559–568.

Dayton, P. K., Robilliard, G. A., & Devries, A. L. (1969). Anchor ice formation in McMurdo Sound, Antarctica, and its biological effects. *Science, 163*, 273–274.

Díaz, A., Gérard, K., González-Wevar, C., Maturana, C., Féral, J.-P., David, B., et al. (2018). Genetic structure and demographic inference of the regular sea urchin *Sterechinus neumayeri* (Meissner, 1900) in the Southern Ocean: The role of the last glaciation. *PLoS One, 13*, e0197611.

Duarte, A. W., Dayo-Owoyemi, I., Nobre, F. S., Pagnocca, F. C., Chaud, L. C., Pessoa, A., Felipe, M. G., & Sette, L. D. (2013). Taxonomic assessment and enzymes production by yeasts isolated from marine and terrestrial Antarctic samples. *Extremophiles, 17*, 1023–1035.

Gaino, E., Bavestrello, G., Cattaneo-Vietti, R., & SaraÁ, M. (1994). Scanning electron microscope evidence for diatom uptake by two Antarctic sponges. *Polar Biology, 14*, 55–58.

Gili, J. M., Alvà, V., Pagès, F., Klöser, H., & Arntz, W. E. (1996). Benthic diatoms as the major food source in the sub-Antarctic marine hydroid *Silicularia rosea*. *Polar Biology, 16*, 507–512.

Gili, J.-M., Arntz, W. E., Palanques, A., et al. (2006). A unique assemblage of epibenthic sessile suspension feeders with archaic features in the high-Antarctic. *Deep Sea Research Part II, 53*, 1029–1052.

González-Aravena, M., Urtubia, R., Del Campo, K., Lavín, P., Wong, C. M. V. L., Cárdenas, C. A., & González-Rocha, G. (2016). Antibiotic and metal resistance of cultivable bacteria in the Antarctic sea urchin. *Antarctic Science, 28*, 261–268.

Hamilton, P. B., Poulin, M., & Yang, J.-R. (1997). A new diatom genus *Porannulus* (Bacillariophyta), associated with marine sponges around King George Island, South Shetland Islands, Antarctica. *Diatomologica Research, 12*, 229–242.

Hentschel, U., Hopke, J., Horn, M., Friedrich, A. B., Wagner, M., Hacker, J., & Moore, B. S. (2002). Molecular evidence for a uniform microbial community in sponges from different oceans. *Applied and Environmental Microbiology, 68*, 4431–4440.

Herrera, L. M., García-Laviña, C. X., Marizcurrena, J. J., Volonterio, O., Ponce de León, R., & Castro-Sowinski, S. (2017). Hydrolytic enzyme-producing microbes in the Antarctic oligochaete *Grania* sp. (Annelida). *Polar Biology, 40*, 947–953.

Kellogg, D. E., Kellogg, T. B., Dearborn, J. H., Edwards, K. C., & Fratt, D. B. (1982). Diatoms from brittle star stomach contents: Implications for sediment reworking. *Antarctic Journal of the United States, 17*, 167–169.

Laich, F., Chávez, R., & Vaca, I. (2014). *Leucosporidium escuderoi* f.a., sp. nov., a basidiomycetous yeast associated with an Antarctic marine sponge. *Antonie Van Leeuwenhoek, 105*, 593–601.

Mangano, S., Michaud, L., Caruso, C., Brilli, M., Bruni, V., Fani, R., & Lo Giudice, A. (2009). Antagonistic interactions among psychrotrophic cultivable bacteria isolated from Antarctic sponges: A preliminary analysis. *Research in Microbiology, 160*, 27–37.

Mangano, S., Michaud, L., Caruso, C., & Lo Giudice, A. (2014). Metal and antibiotic-resistance in psychrotrophic bacteria associated with the Antarctic sponge *Hemigellius pilosus* (Kirkpatrick, 1907). *Polar Biology, 37*, 227–235.

Mangano, S., Caruso, C., Michaud, L., & Lo Giudice, A. (2018). First evidence of quorum sensing activity in bacteria associated with Antarctic sponges. *Polar Biology, 41*, 1435–1445.

Margesin, R., & Miteva, V. (2011). Diversity and ecology of psychrophilic microorganisms. *Research in Microbiology, 162*, 346–361.

McClintock, J. B., Amsler, C. D., Baker, B. J., & van Soest, R. W. (2005). Ecology of Antarctic marine sponges: An overview. *Integrative and Comparative Biology, 45*, 359–368.

Núñez-Pons, L., Carbone, M., Vázquez, J., Gavagnin, M., & Avila, C. (2013). Lipophilic defenses from *Alcyonium* soft corals of Antarctica. *Journal of Chemical Ecology, 39*, 675–685.

Papaleo, M. C., Fondi, M., Maida, I., Perrin, E., Lo Giudice, A., Michaud, L., Mangano, S., Bartolucci, G., Romoli, R., & Fani, R. (2012). Sponge-associated microbial Antarctic communities exhibiting antimicrobial activity *against Burkholderia cepacia* complex bacteria. *Biotechnology Advances, 30*, 272–293.

Rodríguez-Marconi, S., De la Iglesia, R., Díez, B., Fonseca, C. A., Hajdu, E., & Trefault, N. (2015). Characterization of bacterial, archaeal and eukaryote symbionts from Antarctic sponges reveals a high diversity at a three-domain level and a particular signature for this ecosystem. *PLoS One, 10*, e0138837.

Rosenber, E., & Zilber-Rosenberg, I. (2016). Microbes drive evolution of animals and plants: The hologenome concept. *mBio, 7*, e01395–e01315. https://doi.org/10.1128/mBio.01395-15.

Rosenber, E., & Zilber-Rosenberg, I. (2018). The hologenome concept of evolution after 10 years. *Microbiome, 6*, 78. https://doi.org/10.1186/s40168-018-0457-9.

Schiaparelli, S. (2014). Biotic interactions. In C. De Broyer, P. Koubbi, H. Griffiths, B. Raymond, U. d'Acoz C.d', A. Van de Putte, B. Danis, B. David, S. Grant, J. Gutt, C. Held, G. Hosie, F. Huettmann, A. Post, & Y. Ropert-Coudert (Eds.)., Biogeographic Atlas of the Southern Ocean *Biogeographic atlas of the Southern Ocean* (pp. 245–252). Cambridge, UK: The Scientific Committee on Antarctic Research, Scott Polar Research Institute. www.scar.org. ISBN:978-0-948277-28-3., Chapter 5.31.

Schiaparelli, S., Albertelli, G., & Cattaneo-Vietti, R. (2003). The epibiotic assembly on the sponge *Haliclona dancoi* (Topsent, 1901) at Terra Nova Bay (Antarctica, Ross Sea). *Polar Biology, 26*, 342–347.

Soldatou, S., & Baker, B. J. (2017). Cold-water marine natural products, 2006 to 2016. *Natural Product Reports, 34*, 585–626.

Tatián, M., Sahade, R., & Esnal, G. B. (2004). Diet components in the food of Antarctic ascidians living at low levels of primary production. *Antarctic Science, 16*, 123–128.

Totti, C., Calcinai, B., Cerrano, C., Camillo, C., Romagnoli, T., & Bavestrello, G. (2005). Diatom assemblages associated with *Sphaerotylus antarcticus* (Porifera: Demospongiae). *Journal of the Marine Biological Association of the UK, 85*, 795–800.

Vaca, I., Faúndez, C., Maza, F., Paillavil, B., Hernández, V., Acosta, F., Levicán, G., Martínez, C., & Chávez, R. (2013). Cultivable psychrotolerant yeasts associated with Antarctic marine sponges. *World Journal of Microbiology and Biotechnology, 29*, 183–189.

Webster, N., & Bourne, D. (2007). Bacterial community structure associated with the Antarctic soft coral, *Alcyonium antarcticum*. *FEMS Microbiology Ecology, 59*, 81–94.

Webster, N. S., Negri, A. P., Munro, M. M. H. G., & Battershill, C. N. (2004). Diverse microbial communities inhabit Antarctic sponges. *Environmental Microbiology, 6*, 288–300.

Xin, Y., Kanagasabhapathy, M., Janussen, D., Xue, S., & Zhang, W. (2011). Phylogenetic diversity of Gram-positive bacteria cultured from Antarctic deep-sea sponges. *Polar Biology, 34*, 1501–1512.

Index

S. Castro-Sowinski (ed.), *The Ecological Role of Micro-organisms in the Antarctic Environment*, Springer Polar Sciences,
https://doi.org/10.1007/978-3-030-02786-5

MIX
Papier aus verantwortungsvollen Quellen
Paper from responsible sources
FSC® C105338

If you have any concerns about our products,
you can contact us on
ProductSafety@springernature.com

In case Publisher is established outside the EU,
the EU authorized representative is:
Springer Nature Customer Service Center GmbH
Europaplatz 3, 69115 Heidelberg, Germany

Printed by Libri Plureos GmbH
in Hamburg, Germany